Fundamental Toxicology

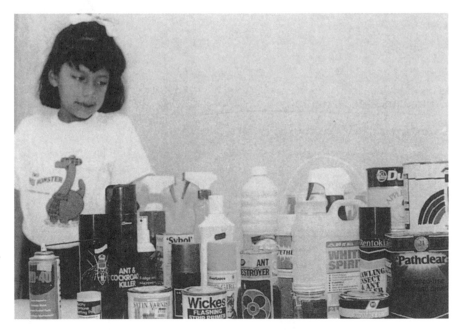

Frontispiece *Potentially toxic and dangerous chemicals are now part of our everyday life, both in our homes and in our places of work.*
(Photo: Courtesy of H.G.J. Worth, The King's Mill Centre for Health Care Services, Sutton-in-Ashfield)

Books are to be returned on or before
the last date below.

7–DAY LOAN

Fundamental Toxicology

Edited by

John H. Duffus
The Edinburgh Centre for Toxicology

Howard G. J. Worth
Healthcare Scientist Consultant

RSCPublishing

ISBN 0-85404-614-3

A catalogue record for this book is available from the British Library

Published by The Royal Society of Chemistry,
Thomas Graham House, Science Park, Milton Road,
Cambridge CB4 0WF, UK

Registered Charity Number 207890

For further information see our web site at www.rsc.org

Typeset by Macmillan India Ltd, Bangalore, India
Printed by Biddles Ltd, King's Lynn, Norfolk, UK

Preface

When the first edition of *Fundamental Toxicology for Chemists* was published in 1996, we recognised the increasing awareness of safety and a growing consciousness of the need for safety standards. This had resulted in legislation concerned with safe practice in the work place, which was led by Europe and North America and other developed countries and which had spread to many other areas of the world.

In the United Kingdom the trend was spearheaded by the Health and Safety at Work Act in 1974, followed by legislation in 1978 concerned with safe practice of work in clinical laboratories and post mortem rooms, and then by regulations for the Control of Substances Hazardous to Health (COSHH). At the international level, the International Programme on Chemical Safety (IPCS), a joint activity of the World Health Organisation (WHO), the United Nations Environmental Programme (UNEP) and the International Labour Organisation (ILO) have published many valuable documents on chemical safety in conjunction with the Commission of the European Communities (CEC). This is merely one example of international collaboration. At present, the European Union is about to introduce a new regulatory framework in the form of the Registration Evaluation and Authorisation of Chemicals (REACH) proposals, which will cover all the constituent countries.

Much safety legislation and safe practice is concerned with the correct handling and use of chemicals. It is expected that chemists should be aware of the dangers of the chemicals that are used in their laboratories, and that there should be documentation and legislation to help this safety process. But the use of chemicals is not confined to the laboratory or the factory. Chemicals are used increasingly in domestic and non-technical environments, where their safe handling is no longer solely the concern of qualified chemists. For instance, consider the use of domestic cleaners, solvents and detergents, weed killers and pesticides and proprietary medicines. The question is asked, therefore, who is the person to whom the public might turn to seek help and advice in the safe handling of these chemicals? As like as not, the answer that comes back is, the chemist. It is not unreasonable that the chemist is seen as the person who can give help and advice on the handling of chemicals, on the toxic effects associated with them, and on how to deal with an incident if and when it occurs. However, the need is still not recognised in the curricula for the training of chemists, and indeed, apart from what they pick up indirectly during their educational progress, there is usually no formal training in toxicology. This makes the chemist very vulnerable as a result of being given new responsibilities without adequate training to handle them. Thus, this book was written originally with the chemist in mind.

The above was the situation when we edited the first edition of *Fundamental Toxicology for Chemists*, but things have moved on. Even my daughter (HGJW) who appears in the Frontispiece of both editions is no longer a little girl! Legislation has increased. It has become more detailed and more complex, and even more widespread across the world. The public are better informed about toxic effects and their rights in relation to any consequential adverse effects. The scientific understanding of toxicology has increased and so, hopefully, has the knowledge of non-toxicologists, but it is unlikely to have kept up with the advances in toxicology. Thus, it has become necessary to produce a second edition of this book, not just for chemists, but for all those scientists who work with chemicals and now have to take the responsibility for any harm that may arise from their use. We are gratified that the Royal Society of Chemistry (RSC) has invited us to do this, and that it is again carried out under the auspices of the International Union of Pure and Applied Chemistry (IUPAC).

Every chapter has been reviewed and updated. As a result many have undergone a major restructuring, and some have been rewritten. Four new chapters have been added namely, 'Introduction to Toxicogenomics', 'Pathways and Behaviour of Chemicals in the Environment', 'Toxicology in the Clinical Laboratory' and 'Pharmaceutical Toxicology'. These have made the text a far more comprehensive guide to current toxicology than it was. The appendices include a 'Curriculum of Fundamental Toxicology for Chemists' and a 'Glossary of Terms used in Toxicology'. These were both in the previous edition, but have been revised. The glossary of terms is based on two IUPAC publications: J.H. Duffus, Glossary for Chemists of Terms Used in Toxicology (IUPAC Recommendations, 1993), *Pure Appl. Chem.*, 1993, **65**, 2003–2122; M. Nordberg, J.H. Duffus and D.M. Templeton, Glossary of Terms Used in Toxicokinetics (IUPAC Recommendations, 2004), *Pure Appl. Chem.*, 2004, **76**, 1033–1082. In addition, we have added a further appendix of commonly used abbreviations. This includes terms that are familiar to toxicologists such as lifetime average daily dose (LADD), for example, but are not so familiar to other scientists. It also includes the names of international bodies and pieces of legislation that are commonly abbreviated and may appear in other textbooks without definition.

Chemistry has had a poor press in recent decades partly because the public has the misconception that manmade chemicals are inherently bad and therefore toxic, while naturally occurring substances are inherently good and healthy. Nothing of course is further from the truth as may be illustrated by a survey of the use of animal and plant extracts over the centuries. It is well known that Cleopatra committed suicide by the administration of snake venom. Roman ladies distilled belladonna, which means beautiful woman, and used it as eye drops to make their pupils dilate. Belladonna is extracted from the plant known as deadly nightshade. Lucrezia Borgia made use of an extract from *Nux vomica* whose active ingredient is strychnine. This is to say nothing of Shakespeare's characters who took or administered an impressive range of animal and plant toxins. Hopefully, the explanation of the science of toxicology in this book will go some way to redressing the balance and putting manmade and natural chemicals into a proper perspective as parts of a total group of substances, and even micro-organisms, which must be considered as a whole in order to ensure their safe use.

Again, we thank IUPAC for their support of this project. In particular, we thank the committee of Division VII, Chemistry and Human Health, and the Subcommittee on Toxicology and Risk Assessment, for their encouragement and assistance. Finally, our thanks go to our team of internationally recognised authors without whose expertise and effort this book could not have been published.

John H. Duffus
Howard G.J. Worth
(Editors)

Contents

Chapter 23 **Toxicology in the Clinical Laboratory** 304
Robin A. Braithwaite

Contributors

A. Adisesh, *Health and Safety Laboratory, Harpur Hill, Buxton, Derbyshire SK17 9JN; e-mail: Anil.Adisesh@hsl.gov.uk.*

R. Agius, *Director, Centre for Occupational and Environmental Health, The University of Manchester, Stopford Building, Oxford Road, Manchester M13 9PL; e-mail: Raymond.Agius@manchester.ac.uk.*

R.D. Aldridge, *Consultant Dermatologist, Department of Dermatology, Royal Infirmary NHS Trust, The Royal Infirmary of Edinburgh, Lauriston Place, Edinburgh EH3 9HA, Scotland.*

H.R. Andersen, *Landesamt fur Natur und Umwelt des Landes Schleswig-Holstein, LGA SH 50, Brunswickerstr. 4, KIEL 24105,Germany.*

D. Boverhof, *Biochemistry and Molecular Biology, Michigan State University, 223 Biochemistry Building, Wilson Road, East Lansing Michigan, MI 48824-1319, USA; e-mail: boverho5@msu.edu.*

R.A. Braithwaite, *Regional Laboratory for Toxicology, Sandwell and West Birmingham NHS Trust, City Hospital, Birmingham B18 7QH.*

J. Burt, *Biochemistry and Molecular Biology, Michigan State University, 223 Biochemistry Building, Wilson Road, East Lansing Michigan, MI 48824-1319, USA; e-mail: burtje@msu.edu.*

J.H. Duffus, *The Edinburgh Centre for Toxicology, 43 Mansionhouse Road, Edinburgh EH9 2JD; e-mail: J.H. Duffus@blueyonder.co.uk.*

J.S.L. Fowler, *Research and Liaison in Toxicology, 66 Meadow Road, Loughton IG10 4HX; e-mail: john.toni@btopenworld.com.*

B. Heinzow, *Landesamt fur Natur und Umwelt des Landes Schleswig-Holstein, LGA SH 50, Brunswickerstr. 4, KIEL 24105, Germany; e-mail: birger.heinzow@lgash-ki.landsh.de.*

R.F.M. Herber, *Tollenslaan 16, Bilthoven, NL-3723 DH, The Netherlands; e-mail: Rob@herber.com.*

M. Herrchen, *Fraunhofer Institute of Molecular Biology and Applied Ecology, Schmallenberg D-57392, Germany; e-mail: Monika.Herrchen@ime.fraunhofer.de.*

H.P.A. Illing, *PICS, Sherwood, 37 Brimstage Road, Heswall, Wirral CH80 1XE; e-mail: paul@sherwood37.demon.co.uk.*

D. McGregor, *Toxicity Evaluation Consultants, 38 Shore Road, Aberdour Fife KY3 0TU, Scotland, UK; e-mail: mcgregortec@btinternet.com.*

A.L. Jones, *Medical Director - National Poisons Information Service, Guy's and St. Thomas' Hospital Trust, Medical Toxicology Unit, Avonley Road, London SE14 5ER, England; e-mail: Alison.Jones@gstt.nhs.uk.*

D.A. Mckay, *Department of Dermatology, Royal Infirmary NHS Trust, The Royal Infirmary of Edinburgh, Lauriston Place, Edinburgh EH3 9HA, Scotland.*

M.V. Park, *28 Coltbridge Terrrace, Edinburgh EH12 6AE, Scotland; e-mail: m.v.park@ntlworld.com.*

A.G. Renwick, *Clinical Pharmacology Group, School of Medicine, University of Southampton, Biomedical Sciences Building, Bassett Crescent East, Southampton SO16 7PX; e-mail: A.G.Renwick@soton.ac.uk.*

D.M. Templeton, *Department of Laboratory Medicine and Pathobiology, University of Toronto, Medical Sciences Building, 1 King's College Circle, Toronto, Ontario M5S 1A8, Canada; e-mail: doug.templeton@utoronto.ca.*

F.M. Sullivan, *Consultant in Toxicology, Harrington House, 8 Harrington Road, Brighton BN1 6RE, England; e-mail: fsullivan@mistral.co.uk.*

G. Wild, *Department of Immunology, Sheffield Teaching Hospitals, NHS Trust, Sheffield S5 7YT; e-mail: graeme.wild@sth.nhs.uk.*

M. Wilkinson, *Biological Sciences (T8), Heriot-Watt University, Riccarton, Edinburgh EH14 4AS, Scotland; e-mail: M.Wilkinson@hw.ac.uk.*

G. Winneke, *Head of Department–Medical Psychology, Medizinisches Institut Fur Umwelthygiene, Heinrich-Heine-Universitat Dusseldorf, Auf'm Hennekamp 50, 40225 Dusseldorf, Germany; e-mail: gerhard.winneke@uniduesseldorf.de.*

H.G.J. Worth, *1 Park Court, Mansfield, Nottinghamshire NG18 2AX; e-mail: howard.worth@btopenworld.com.*

T. Zacharewski, *Biochemistry and Molecular Biology, Michigan State University, 223 Biochemistry Building, Wilson Road, East Lansing Michigan, MI 48824-1319, USA; e-mail: tzachare@pilot.msu.edu.*

Chapter 1

Introduction to Toxicology

JOHN H. DUFFUS

1.1 INTRODUCTION

Toxicology is the fundamental science of poisons. A poison is generally considered to be any substance that can cause severe injury or death as a result of a physicochemical interaction with living tissue. However, all substances are potential poisons since all of them can cause injury or death following excessive exposure. On the other hand, all chemicals can be used safely if exposure of people or susceptible organisms to chemicals is kept below defined tolerable limits, *i.e.* if handled with appropriate precautions. If no tolerable limit can be defined, zero exposure methods must be used.

Exposure is a function of the amount (or concentration) of the chemical involved, and the time and frequency of its interaction with people or other organisms at risk. For very highly toxic substances, the tolerable exposure may be close to zero. In deciding what constitutes a tolerable exposure, it is essential to have data relating exposure to the production of injury or adverse effect. A problem often arises in deciding what constitutes an injury or adverse effect.

An adverse effect is defined as an abnormal, undesirable or harmful change following exposure to the potentially toxic substance. The ultimate adverse effect is death but less severe adverse effects may include altered food consumption, altered body and organ weights, visible pathological changes or simply altered enzyme levels. A statistically significant change from the normal state of the person at risk is not necessarily an adverse effect. The extent of the difference from normal, the consistency of the altered property and the relation of the altered property to the total well-being of the person affected have to be considered.

An effect may be considered harmful if it causes functional or anatomical damage, irreversible change in homeostasis or increased susceptibility to other chemical or biological stress, including infectious disease. The degree of harm of the effect can be influenced by the state of health of the organism. Reversible changes may also be harmful, but often they are essentially harmless. An effect which is not harmful is usually reversed when exposure to the potentially toxic chemical ceases. Adaptation of the exposed organism may occur so that it can live normally in spite of an irreversible effect.

In immune reactions leading to hypersensitivity or allergic effects, the first exposure to the causative agent may produce no adverse effect, although it sensitizes the organism to respond adversely to future exposures, often at a very low level.

The amount of exposure to a chemical required to produce injury varies over a very wide range depending on the chemical and the form in which it occurs. The extent of possible variation in harmful exposure levels is indicated in Table 1.1, which compares median lethal dose (LD_{50}) values for a number of potentially toxic chemicals. The LD_{50} value is more descriptively called the median lethal dose and is defined below.

The LD_{50} is the statistically derived single dose of a chemical that can be expected to cause death in 50% of a given population of organisms under a defined set of experimental conditions. Where LD_{50} values are quoted for human beings, they are derived by extrapolation from studies with mammals or from observations following accidental or suicidal exposures.

The LD_{50} has often been used to classify and compare toxicity among chemicals but its value for this purpose is limited. A commonly used classification of this kind is shown in Table 1.2. Such a classification is entirely arbitrary and has some intrinsic weaknesses. For example, it is difficult to see why a substance with an LD_{50} of 200 mg kg^{-1} body weight should be regarded only as harmful while one with an LD_{50} of 199 mg kg^{-1} body weight is said to be toxic, when the difference in values is minimal. Further, there is no simple relationship between lethality and sublethal toxic effects. In particular, there is no simple relationship between lethality and effects of great concern, such as cancer or abnormal development of the human

Table 1.1 *Approximate acute LD_{50} values for some potentially hazardous substances*

Substance	LD_{50} male rat (mg kg^{-1} body weight) oral administration
Ethanol	7000
Sodium chloride	3000
Cupric sulphate	1500
DDT	100
Nicotine	60
Tetrodotoxin	0.02
Dioxin (TCDD)	0.02

Notes: Values obtained from the Merck Index, Sigma-Aldrich Material Safety Data Sheets (Sigma-Aldrich Library of Chemical Safety Data), and Casarett and Doull's Toxicology. DDT, (1,1,1-trichloro-2,2-bi~chlorophenyl) ethane; TCDD, 2,3,7,8-tetrachloro- dibenzo-p-dioxin.

Table 1.2 *An example of a classification of toxicity based on acute LD_{50} values (used in EC directives on classification, packaging and labelling of chemicals)*

Category	LD_{50} orally to rat (mg kg^{-1} body weight)
Very toxic	Less than 25
Toxic	From 25 to 200
Harmful	From 200 to 2000

embryo. Even in relation to lethality, it is not helpful because it gives no measure of the minimum dose that can be lethal and thus no guide to what might be a 'safe' exposure level.

In decisions relating to chemical safety, the toxicity (hazard) of a substance is less important than the risk associated with its use. Risk is the predicted or actual frequency (probability) of a chemical causing unacceptable harm or effects as a result of exposure of susceptible organisms or ecosystems. Assessment of risk is often assessment of the probability and likely degree of exposure.

By comparison with risk, safety is the practical certainty that injury will not result from exposure to a hazard under defined conditions; in other words, the high probability that injury will *not* result. Practical certainty is defined as a numerically specified low risk or socially acceptable risk applied in decision making for risk management.

In assessing permissible exposure conditions for chemicals, uncertainty factors are applied. A threshold of exposure above which an adverse effect can occur (and below which no such effect is observed) is defined from the available data, and this is divided by an uncertainty factor to lower it to a value that regulatory toxicologists can regard as safe beyond doubt. An uncertainty factor may be defined as a mathematical expression of uncertainty that is used to protect populations from hazards that cannot be assessed with high precision. For example, the 1977 report of the US National Academy of Sciences Safe Drinking Water Committee proposed the following guidelines for selecting uncertainty (safety) factors to be used in conjunction with no observed effect level (NOEL) data. The NOEL should be divided by the following uncertainty factors:

1. An uncertainty factor of 10 should be used when valid human data based on chronic exposure are available.
2. An uncertainty factor of 100 should be used when human data are inconclusive, *e.g.* limited to acute exposure histories, or absent, but when reliable animal data are available for one or more species.
3. An uncertainty factor of 1000 should be used when no long-term, or acute human data are available and experimental animal data are scanty.

This approach is subjective and is being continually updated.

Safety control often involves the assessment of 'acceptable' risk since total elimination of risk is often impossible. 'Acceptable' risk is the probability of suffering disease or injury that will be tolerated by an individual, group, or society. Assessment of risk depends on scientific data but its 'acceptability' is influenced by social, economic and political factors, and by the perceived benefits arising from a chemical or process.

1.2 EXPOSURE TO POTENTIALLY TOXIC SUBSTANCES

Injury can be caused by chemicals only if they reach sensitive parts of a person or other living organism at a sufficiently high concentration and for a sufficient length of time. Thus, injury depends upon the physicochemical properties of the potentially toxic substances, the exact nature of the exposure circumstances, and the health and developmental state of the person or organism at risk.

Major routes of exposure are through the skin (topical), through the lungs (inhalation) or through the gastrointestinal tract (ingestion). In general, for exposure to any given concentration of a substance for a given time, inhalation is likely to cause more harm than ingestion, which, in turn, will be more harmful than topical exposure. Exposure of the eye can have serious consequences and must also be given due consideration.

1.2.1   Skin (Dermal or Percutaneous) Absorption

Many people do not realize that chemicals can penetrate healthy intact skin and so this fact must be emphasized. Among the chemicals that are absorbed through the skin are aniline, hydrogen cyanide, some steroid hormones, organic mercury compounds, nitrobenzene, organophosphate compounds and phenol. Some chemicals, such as phenol or methylmercury chloride, can be lethal if absorbed from a fairly small area (a few square centimetres) of skin. If protective clothing is being worn, it must be remembered that absorption through the skin of any chemical that gets inside the clothing will be even faster than through unprotected skin because the chemical cannot escape and contact is maintained over a longer time.

1.2.2   Inhalation

Gases and vapours are easily inhaled but inhalation of particles depends upon their size and shape. The smaller the particle, the further into the respiratory tract it can go. Dusts with an effective aerodynamic diameter of between 0.5 and 7 μm (the respirable fraction) can persist in the alveoli and respiratory bronchioles after deposition. Peak retention depends upon the aerodynamic shape but is mainly of those particles with an effective aerodynamic diameter of between 1 and 2 μm. Particles of effective aerodynamic diameter less than 1 μm tend to be breathed out again but a significant fraction may persist in the alveoli and cause harmful effects. Inhaled particles may also enter the gut (see below).

The effective aerodynamic diameter is defined as the diameter in micrometers of a spherical particle of unit density that falls at the same speed as the particle under consideration. Dusts of larger diameter than 10 μm either do not penetrate the lungs or lodge further up in the bronchioles and bronchi where cilia create a flow of mucus, which returns them to the pharynx and from there they go to the oesophagus. This process is known as the mucociliary clearance mechanism.

From the oesophagus dusts pass through the gut in the normal way. Particles entering the gut in this way may cause poisoning just as though they had been ingested in the food. A large proportion of inhaled dust enters the gut and so its effects through this route must be assessed. A significant portion of inhaled dust consists of microorganisms. There is thus the possibility of bacterial infection. Presence of fungi and their spores may be associated with hypersensitivity responses or with mycotoxins, which may have a range of effects including cancer or even endocrine effects. As with any foodstuff, constituents of dust may affect the gut directly or be absorbed and cause systemic effects.

Physical irritation by dust particles or fibres can cause very serious adverse health effects but most effects depend upon the solids being dissolved. Special consideration should be given to asbestos fibres which may lodge in the lung and cause fibrosis and cancer even though they are mostly insoluble and therefore do not act like classical toxicants: care should also be taken in assessing possible harm from manmade mineral fibres that have similar properties. The macrophage cells in the lung that normally remove invading bacteria and organic matter may take in insoluble particles. This is called phagocytosis.

Phagocytosis is the process whereby certain body cells, notably macrophages and neutrophils, engulf and destroy invading foreign particles. The cell membrane of the phagocytosing cell (phagocyte) invaginates to capture and engulf the particle. Hydrolytic enzymes are secreted round the particle to digest it and may leak from the phagocyte and cause local tissue destruction if the particle damages the phagocyte. If phagocytic cells are adversely affected by ingestion of insoluble particles, their ability to protect against infectious organisms may be reduced and infectious diseases may follow.

Some insoluble particles such as coal dust and silica dust will readily cause fibrosis of the lung. Others, such as asbestos, may also cause fibrosis depending on the exposure conditions. Fibrosis of the lung leads to breathing problems such as emphysema. It may follow from any chronic inflammation of the lungs and thus be caused even by soluble irritants such as certain metal salts.

Remember that tidal volume (the volume of air inspired and expired with each normal breath) increases with physical exertion; thus absorption of a chemical as a result of inhalation is directly related to the rate of physical work. This is why people living in cities subject to severe air pollution may sometimes be advised to avoid physical activity as far as possible.

1.2.3 Ingestion

As mentioned above, airborne particles breathed through the mouth or cleared by the cilia of the lungs are ingested and, outwith the workplace, we may have little control over this apart from avoiding heavily contaminated air, for example by avoiding active or passive smoking. We can keep our homes clean but air pollution in our immediate environment is otherwise beyond individual control. On the other hand, ingestion of potentially toxic substances in our food and drink, or as medication, is under individual control and, by using common sense, we can minimize any associated risks. The nature of the absorption processes following ingestion is discussed elsewhere.

The importance of concentration and time of exposure has already been pointed out. It should be remembered that exposure may be continuous or it may be repeated at intervals over a period of time; the consequences of different patterns of exposure to the same amount of a potentially toxic substance may vary considerably in their seriousness. In most cases, the consequences of continuous exposure to a given concentration of a chemical will be worse than those of intermittent exposures to the same concentration of the chemical at intervals separated by sufficient time to permit a degree of recovery. However, repeated or continuous exposure to very small amounts of potentially toxic chemicals may be a matter for serious concern if either the chemical or its effects, *e.g.* decreasing numbers of nerve cells, have a tendency to accumulate in the person or organism at risk.

A chemical may accumulate if absorption exceeds excretion; this is particularly likely with substances that combine a fairly high degree of lipid solubility with chemical stability. Such chemicals are found in the group of persistent organic pollutants (POPS), including several organochlorine pesticides, which are now largely, but not entirely, banned from use. Accumulation of water-soluble ions may also be a problem. Divalent lead ions accumulate in bone where they replace the chemically similar calcium ions. While in the bone, they cause little harm but when bone breaks down, during pregnancy or illness, the lead ions enter the blood and may poison the person who has accumulated them or, in the case of pregnancy, the unborn child. Fluoride ions also accumulate in bone and this is of concern in regard to schemes for water fluoridation.

1.3   ADVERSE EFFECTS

Adverse effects may be local or systemic. Local effects occur at the site of exposure of the organism to the potentially toxic substance. Corrosives always act locally. Irritants frequently act locally.

Most substances that are not highly reactive are absorbed and distributed around the affected organism causing systemic injury at a target organ or tissue distinct from the absorption site. The target organ is not necessarily the organ of greatest accumulation. For example, adipose (fatty) tissue accumulates organochlorine pesticides to very high levels but does not appear to be affected by them.

Some substances produce both local and systemic effects. For example, tetraethyl lead damages the skin on contact, and is then absorbed and transported to the central nervous system where it causes further damage.

Effects of a chemical can accumulate even if the chemical itself does not. There is some evidence that this is true of the effects of organophosphate pesticides and other neurotoxins on the nervous system. This may lead to poor functioning of the nervous system in humans in old age. Because of the time difference between exposure and effect, establishing the relationship between such delayed effects and the possible cause, no longer present in the body, is often difficult.

A particularly harmful effect that may accumulate is death of nerve cells, since nerve cells cannot be replaced, though damaged nerve fibres can be regenerated.

It will be clear that the balances between absorption and excretion of a potentially toxic substance and between injury produced and repair are the key factors in determining whether any injury follows exposure. All of the possible adverse effects cannot be discussed here but some aspects should be mentioned specifically.

Production of mutations, tumours and cancer and defects of embryonic and fetal development are of particular concern.

Adverse effects related to allergies appear to be increasing. Allergy (hypersensitivity) is the name given to disease symptoms following exposure to a previously encountered substance (allergen) that would otherwise be classified as harmless. Essentially, an allergy is an adverse reaction of the altered immune system. The process, which leads to the disease response on subsequent exposure to the allergen, is called sensitization. Allergic reactions may be very severe and fatal.

To produce an allergic reaction, most chemicals must act as haptens, *i.e.* combine with proteins to form antigens. Antigens entering the human body or produced within it cause the production of antibodies. Usually at least a week is needed before appreciable amounts of antibodies can be detected and further exposure to the allergen can produce disease symptoms. The most common symptoms are skin ailments such as dermatitis and urticaria, or eye problems such as conjunctivitis. The worst may be death resulting from anaphylactic shock.

Of particular importance in considering the safety of individuals is the possibility of idiosyncratic reactions. An idiosyncratic reaction is an excessive reactivity of an individual to a chemical, for example, an extreme sensitivity to low doses as compared with the average member of the population. There is also the possibility of an abnormally low reactivity to high doses. An example of a group of people with an idiosyncrasy is the group that has a deficiency in the enzyme required to convert methaemoglobin (which cannot carry oxygen) back into haemoglobin; this group is exceptionally sensitive to chemicals like nitrites, which produce methaemoglobin. Idiosyncratic reactions may occur to foodstuffs and to prescription drugs, giving harmful effects, which may be wrongly ascribed to chemicals in the environment or in the workplace.

Another factor to be considered is whether the adverse effects produced by a potentially toxic chemical are likely to be immediate or delayed. Immediate effects appear rapidly after exposure to a chemical while delayed effects appear only after a considerable lapse of time.

Among the most serious delayed effects are cancers; carcinogenesis may take 20 or more years before tumours are seen in humans.

Perhaps the most difficult adverse effects to detect are those that follow years after exposure in the womb: a well-established example of such an effect is the vaginal cancer produced in young women whose mothers were prescribed diethylstilbestrol during pregnancy in order to prevent a miscarriage.

Another important aspect of adverse effects to be considered is whether they are reversible or irreversible. For the liver, which has a great capacity for regeneration, many adverse effects are reversible, and complete recovery can occur. For the central nervous system, in which regeneration of tissue is severely limited, most adverse effects leading to morphological changes are irreversible and recovery is, at best, limited. Carcinogenic and teratogenic effects are also irreversible, but suitable treatment may reduce the severity of such effects.

1.4 CHEMICAL INTERACTIONS

A major problem in assessing the likely effect of exposure to a chemical is that of assessing possible interactions. The simplest interaction is an additive effect: this is an effect, which is the result of two or more chemicals acting together and is the simple sum of their effects when acting independently. In mathematical terms,

$$1 + 1 = 2, \quad 1 + 5 = 6, etc.$$

The effects of organochlorine pesticides are usually additive.

More complex is a synergistic (multiplicative) effect: this is an effect of two chemicals acting together, which is greater than the simple sum of their effects when acting alone; it may be called synergism. In mathematical terms,

$$1 + 1 = 4, \quad 1 + 5 = 10, \textit{ etc.}$$

Asbestos fibres and cigarette smoking act together to increase the risk of lung cancer by a factor of 40, taking it well beyond the risk associated with independent exposure to either of these agents.

Another possible form of interaction is potentiation. In potentiation, a substance that on its own causes no harm makes the effects of another chemical much worse. This may be considered to be a form of synergism. In mathematical terms,

$$0 + 1 = 5, \quad 0 + 5 = 20, \textit{ etc.}$$

For example, isopropanol, at concentrations that are not harmful to the liver, increases (potentiates) the liver damage caused by a given concentration of carbon tetrachloride.

The opposite of synergism is antagonism: an antagonistic effect is the result of a chemical counteracting the adverse effect of another; in other words, the situation where exposure to two chemicals together has less effect than the simple sum of their independent effects. Such chemicals are said to show antagonism. In mathematical terms,

$$1 + 1 = 0, \quad 1 + 5 = 2, \textit{ etc.}$$

1.5   TOLERANCE AND RESISTANCE

Tolerance is a decrease in sensitivity to a chemical following exposure to it or to a structurally related substance. For example, cadmium causes tolerance to itself in some tissues by inducing the synthesis of the metal-binding protein, metallothionein. However, it should be noted that cadmium–metallothionein accumulates in the kidney where it causes nephrotoxicity.

Resistance is almost complete insensitivity to a chemical. It usually reflects the metabolic capacity to inactivate and eliminate the chemical and its metabolites rapidly.

1.6   TOXICITY TESTING

1.6.1   Dose–Response and Concentration–Response

A dose–response (concentration–response) relationship is defined as the association between dose (concentration) and the incidence of a defined biological effect in an exposed population, usually expressed as percentage. Historically the defined effect was death. The classic dose–response or concentration–response relationship is shown in Figure 1.1. This is a theoretical curve and in practice such a Gaussian curve is rarely found. Curves of this kind form the basis for the determination of the LD_{50} or the LC_{50} (the median lethal concentration). The LD_{50} and LC_{50} are specific cases of the generalized values LDn and LCn. The LDn is the dose of a toxicant lethal to $n\%$ of a test population. The LCn is the exposure concentration of a toxicant lethal to $n\%$ of a test population. Thus, the LD_{50} is the statistically derived single dose of

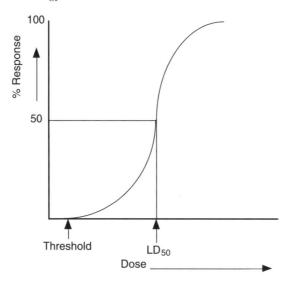

Figure 1.1 *The classic dose–response curve*

a chemical that can be expected to cause death in 50% of a given population of organisms under a defined set of experimental conditions. Similarly, the LC_{50} is the statistically derived exposure concentration of a chemical that can be expected to cause death in 50% of a given population of organisms under a defined set of experimental conditions.

Another important value that may be derived from the relationship shown is the threshold dose or concentration, the minimum dose or concentration required to produce a detectable response in the test population. The threshold value can never be derived with absolute certainty and therefore the lowest observed effect level (LOEL) or the NOEL have normally been used instead of the threshold value in deriving regulatory standards. There is a move to replace these values by the benchmark dose (BMD). This is defined as the statistical lower confidence limit on the dose that produces a defined response (called the benchmark response or BMR, usually 5 or 10%) in a given population under defined conditions for an adverse effect compared to background, defined as 0%.

The use of the LD_{50} in the classification of potentially toxic chemicals has been described; it must be emphasized that such a classification is only a very rough guide to relative toxicity. The LD_{50} tells us nothing about sublethal toxicity. Any classification based on the LD_{50} is strictly valid only for the test population and conditions on which it is based and on the related route of exposure. The LD_{50} tells us nothing about the shape of the dose–response curve on which it is based. Thus, two chemicals may appear to be equally toxic since they have the same LD_{50}, but one may have a much lower lethal threshold and kill members of an exposed population at concentrations where the other has no effect (Figure 1.2). Remember, these are theoretical curves and in practice Gaussian curves of this sort are rarely, if ever, found.

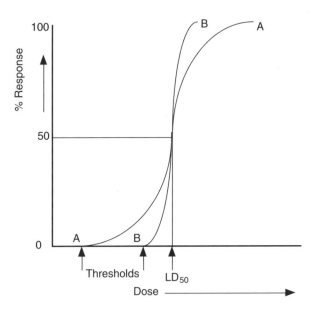

Figure 1.2 *Two substances with the same LD$_{50}$ but different lower lethal thresholds*

The determination and use of the LD$_{50}$ are likely to decline in future as fixed dose testing becomes more widely used. In fixed dose testing, the test substance may be administered to rats or other test species at no more than three dose levels: the possible dose levels are preset legally to equate with a regulatory classification or ranking system. Dosing is followed by an observation period of 14 days. The dose at which toxic signs are detected is used to rank or classify the test materials.

A retrospective study of LD$_{50}$ values showed that between 80 and 90% of those compounds which produced signs of toxicity but no deaths at dose levels of 5, 50 or 500 mg kg^{-1} body weight oral administration had LD$_{50}$ values from the same studies of more than 25, from 25 to 200, or from 200 to 2000 mg kg^{-1} body weight, corresponding to the European Union classification for very toxic, toxic and harmful.

The initial test dose level is selected with a view to identifying toxicity without mortality occurring. Thus, if a group of five male and five female rats is tested with an oral dose of 500 mg kg^{-1} body weight and no clear signs of toxicity appear, the substance should not be classified in any of the defined categories of toxicity.

If toxicity is seen but no mortality, the substance can be classed as 'harmful'. If mortality occurs, retesting with a dose of 50 mg kg^{-1} body weight is required. If no mortality occurs at the lower dose but signs of toxicity are detected, the substance would be classified as 'toxic'. If mortality occurs at the lower dose, retesting at 5 mg kg^{-1} body weight would be carried out and if signs of toxicity were detected and mortality occurred, the substance would be classified as 'very toxic'.

For a full risk assessment, testing at 2000 mg g^{-1} body weight is also required if no signs of toxicity are seen at 500 mg kg^{-1} body weight.

Fixed dose testing reduces the number of animals required and, because mortality need not occur, also greatly reduces possible animal suffering. Fixed dose testing can also identify substances that have high LD_{50} values but still cause acute toxic affects at relatively low doses or exposures.

In assessing the significance of LD_{50} or other toxicological values, it is necessary to note the units used in expressing dosage. Normally dosage is expressed in mg kg^{-1} body weight, but it may be expressed as mg cm^{-2} body surface area as this has been shown in a number of cases to permit more accurate extrapolation between animals of different sizes and from test mammalian species to humans.

For biocides, selective toxicity is the key property, since they are to be used to kill pests with minimal harm to other organisms. Selective toxicity depends upon differences in biological characteristics that may be either quantitative or qualitative. Minimizing the amount of pesticide used and targeting its application is crucial to avoid harm to non-target organisms.

Although now applied to many species, toxicity testing was originally aimed at establishing, by tests on laboratory animals, what effects chemicals are likely to have on human beings who may be exposed to them and the shape of the dose–response relationship. On a body weight basis, it is assumed for toxicity data extrapolation that humans are usually about 10 times more sensitive than rodents. On a body surface–area basis, humans usually show about the same sensitivity as test mammals, *i.e.* the same dose per unit of body surface area will give the same given defined effect, in about the same percentage of the population. Knowing the above relationships, it is possible to estimate the exposure to a chemical that humans should be able to tolerate.

In many countries there is now a defined set of tests that must be carried out on every new chemical that is to be used or produced in an appreciable quantity, usually above 1 tonne/year. Table 1.3 gives an example of test requirements applicable in a number of such countries.

1.7 EPIDEMIOLOGY AND HUMAN TOXICOLOGY

Epidemiology is the analysis of the distribution and determinants of health-related states or events in human populations and the application of this study to the control of health problems. It is the only ethical way to obtain data about the effects of chemicals other than drugs on human beings and hence to establish beyond doubt that toxicity to humans exists. The following are the main approaches that have been used in epidemiology.

1.7.1 Cohort Study

A cohort is a component of the population born during a particular period and identified by the period of birth so that its characteristics (such as causes of death and numbers still living) can be ascertained as it enters successive time and age periods. The term 'cohort' has broadened to describe any designated group of persons followed or traced over a period of time.

In a cohort study, one identifies cohorts of people who are, have been, or in the future may be exposed or not exposed, or exposed in different degrees, to a factor or

Table 1.3　*Example of the information required in some countries for notification and hazard assessment of new chemicals*

Base set information

1　Identity of the substance

　1.1　Name

　　　1.1.1　Names in the IUPAC nomenclature
　　　1.1.2　Other names (usual name, trade name, abbreviation)
　　　1.1.3　CAS number and CAS name (if available)

　1.2　Empirical and structural formula
　1.3　Composition of the substance

　　　1.3.1　Degree of purity (%)
　　　1.3.2　Nature of impurities, including isomers and by-products
　　　1.3.3　Percentage of (significant) main impurities
　　　1.3.4　If the substance contains a stabilizing agent or an inhibitor or other
　　　　　　additives, specify: nature, order of magnitude: ... ppm; ...%
　　　1.3.5　Spectral data (UV, IR, NMR or mass spectrum)
　　　1.3.6　Chromatographic data (HPLC, GC)

　1.4　Methods of detection and determination
　　　A full description of the methods used or the appropriate bibliographical references

2　Information on the substance

　2.0　Production (process, quantity and estimate of resultant exposure)

　2.1　Proposed uses

　　　2.1.1　Types of use
　　　　　　Describe: the function of the substance and the desired effects (including
　　　　　　processes, form marketed, quantity and exposure estimate)
　　　2.1.2　Fields of application with approximate breakdown

　　　　　　– Industries
　　　　　　– Farmers and skilled trades
　　　　　　– Use by the public at large

　　　2.1.3　Waste quantities and composition of waste

　2.2　Estimated production and imports for each of the anticipated uses or fields of
　　　application

　　　2.2.1　Overall production and/or imports in tonnes per year

　　　　　　– First 12 months
　　　　　　– Thereafter

　　　2.2.2　Production and/or imports, broken down in accordance with 2.1.1 and 2.1.2,
　　　　　　expressed as a percentage

　　　　　　– First calendar year
　　　　　　– The following calendar years

　2.3　Recommended methods and precautions concerning:

　　　2.3.1　Handling
　　　2.3.2　Storage
　　　2.3.3　Transport
　　　2.3.4　Fire (nature of combustion gases or pyrolysis, where proposed uses justify
　　　2.3.5　Other dangers, particularly chemical reaction with water or tendency to
　　　　　　explode as a dust

Table 1.3 *(Continued)*

Base set information

2.4 Emergency measures in the case of accidental spillage

2.5 Emergency measures in the case of injury to persons (*e.g.* poisoning)

3 Physicochemical properties of the substance

3.0 State of the substance at 20 °C and 101.3 kPa

3.1 Melting point

3.2 Boiling point °C at ... Pa

3.3 Relative density (D_4^{20})

3.4 Vapour pressure Pa at ... °C

3.5 Surface tension N m^{-1} (...°C)

3.6 Water solubility mg l^{-1} (...°C)

3.7 Fat solubility
Solvent oil (to be specified) mg 100 g^{-1} solvent (...°C)

3.8 Partition coefficient *n*-Octanol/water

3.9 Flashpoint ... °C. Open cup and closed cup

3.10 Flammability

3.11 Explosive properties

3.12 Self ignition temperature ... °C

3.13 Oxidizing properties

3.14 Granulometry (particle size distribution)

4 Toxicological studies

4.1 Acute toxicity
Substances other than gases shall be administered *via* two routes at least one of which should be the oral route. The other route will depend on the intended use and on the physical properties of the substance. Gases and volatile liquids should be administered by inhalation. In all cases, observation of the animals should be carried out for at least 14 days. Unless there are contraindications, the rat is the preferred species for oral and inhalation experiments. The experiments in 4.1.1, 4.1.2, and 4.1.3 shall be carried out on both male and female subjects.

4.1.1 Administered orally
LD_{50} (mg kg^{-1}) or acceptable alternative
Effects observed, including in the organs

4.1.2 Administered by inhalation
LC_{50} (ppm) or acceptable alternative
Duration of exposure in hours
Effects observed, including in the organs

4.1.3 Administered cutaneously (percutaneous absorption)
LD_{50} (mg l^{-1}) or acceptable alternative
Effects observed, including in the organs

4.1.4 Skin irritation
The substance should be applied to the shaved skin of an animal, preferably an albino rabbit.
Duration of exposure in hours

4.1.5 Eye irritation. The rabbit is the preferred animal. Duration of exposure in hours

4.1.6 Skin sensitization. To be determined by a recognized method using a guinea pig.

Table 1.3   *(Continued)*

Base set information

4.2   Repeated dose toxicity

The route of administration should be the most appropriate considering the intended use, the acute toxicity and the physical and chemical properties of the substance. Unless there are contraindications, the rat is the preferred species for oral and inhalation experiments.

4.2.1   Repeated dose toxicity (28 days)

Effects observed on the animal and organs according to the concentrations used, including clinical and laboratory investigations.

Dose for which no toxic effect is observed.

4.3   Other effects

4.3.1   Mutagenicity (including carcinogenic pre-screening test)
The substance should be examined during a series of two tests, one of which should be bacteriological, with and without metabolic activation, and one non-bacteriological with and without metabolic activation

4.3.2   Screening for toxicity related to reproduction

4.3.3   Assessment of toxicokinetic behaviour of the substance

5   Ecotoxicological studies

5.1   Effects on organisms

5.1.1   Acute toxicity for fish

5.1.2   Acute toxicity for Daphnia LC_{50}

5.1.3   Growth inhibition test on algae

5.1.4   Bacteriological inhibition

5.2   Degradation: biotic and abiotic

5.3   Absorption/desorption screening test

6   Possibility of rendering the substance harmless

6.1   For industry/skilled trades

6.1.1   Possibility of recycling

6.1.2   Possibility of neutralization of harmful effects

6.1.3   Possibility of destruction:

– Controlled discharge

– Incineration

– Water purification station

– Others

6.2   For the public at large

6.2.1   Possibility of recycling

6.2.2   Possibility of neutralization of harmful effects

6.2.3   Possibility of destruction:

– Controlled discharge

– Incineration

– Water purification station

– Others

---

factors hypothesized to influence the probability of occurrence of a given disease or other outcome. Alternative terms for such a study – follow-up, longitudinal and prospective study – describe an essential feature of the method, observation of the population for a sufficient number of person-years to generate reliable incidence or mortality rates in the population subsets. This generally means studying a large population, studying for a prolonged period (years), or both.

Cohort studies involve cohort analysis. Cohort analysis is the tabulation and analysis of morbidity or mortality rates in relationship to the ages of the members of the cohort, identified by their birth period, and followed as they pass through different ages during part or all of their life span. Under certain circumstances, such as studies of migrant populations, cohort analysis may be performed according to duration of residence in a country rather than year of birth, in order to relate health or mortality experience to duration of exposure.

1.7.2 Retrospective Study

A retrospective study is used to test hypotheses of cause (aetiological hypotheses) in which inferences about exposure to the putative causal factor(s) are derived from data relating to characteristics of the persons or organisms under study or to events or experiences in their past: the essential feature is that some of the persons under study have the disease or other outcome condition of interest, and their characteristics and past experiences are compared with those of other, unaffected persons. Persons who differ in the severity of the disease may also be compared.

1.7.3 Case Control Study

A case control study starts with the identification of persons with the disease (or other outcome variable) of interest, and a suitable control (comparison and reference) group of persons without the disease. The relationship of an attribute to the disease is examined by comparing the diseased and non-diseased with regard to how frequently the attribute is present or, if quantitative, the levels of the attribute, in the two groups.

1.7.4 Cross-Sectional Study (of Disease Prevalence and Associations)

A cross-sectional study examines the relationship between diseases (or other health-related characteristics) and other variables of interest as they exist in a defined population at one particular time. Disease prevalence rather than incidence is normally recorded in a cross-sectional study and the temporal sequence of cause and effect cannot necessarily be determined.

1.7.5 Confounding

Confounding is one of the biggest difficulties in carrying out a successful epidemiological investigation.

Confounding can occur in a number of different ways. First, there is the situation in which the effects of two processes are not distinguishable from one another: this leads to the situation where the distortion of the apparent effect of an exposure on risk is brought about by the association of other factors that can influence the outcome. Secondly, there is the possibility of a relationship between the effects of two or more causal factors as observed in a set of data, such that it is not logically possible to separate the contribution that any single causal factor has made to an effect. Finally, there is the situation in which a measure of the effect of an exposure on risk is distorted because of the association of exposure with other factor(s) that influence the outcome under study.

A confounding variable (confounder) is defined as a changing factor that can cause or prevent the outcome of interest, is not an intermediate variable, and is not associated with the factor under investigation, such a variable must be controlled in order to obtain an undistorted estimate of the effect of the study factor on risk.

BIBLIOGRAPHY

B. Ballantyne, T. Marrs and T. Syverson (eds), *General and Applied Toxicology*, Macmillan Reference & Grove's Dictionaries, London, New York, 1999, 2199 (text), 2154 (indexes).

A.W. Hayes (ed), *Principals and Methods in Toxicology*, Taylor and Francis, Philadelphia, 2001, 1887.

H.P.A. Illing, *Toxicity and Risk – Context, Principles and Practice*, Taylor & Francis, Basingstoke, 2001, 144.

C.D. Klaasen (ed), *Casarett and Doull's Toxicology – The Basic Science of Poisons*, McGraw-Hill, New York, 2001, 1236.

J.M. Last (ed), *A Dictionary of Epidemiology*, Oxford University Press, New York, 2001.

N.H. Stacey and C. Winder (eds), *Occupational Toxicology*, Taylor and Francis, Basingstoke, 2001, 400.

J. Timbrell, *Introduction to Toxicology*, Vol 3, Taylor and Francis, Basingstoke, 2001, 192.

P. Wexler (ed), *Encyclopedia of Toxicology*, 2nd edition, Academic Press, New York, 2005, 1500 (approx.).

Chapter 2

Introduction to Toxicodynamics

ROBERT F.M. HERBER

2.1 INTRODUCTION

Toxicodynamics is the study of toxic actions on living systems, including the reactions with and binding to cell constituents, and the biochemical and physiological consequences of these actions. The following chapter is devoted to toxicokinetics. What is the difference between toxicodynamics and toxicokinetics? When your friend drinks a few glasses of whisky and he or she breathes over you the next day with a foul smelling odour, you are dealing with toxicokinetics. The ethanol from the whisky is transformed within the body to acetaldehyde (and later to acetic acid) and this gives a nasty odour. This transformation needs water, and this again leads to the development of a thirst. These phenomena are not in themselves toxic to your friend but are manifestations of toxicokinetics, the way in which the body handles potentially toxic substances.

If your friend continues to drink whisky or other alcoholic beverages excessively on a regular basis, there is a chance that he or she will develop cirrhosis of the liver, a toxic effect. Excessive ethanol injures the liver by blocking the normal metabolism of protein, fats, and carbohydrates. In cirrhosis of the liver, scar tissue replaces normal, healthy tissue, blocking the flow of blood through the organ and preventing it from working as it should. In this situation you are dealing with toxicodynamics, the mechanism by which a toxic effect is produced.

The father of toxicology, Philippus Aureolus Theophrastus Bombastus von Hohenheim (better known as Paracelsus), formulated in early 1538 the most important thesis in toxicology, which may be stated as follows: "What is it that is not a poison? All things are poisons and there are none that are not. Only the dose decides that a thing is not poisonous."

Since all chemicals can produce injury or death under some exposure conditions, there is no such thing as a 'safe' chemical in the sense that it will be free of injurious effects under all circumstances of exposure. However, it is also true that there is no chemical that cannot be used safely by limiting the dose or exposure. Thus, reference should not be made to toxic and non-toxic substances or compounds, but rather to a toxic or non-toxic dose. A dose is the total amount of a substance administered to, taken, or absorbed by, an organism. A well-known example is sodium chloride, which is in small quantities necessary (essential) for human life, in larger quantities it is used as a flavour in foods, and, in some parts of Asia, in the past, used as a suicide agent. In the case of liver cirrhosis a certain dose, several glasses of alcohol everyday, over at least 10 years of drinking, will cause the effect. There is a difference in metabolism between women and men. For women, two to three glasses a day may be sufficient to cause the effect, while for men three to four glasses a day are required over a decade.

In food toxicology, substances are sometimes defined as essential or non-essential for humans. The distribution, with time, of concentration of non-essential elements in different tissues (and blood) follows a non-Gaussian pattern, as these elements are not present unless there has been exposure to the compound. The normal concentration will be zero or close to zero. The concentration of essential elements, however, should be around the optimal concentration, and the body tends to keep the concentration of such elements in a steady state. This is a reflection of the homeostasis (metabolic stability) of the body.

Adverse or toxic effects in a biological system are not produced by a chemical agent unless that agent or its biotransformation products reach appropriate sites in the body at a concentration and for a period of time sufficient to produce toxic effects. The toxic effect is thus dependent on the chemical and on its properties, the exposure situation, and the susceptibility of the cell, the biological system, or the subject. Adverse effects may differ from undesirable to death, with all possible effects between, dependent on the compound and dose.

When a subject is exposed *via* the mouth (oral), the lungs, or other tissues, intake occurs. Following intake, a proportion of the compound (up to 100%) will be absorbed by the body. This is called the uptake. The difference between the intake and the uptake is the amount of the substance that leaves the body *via* the urine, faeces, exhaled air, hair, nails, sweat, *etc.* Compounds retained in the body will be metabolized or remain unchanged and be transported through the body. Toxic effects may appear during metabolism (often in the liver) and by toxic (un)metabolized compounds retained in different organs. It is important to be aware of the possible different chemical forms (speciation) of a substance, and also of the different effects the substance may have when it is outside or inside the body. As an example, let us consider the element chromium. Chromium(III) has been said to be essential and may be involved in glucose metabolism. High concentrations of Cr(III) are not absorbed, but are excreted (uptake is limited). Cr(III) in high concentrations externally, however, may cause dermal problems. Non-essential Cr(VI) compounds can be absorbed if they are water-soluble and are metabolized to Cr(III) in the body. Thus, a high concentration of Cr(III) may occur which will cause renal toxicity. Soluble Cr(VI) compounds have an irritant and corrosive effect on the skin, eyes, and lungs. However, insoluble Cr(VI) cannot enter the body and if inhaled as an aerosol, remains in the lungs. Long-term exposure of the lungs in this manner (at least 6 months) may result in bronchial cancer. This example makes it clear that both physical and chemical properties can contribute to toxicity.

2.2 DOSE–TOXICITY RELATIONSHIPS

2.2.1 Dose–Effect Relationship

In toxicology, the most fundamental concept is that of dose. It is important to recognize the difference between dose and concentration. Concentration is the dose of a toxicant expressed per unit volume of the medium in which the toxicant occurs.

Thus, if we know the concentration of the toxicant and the amount of medium administered to an organism, we can calculate the total dose received by the organism.

In the case of essential substances, there is a gradual transition from deficiency to effectiveness depending on the dose given. At low dose, when deficiency occurs, the metabolized or excreted amount of a substance is greater than the uptake, leading to a shortage of the substance in the cell or organ. A well-known example is vitamin deficiency, leading, in the case of vitamin C, to scurvy. An intake of vitamin C will redress the balance and scurvy disappears. Thus, a certain minimum amount is necessary to prevent deficiency. This is illustrated in Figure 2.1, where the effects of essential and non-essential substances are shown as a function of dose. For essential nutrients, there is a gradual change in effects from deficiency to health and then, at higher doses, to fatal toxicity.

Another example of an essential nutrient is zinc. There is a homeostatic control of zinc in blood serum, but deficiency may lead to skin-healing problems, which are overcome by zinc treatment at the wound site. In excess, oral intake of zinc may cause gastrointestinal problems (1 g or more).

The dose-effect (D-E) relationship was originally studied in detail by pharmacologists to establish the dose levels of drugs, which produced beneficial effects and those which produced harmful effects on the body. This relationship is also important in toxicology. Figure 2.1 shows the relationship between the dose and different effects of both essential nutrients and non-essential substances. The difference between these groups is that no benefit can be expected from a non-essential substance and thus such substances may be classified directly as toxic if they cause harm at low doses and banned. This is not possible for essential nutrients and it is only in recent times that the problem of protecting people from the risks of high doses of such substances as vitamins has begun to be tackled seriously.

Figure 2.1 refers to no effects, early effects, and clinical effects. The maximum dose that produces no detectable change under defined conditions of exposure is referred to as the no effect level (NEL). It is not always easy to determine whether an effect should be detectable under the test conditions applied, and, even then, the significance of any observed effect. Thus a more practical measure is the no observed adverse effect level (NOAEL). An adverse effect is a change in morphology, physiology, growth, development, or life span of an organism that results in impairment of functional capacity, or impairment of capacity to compensate for additional stress, or increase in susceptibility to the harmful effects of other environmental influences. Early effects, or low-dose effects, are not necessarily

(a) Ill Health → Good Health → Early Effects → Clinical Effects → Death

Increasing Dose → → → → → → → → → → → → → → → → → →

(b) No Effects → Early Effects → Clinical Effects → Death

Increasing Dose → → → → → → → → → → → → → → → → → →

Figure 2.1 *The simple relationship between dose and effects for (a) essential nutrient substances, (b) non-essential substances*

adverse effects. Thus, an NOAEL may be above the level causing an early (or low dose) effect, but below the level causing a clinical effect.

Clinical effects are generally adverse effects, *e.g.* occupational asthma (which may be caused by many compounds). Other examples of clinical effects are disturbances in the peripheral and central nervous system (*e.g.* tremor, polyneuropathy), and disturbances of liver and kidney function. The body has relatively few organs and the number of industrially used chemicals is about 70,000. Very few early effects or clinical effects are caused specifically by only one substance. Thus, any given effect may be caused by one or a number of substances, and there are many possibilities of interactions in the production of a toxic effect.

The branch of science relating the production of effects to the structure of a substance is called the study of quantitative structure–activity relationship (QSAR). QSAR has been used to predict toxic properties in the environment of many organic fat-soluble compounds, especially drugs.

Dose–effect relationships are generally S-shaped, as can be seen in Figure 2.2. In this example, there is an exponential rise above the baseline in urinary β_2-microglobulin, a parameter for assessing renal tubular damage, as a function of increasing dose of cadmium in urine above a threshold value. Beyond the exponential rise to a plateau level, enhancing the dose gives only a moderate rise in the effect.

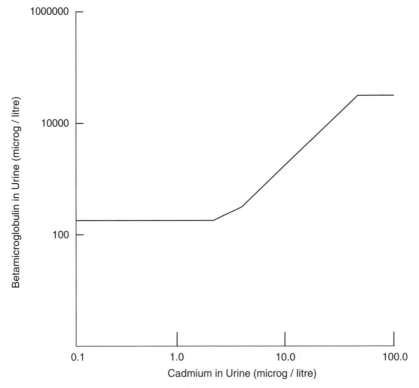

Figure 2.2 *Dose–effect curve for cadmium in urine and β_2-microglobulin excretion, an early measure of renal tubule damage. Note the exponential rise after no change occurs at low exposures to cadmium*

The steepness of the D–E relationship determines the value at which the NOAEL must be set. The NOAEL has to be set at a lower dose when the curve rises steeply than when the slope is shallow.

Early effects are those that can be observed before adverse change has occurred. Examples are compounds that smell offensive at an NEL. Mercaptans, for example, are used as a warning odour in natural gas. At higher levels mercaptans are toxic. Thus, an effect occurring at low concentrations where it is not adverse may be used for a beneficial purpose. Other examples of early effects are the excretion of certain enzymes and specific proteins into urine as a consequence of low exposure to certain compounds such as metals, solvents, pesticides, and drugs.

A D–E relationship exists for nearly all organ systems. But there are two exceptions, cancer and the immune system. These are dealt with in Chapters 9 and 12.

2.2.2 Biological Effect Monitoring

Biological effect monitoring (BEM) is defined as the continuous or repeated measurement and assessment of early biological effects in exposed humans, carried out in order to evaluate ambient exposure and health risk by comparison with appropriate reference values based on knowledge of the probable relationship between ambient exposure and biological effects. BEM is used in environmental and occupational toxicology, and sometimes in forensic toxicology to monitor possible problems at an early stage. Mostly, BEM is carried out using readily available specimens such as urine, blood, and faeces.

When a BEM parameter is exceeded, the signs and symptoms measured are not necessarily related to exposure to specific agents, nor to conditions of work or environment. BEM parameters are sensitive, but non-selective, and thus may serve as a safety net when the exposure is not known (*e.g.* exposure to many different solvents or a mixture of solvents).

2.2.3 Dose–Response Relationship

In Figure 2.2 if effect is replaced by response, a similar dose–response (D–R) curve is obtained. A response may be the reaction of an organism or part of an organism (such as a muscle) to a stimulus, and as such the relevant curve is similar to a D–E curve. An alternative definition of response is the proportion of a group of individuals that demonstrates a defined effect in a given time at a given dose rate. Such D–R relationships are quite different from the D–E relationships mentioned earlier. D–R relationships in this sense can be used to differentiate between groups. If a D–R relationship for one group has a steeper slope than for another, the first group will be more sensitive to the dose than the latter. Consequently, D–R relationships can be used to detect sensitive groups.

2.2.4 Acute and Chronic Effects

As mentioned earlier, dose is characterized by the integration of exposure concentration over time, and this produces a certain effect. This is a useful approach in assessing chronic effects, where the time scale is long, *e.g.* days or even years. It is also useful where there is a wide range of possible exposure concentrations. Chronic effects are thus dependent on two parameters, time and concentration. It follows, therefore, that an effect will be chronic if the exposure and hence the dose is persistent. Occasionally, there may be a time shift between dose and effect; this is called a latent period. Generally, not only is a minimum dose needed to cause an effect (a dose above the NOAEL), but also a minimum exposure time. This minimum time is called the minimum duration of exposure. In cases of chronic exposure, monitoring can take place by BEM or health surveillance, whenever a D–E relationship exists. It should be noted that no simple D–E relationship exists for immunological effects.

For acute effects, the time component of the dose is not important, as a high dose or exposure concentration, sometimes instantaneous, is responsible for these effects. For very brief exposure to very high concentration of a toxic substance, a concentration–effect (or D–E) relationship may exist, but this is not generally relevant. Acute effects are almost always the result of accidents (both in the home and at the workplace). Otherwise, they may result from attempted criminal poisoning (*e.g.* with arsenic or thallium) or self-poisoning (suicide).

Effects may be local or systemic. Local effects are mainly to the skin, eyes, and respiratory tract, and it is often difficult to find a dose or exposure concentration–effect relationship. With systemic effects, the whole body or a number of organs may be affected and usually a D–E relationship will exist.

2.3   TOXICITY TESTING AND HEALTH RISK

To test the toxicity of newly developed compounds, a complex testing system has been developed. The first step is to select a critical effect of the compound and an appropriate study protocol (see Chapter 4). The adequacy of the study protocol must then be ascertained before proceeding. Following the study or studies, NOAELs, using all available data, are determined. Knowledge of interspecies variability in toxicokinetics and toxicodynamics can lead to new insights and improve the derivation of human NOAELs. As we learn more, we can make better allowance for human variability and this may require an ongoing re-assessment of the available data (see Chapter 5).

In some countries, a three-step procedure has been developed in order to derive permissible exposure levels (PELs). After the first step, the scientific process above, pressure groups (mostly companies, trade unions, and the government) decide in a second step whether a suggested PEL based on the NOAEL should be modified, and finally in a third, administrative, step the agreed PEL is published. It should be emphasized that PEL values are chosen to minimize the risk of adverse effects, the health risk (see Chapter 5).

BIBLIOGRAPHY

G.D. Clayton and F.E. Clayton (eds), *Patty's Industrial Hygiene and Toxicology*, Wiley, New York, 1991 (10 vols.).

J.H. Duffus, S.S. Brown, N. de Fernicola, P. Grandjean, R.F. Herber, C.R. Morris and J.A. Sokal, Glossary for chemists of terms used in toxicology, *Pure Appl. Chem.*, 1993, **65**, 2003–2122.

WHO, *Environmental Health Criteria, International Programme on Chemical Safety*, World Health Organization, Geneva. (Currently there are more than 230 volumes. Red books deal with substances and yellow books with methods or techniques such as epidemiology, quality control, *etc.*) See http://www.who.int/ipcs/publications/ehc/en/.

C.D. Klaassen (ed), *Casarett and Doull's Toxicology*, 6th edn. McGraw-Hill, New York, 2001.

A.D. McNaught and A. Wilkinson, *Compendium of Chemical Terminology The Gold Book*, 2nd edn. Blackwell Science, Cambridge, 1997. Available on the IUPAC website at http://www.iupac.org/publications/books/author/mcnaught. html.

Chapter 3

Toxicokinetics

ANDREW G. RENWICK

3.1 INTRODUCTION

The sequence between exposure to a chemical and the generation of an adverse effect can be divided into two aspects (Figure 3.1); toxicokinetics or the delivery of the compound to its site of action and toxicodynamics or the response at the site of action. This subdivision is particularly useful in risk assessment (see later).

Toxicokinetics is the study of the movement of chemicals around the body. It includes absorption (transfer from the site of administration into the general circulation), distribution (*via* the general circulation into and out of the tissues), and elimination (from the general circulation by metabolism or excretion). The term toxicokinetics has useful connotations with respect to the high doses used in toxicity studies, but it may be misleading if interpreted as the 'movement of toxicants around the body' since, as all toxicologists agree, 'all things are toxic and it is only the dose which renders a compound toxic'. Toxicodynamics relates to the processes and changes that occur in the target tissue, such as metabolic bioactivation and covalent binding, and result in an adverse effect.

Useful toxicokinetic data may be derived using a radiolabelled dose of the chemical, *i.e.* in which a proton in the molecule is replaced by a tritium atom or a carbon or sulfur atom is replaced by the radioactive equivalent (^{14}C or ^{35}S). Such studies are invaluable in following the fate of the chemical skeleton as it is transferred from the site of administration into the blood, is distributed to the tissues, and is eliminated as carbon dioxide or more likely as metabolites in air, urine, or bile. The advantage of using the radiolabelled chemical is that measured radioactivity reflects both the chemical and its metabolites, and this allows quantitative balance studies to be performed, *e.g.* to determine how much of the dose is absorbed, which organs accumulate the compound, and the pathways of metabolism. However, such simple radioactive absorption, distribution, metabolism, and excretion (ADME) studies provide only a part of the total picture, because the lack of chemical specificity in the methods does not allow an assessment of how much of the chemical is absorbed intact and how much is distributed around the body as the parent chemical. A further advantage of radiolabelling studies is that radiochromatographic methods can be invaluable in the separation and identification of metabolites, which is an important aspect of the fate of the chemical in the body. Thus, initial ADME studies define the

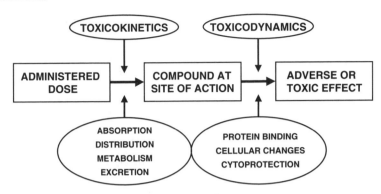

Figure 3.1 *The relationship between delivery of the administered dose to the target site and the generation of the adverse or toxic response*

overall fate of the chemical in the body and recognize the main chemical species (parent compound and/or metabolites) that are present in the circulation and in the urine and faeces following metabolism and excretion.

In recent years, it has been recognized that measurement of the circulating concentrations of the chemical and/or its metabolites can provide useful information on both the magnitude and the duration of exposure of targets for toxicity. The term toxicokinetics is sometimes restricted to studies based on measurements of blood or plasma concentrations, since these provide a vital link between the dosing of experimental animals and the amounts of the chemical in the general circulation (Figure 3.2). Such information can be of great value in the interpretation of species differences in toxic response, and in estimating the possible risk to humans of hazards identified in animal experiments. Toxicokinetic data are also useful in extrapolating across different routes of exposure or administration, as well as from single doses to chronic administration. Chemical-specific toxicokinetic measurements are essential if the results of *in vitro* toxicity tests are to be interpreted logically.

The ever increasing sensitivity of modern analytical techniques should allow the measurements of 'toxicokinetics' in humans receiving the compound at safe exposure levels. Thus, toxicokinetic differences between test animals and humans are open to direct measurement, and such data should increase confidence in the extrapolation process. In contrast, it is unethical intentionally to generate potentially adverse effects in humans and therefore data on inter-species differences in toxicodynamics are limited to observations following accidental poisonings, mild and reversible biomarkers of the potential adverse effect, and *in vitro* studies related to the mode of action of the chemical in animals.

The toxicokinetics of a chemical are determined by measuring the concentrations of the chemical in plasma (usually) or blood at various times following a single dose. The fundamental parameters that define the rates and extents of distribution and elimination are derived from data following an intravenous dose (Figure 3.2). The parameters relating to absorption from an extravascular site of administration, such as gut, lungs, *etc.*, are derived from comparisons of data following an extravascular dose with an intravenous dose. Additional useful information can be obtained from

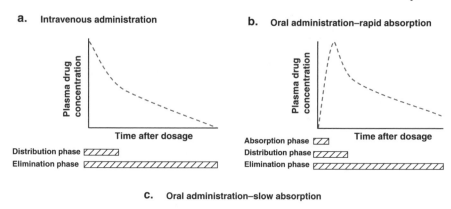

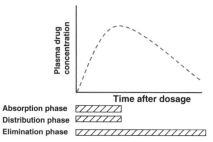

Figure 3.2 *The plasma concentration–time profiles of a chemical following intravenous and oral dosage*

measurements of the concentrations in plasma (or blood) over a period of 24 h in animals treated chronically with the chemical since the area under the plasma concentration–time curve often referred to as 'area under the curve' (AUC) is the best indication of exposure.

The interpretation of toxicokinetic data requires an understanding of both the biological basis of the processes of absorption, distribution, and elimination and the way that simple measurements of plasma or blood concentrations can be converted into useful quantitative kinetic parameters that describe these processes. The mathematics used to define and describe the movement of a chemical around the body can display various levels of sophistication and complexity. Compartmental analysis (Figure 3.3) allows the derivation of a mathematical equation which fits the data and allows the prediction of plasma concentrations at time points that were not measured directly and also outside the confines of the period of experimental observations. Physiologically based pharmacokinetic (PBPK) modelling (Figure 3.4) allows a greater interpretation of the data in biologically relevant terms but requires a sophisticated database to produce valid results. PBPK models (see below) can be used to bridge the gap between species, based on physiological differences and *in vitro* metabolic data, and extended to a biologically based dose–response model by the incorporation of *in vitro* response data.

This chapter will consider the biological basis of the processes of absorption, distribution, and elimination and describe the basic parameters, *e.g.* bioavailability, apparent volume of distribution, clearance, and half-life, which are most valuable

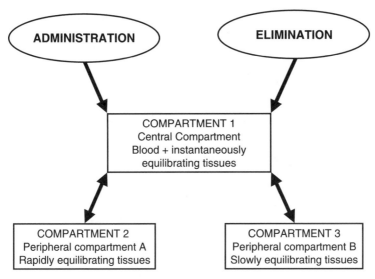

Figure 3.3 *Compartmental analysis. In the example shown, the body is considered to consist of two peripheral compartments that equilibrate with the central compartment. Strictly speaking the only property that links tissues that are part of the same "compartment" is the rate of transfer into and out of the tissue. The central compartment usually comprises blood and well-perfused tissues and equilibrates instantaneously. In the example shown, the compound is eliminated from the central compartment, for example by extraction by the liver or kidneys. The number of compartments necessary in the mathematical model fitted to the data depends on the number of exponential terms necessary to describe the plasma concentration–time curve. The mathematical model can be used to estimate the concentration in plasma or blood at any time after dosage*

because they are open to physiological interpretation. Absorption, distribution, and elimination can be considered in terms of the rate and the extent of the process.

3.2 ABSORPTION

The term absorption describes the process of the transfer of the parent chemical from the site of administration into the general circulation, and applies whenever the chemical is administered *via* an extravascular route (*i.e.* not by direct intravascular injection). The term 'absorption' is also used to describe the extent to which the radioactivity from a radiolabelled chemical is transferred from the site of administration into the excreta and/or expired air. However, many chemicals will be metabolized or transformed during their passage from the site of administration into the general circulation, so that little parent chemical may reach the general circulation, despite the fact that all of the radiolabel may leave the site of administration and be eliminated in the urine. This raises the possibility of confusion in discussing the 'extent of absorption' depending on whether the data refer to the parent chemical *per se*, or to radiolabel (which will include the chemical plus metabolites). This confusion is resolved by the proper use of the term bioavailability given below to describe the extent of absorption.

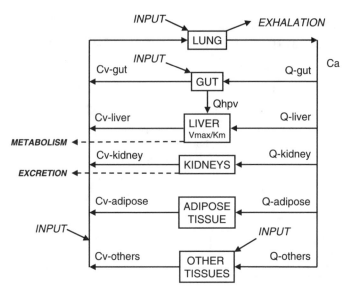

Figure 3.4 *Physiologically based pharmacokinetic model (PBPK). The PBPK model is derived from known rates of organ blood flow, the partition coefficient of the chemical between blood and the tissue, and the rates of the process of elimination, such as V_{max} and K_m for enzymes. PBPK modelling represents a powerful technique for estimating the dose delivered to specific tissues and can facilitate inter-species extrapolation by replacing animal blood flows and enzyme kinetic constants with human data. Removal across an organ equals $(C_a–C_v)$ times the organ blood flow (Q)*

3.2.1 Rate of Absorption

The rate of absorption may be of toxicological importance because it is a major determinant of the peak plasma concentration and, therefore, the likelihood of acute toxic effects. Transfer of chemicals from the gut lumen, lungs, or skin into the general circulation involves movement across cell membranes, and simple passive diffusion of the unionized molecule down a concentration gradient is the most important mechanism. Lipid-soluble molecules tend to cross cell membranes easily and are absorbed more rapidly than water-soluble ones. The gut wall and lungs provide a large and permeable surface area and allow rapid absorption; in contrast the skin is relatively impermeable and even highly lipid-soluble chemicals can enter only slowly. The lipid solubility and rate of absorption depend on the extent of ionization of the chemical. Compounds are most absorbed from regions of the gastrointestinal tract at which they are least ionized. Weak bases are not absorbed from the stomach, but are absorbed from the duodenum which has a higher luminal pH, whereas weak acids are absorbed from the stomach. The rate of absorption can be affected by the vehicle in which the compound is given, because rapid absorption requires the establishment of a molecular solution of the chemical in the gut lumen. Extremely lipid-soluble compounds, such as dioxins, may be only partially absorbed, because they do not form a molecular solution in the aqueous phase of the intestinal contents. There are few membrane barriers to absorption following subcutaneous or intramuscular dosage, and the absorption rate may be limited by the water solubility of the injected materials; slow absorption occurs with lipid-soluble compounds injected in an oily vehicle (which contrasts with the rapid absorption possible if such a dose is given *via* the gastrointestinal tract). Irrespective of the route of administration, the rate of absorption is determined from the early time points after dosing (Figure 3.5), and is usually described by an absorption rate constant or absorption half-life.

3.2.2 Extent of Absorption

The extent of absorption is important in determining the total body exposure or internal dose, and therefore is an important variable during chronic toxicity studies and/or chronic human exposure. The extent of absorption depends on the extent to which the chemical is transferred from the site of administration, such as the gut lumen, into the local tissue, and the extent to which it is metabolized or broken down by local tissues prior to reaching the general circulation. An additional variable affecting the extent of absorption is the rate of removal from the site of administration by other processes compared with the rate of absorption (see below).

Chemicals given *via* the gastrointestinal tract may be subject to a wide range of pH values and metabolizing enzymes in the gut lumen, gut wall, and liver before they reach the general circulation. The initial loss of chemical prior to it ever entering the blood is termed first-pass metabolism or pre-systemic metabolism; it may in some cases remove up to 100% of the administered dose so that none of the parent chemical reaches the general circulation. The intestinal lumen contains a range of hydrolytic enzymes involved in the digestion of nutrients. The gut wall can perform similar hydrolytic reactions and contains enzymes involved in oxidation, such as

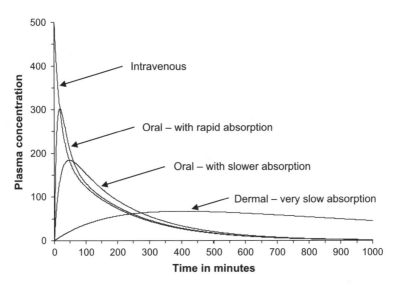

Figure 3.5 *The influence of the rate of absorption of a chemical on the plasma concentration–time curve. A relatively flat low profile is obtained when the rate of absorption is less than the rate of elimination, and this pattern is normally seen with transdermal absorption*

cytochrome P450 3A4, and conjugation of foreign chemicals. Enterocytes contain P-glycoprotein (PGP) which transports a range of absorbed complex foreign chemicals from the cytosol back into the gut lumen, which can increase the likelihood of first-pass metabolism in the gut lumen or gut wall, or incomplete absorption from the gut lumen. The portal circulation drains into the hepatic portal vein which carries compounds absorbed across the gut wall to the liver, which is the main site of foreign compound metabolism, and is responsible for most first-pass metabolism. The other main reason for incomplete absorption of the parent chemical occurs when the rate of absorption is so slow that the chemical is lost from the body before absorption is complete. Examples of this include incomplete absorption of very water-soluble chemicals from the gut and their loss in the faeces, or incomplete dermal absorption, before the chemical is removed from the skin by washing.

Irrespective of the reason that is responsible for the incomplete absorption of the chemical as the parent compound, it is essential that there is a parameter which defines the extent of transfer of the intact chemical from the site of administration into the general circulation. This parameter is the bioavailability, which is simply the fraction of the dose administered that reaches the general circulation as the parent compound. (The term bioavailability is perhaps the most misused of all kinetic parameters and is sometimes used incorrectly in a general sense as the amount available specifically to the site of toxicity.)

The fraction absorbed or bioavailability (F) is determined by comparison with intravenous (i.v.) dosing (where $F = 1$ by definition). The bioavailability can be determined from the area under the plasma concentration–time curve (AUC) of the parent compound (see Figure 3.6), or the percentage dose excreted in urine as the parent compound, *i.e.* for an oral dose:

$$F = \frac{\text{AUC oral}}{\text{AUC}} \times \frac{\text{dose i.v.}}{\text{dose oral}}$$

$$F = \frac{\%\ \text{in urine as parent compound after oral dosing}}{\%\ \text{in urine as parent compound after intravenous dosing}}$$

3.3 DISTRIBUTION

Distribution is the reversible transfer of the chemical between the general circulation and the tissues. Irreversible processes such as excretion, metabolism, or covalent binding are part of elimination and do not contribute to distribution parameters. The important distribution parameters relate to the rate and extent of distribution.

3.3.1 Rate of Distribution

The rate at which a chemical may enter or leave a tissue may be limited by two factors:

   (i)   the ability of the compound to cross cell membranes and
  (ii)   the blood flow to the tissues in which the chemical accumulates.

The rate of distribution of highly water-soluble compounds may be slow due to their slow transfer from plasma into body tissues such as liver and muscle; water-soluble

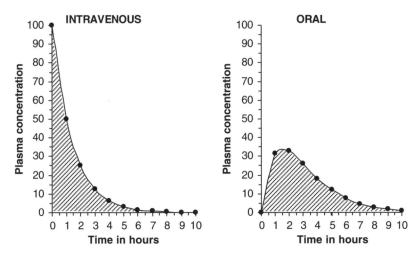

Figure 3.6 *The relationship between the area under the plasma concentration–time curve (AUC) and bioavailability. By definition, the bioavailability (fraction absorbed as the parent compound) is 1 for an intravenous dose. For other routes the bioavailability is given by the AUC for that route divided by the AUC after an intravenous dose (normalized to the same dose in mg kg^{-1})*

compounds do not accumulate in adipose tissue. In contrast, very lipid-soluble chemicals may rapidly cross cell membranes but the rate of distribution may be slow because they accumulate in adipose tissue, and their overall distribution rate may be limited by blood flow to adipose tissue.

Highly lipid-soluble chemicals may show two distribution phases: a rapid initial equilibration between blood and well perfused tissues, and a slower equilibration between blood and poorly perfused tissues (Figure 3.7). The rate of distribution is indicated by the distribution rate constant(s), which is(are) determined from the decrease in plasma concentrations in early time points after an intravenous dose. The rate constants refer to a mean rate of removal from the circulation and may not correlate with uptake into a specific tissue (for which the PBPK approach is more appropriate; see Figure 3.4). Once an equilibrium has been reached between the general circulation and a tissue, any process which lowers the blood (plasma) concentration will cause a parallel decrease in the tissue concentration (see Figure 3.8). Thus the elimination half-life measured from plasma or blood samples is also the elimination half-life from tissues.

3.3.2 Extent of Distribution

The extent of tissue distribution of a chemical depends on the relative affinity of the blood or plasma compared with the tissues. Highly water-soluble compounds that are unable to cross cell membranes readily (*e.g.* tubocurarine) are largely restricted to extracellular fluid (about 13 L per 70 kg body weight). Water-soluble compounds capable of crossing cell membranes (*e.g.* caffeine, ethanol) are largely present in total body water (about 41 L per 70 kg body weight). When one or more body tissues

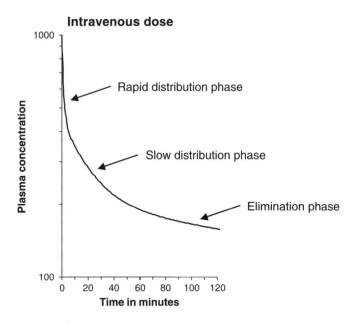

Figure 3.7 *The plasma concentration–time curve for a chemical that requires a three-compartment model (see Figure 3.3)*

has an affinity for the chemical, such as reversible tissue binding, then the blood (plasma) concentration will be lower than if the compound was evenly distributed through body water. Lipid-soluble compounds frequently show extensive uptake into tissues and may be present in the lipids of cell membranes, adipocytes, central nervous system (CNS), *etc.*; the partitioning between circulating lipoproteins and tissue constituents is complex and may result in extremely low plasma concentrations. A factor which may further complicate the plasma/tissue partitioning is that some chemicals bind reversibly to circulating proteins such as albumin (for acid molecules) and α_1-acid glycoprotein (for basic molecules).

The internal environment of the brain is controlled by the endothelial cells of the blood capillaries to the brain which have tight junctions between adjacent cells, fewer and smaller pores, little endocytosis, and the presence of transporters such as PGP which can extrude chemicals that diffuse across the blood brain barrier. In consequence, water-soluble molecules cannot 'leak' into the brain between endothelial cells (as could happen, for example, in muscle capillaries) and are excluded from the brain. The endothelial membranes have specific transporters for the uptake of essential water-soluble nutrients and some ions and also for the exclusion of organic acids. This so-called blood–brain barrier serves to exclude most water-soluble compounds, so that CNS toxicity may be limited. In contrast, lipid-soluble chemicals readily cross the blood–brain barrier and the CNS is a common site for toxicity (*e.g.* organic solvents). Similar permeability barriers are present in the choroid plexus, retina, and testes.

The extent and pattern of tissue distribution can be investigated by direct measurement of tissue concentrations in animals. Tissue concentrations cannot be

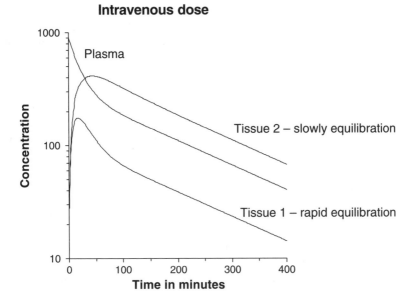

Figure 3.8 *Tissue distribution of a chemical after an intravenous bolus dose. Tissue 1 shows a greater rate of uptake and reaches equilibrium before tissue 2. Tissue 1 shows a lower affinity than tissue 2, so that the concentrations are lower. The concentrations measured in toxicokinetic studies are usually the total concentration (free + bound to proteins or present in cellular lipids) and tissue 2 may show greater tissue binding than tissue 1. The concentrations in all tissues decrease in parallel once all tissues have reached equilibrium with plasma*

measured in human studies and, therefore, the extent of distribution in humans has to be determined based solely on the concentrations remaining in plasma or blood after distribution is complete. The parameter used to reflect the extent of distribution is the apparent volume of distribution (V), which relates the total amount of the chemical in the body (Ab) to the circulating concentration (C) at any time after distribution is complete:

$$V = \frac{Ab}{C}$$

V may be regarded as the volume of plasma in which the body load appears to have been dissolved and simply represents a dilution factor. The volumes of distribution of tubocurarine and caffeine are about 13 and 41 L per 70 kg because of their restricted distribution (see above). However, when a chemical shows a more extensive reversible uptake into one or more tissues the plasma concentration will be lowered and the value of V will increase. For highly lipid-soluble chemicals, such as organochlorine pesticides, which accumulate in adipose tissue, the plasma concentration may be so low that the value of V may be many litres for each kilogram of body weight. This is not a real volume of plasma and therefore V is called the apparent volume of distribution. It is an important parameter because extensive reversible distribution into tissues, which will give a high value of V, is associated with a low elimination rate and a long half-life (see below). It must be emphasized that the apparent volume of distribution simply reflects the extent to which the chemical has moved out of the site of measurement (the general circulation) into tissues, and it does not reflect uptake into any specific tissue(s).

Information on the uptake into specific tissues requires sampling of that specific tissue, although PBPK modelling can provide useful estimates of tissue concentrates based on *in vitro* partition coefficients and organ blood flows. Once equilibrium has been reached for a tissue, the tissue/plasma ratio will remain constant, so that as the chemical is eliminated from the plasma, the chemical will leave the tissue, maintaining the same ratio (Figure 3.8).

3.4 ELIMINATION

The parameter most commonly used to describe the rate of elimination of a chemical is the half-life (Figure 3.9). Most toxicokinetic processes are first-order reactions, *i.e.* the rate at which the process occurs is proportional to the amount of chemical present. High rates (expressed as mass/time) occur at high concentrations and the rate decreases as the concentration decreases; in consequence the decrease is an exponential curve. The usual way to analyse exponential changes is to use logarithmically transformed data which converts an exponential into a straight line. The slope of the line is the rate constant (k) for the process and the half-life for the

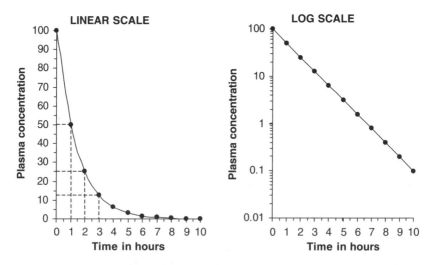

Figure 3.9 *The half-life of a chemical and its determination from plasma data. In the example in this figure the half-life is 1 h. Logarithmic conversion allows the concentration data to be fitted by linear regression analysis; the half-life is calculated as 0.693/slope. Plasma kinetic data are usually fitted by a non-linear least-squares method and there are various programmes available, such as Win-Nonlin*

process is calculated as $0.693/k$. Rate constants and half-lives can be determined for absorption, distribution, and elimination processes.

There are two important biological variables that determine the rate at which a chemical can be eliminated from the body: (i) the functional capacity/ability of the organs of elimination to remove the chemical from the body (the clearance) and (ii) the extent of distribution of the chemical from the general circulation into tissues.

The clearance of a chemical is determined by the ability of the organs of elimination (*e.g.* the liver, kidney, or lungs) to extract the chemical from the plasma or blood and permanently remove it by metabolism or excretion. (Note that this is different from distribution in which the chemical is free to leave the tissue and re-enter the blood when the concentration in the general circulation decreases.)

The mechanisms of elimination depend on the chemical characteristics of the compound:

- volatile chemicals are exhaled,
- water-soluble chemicals are eliminated in the urine and/or bile and
- lipid-soluble chemicals are eliminated by metabolism to more water-soluble molecules, which are then eliminated in the urine and/or bile.

Foreign compound metabolism is an enormous subject and involves a wide range of enzyme systems. Foreign chemicals (xenobiotics) may be metabolized by the enzymes of normal intermediary metabolism, *e.g.* esterases will hydrolyse ester groups. Alternatively, chemicals may be metabolized by enzymes such as cytochrome P450, a primary function of which is xenobiotic metabolism. Species differences in metabolism can be a major source of differences in toxic response. The usual consequence of metabolism is the formation of an inactive excretory product so that species with low metabolizing ability will be likely to show greater toxicity. However, for many compounds, metabolism is a critical step in the generation of a toxic or reactive chemical entity (bioactivation), and for such compounds high rates of metabolism will be linked with greater toxicity. If a chemical undergoes metabolic activation then toxicokinetic studies should measure both the parent chemical and the active metabolite. If the metabolite is so reactive that it does not leave the tissue in which it is produced (*e.g.* alkylating metabolites of chemical carcinogens), then toxicokinetic studies should define the delivery of the parent chemical to the tissues, and the process of local activation should be regarded as part of tissue sensitivity (toxicodynamics) because it is not strictly speaking part of toxicokinetics, *i.e.* the movement of the chemical and/or metabolites around the body.

The best measure of the ability of the organs of elimination to remove the compound from the body is the clearance (*CL*):

$$CL = \frac{\text{rate of elimination}}{\text{plasma concentration}}$$

Because the rate of elimination is proportional to the concentration (see Figure 3.9), clearance is a constant for first-order processes and is independent of dose. It can be regarded as the volume of plasma (or blood) cleared of compound within a unit of time (*e.g.* mL min^{-1}).

Renal clearance depends on the extent of protein binding, tubular secretion and passive reabsorption in the renal tubule; it can be measured directly from the concentrations present in plasma and urine:

$$CL = \frac{\text{rate of elimination in urine}}{\text{plasma concentration}}$$

The total clearance or plasma clearance (which is the sum of all elimination processes, *i.e.* renal + metabolic, *etc.*) is possibly the most important toxicokinetic parameter. It is measured from the total amount of compound available for removal (*i.e.* an intravenous dose) and the total area under the plasma concentration–time curve (AUC) extrapolated to infinity.

$$CL = \frac{\text{Dose i.v.}}{\text{AUC i.v.}}$$

Plasma clearance reflects the overall ability of the body to remove permanently the chemical from the plasma. Plasma clearance is the parameter that is altered by factors such as enzyme induction, liver disease, kidney disease, inter-individual or inter-species differences in hepatic enzymes or in some cases organ blood flow. Once the chemical is in the general circulation, the same volume of plasma will be cleared of chemical per minute (*i.e.* the clearance value) applies irrespective of the route of delivery of chemical into the circulation. However, the bioavailability (*F*) will determine the proportion of the dose reaching the general circulation. Therefore, bioavailability has to be taken into account if clearance is calculated from data from a non-intravenous route (*e.g.* oral):

$$CL = \frac{\text{dose oral} \times F}{\text{AUC oral}}$$

Measurement of dose/AUC for an oral dose determines *CL/F*, which contains two potentially independent variables – the amount of chemical delivered to the blood from the site of administration and the clearance of chemical present in the blood.

The overall rate of elimination, as indicated by the terminal half-life ($t^{\frac{1}{2}}$), is dependent on two physiologically related and independent variables:

$$t^{\frac{1}{2}} = \frac{0.693V}{CL}$$

where *CL* is the ability to extract and remove irreversibly the compound from the general circulation, and *V* the extent to which the compound has left the general circulation in a reversible equilibrium with tissues.

Therefore, a chemical may have a long half-life because the organs of elimination have a low ability to remove it from plasma and/or because it is extensively distributed to body tissues and only a small proportion of the total body burden remains in the plasma and is available for elimination. Chemicals that are extremely lipid-soluble and are sequestered in adipose tissue are eliminated slowly. Lipid-soluble organochlorine compounds, which are not substrates for P450 oxidation, due to the blocking of possible sites of oxidation by chloro-substituents, are eliminated extremely slowly: for example, the half-life of 2,3,7,8-tetrachlorodibenzodioxin (TCDD) is about 8 years in humans.

3.5   CHRONIC ADMINISTRATION

Most toxicity studies involve continuous, or chronic, administration of the chemical either *via* incorporation into the diet or by daily gavage doses. The kinetic concepts and parameters of a single dose (as discussed above) apply to chronic administration, but the exposure has to allow for the fact that not all of the previous dose(s) may have been eliminated when the subsequent dose is given. Therefore, there may be an increase in plasma concentration (and body load) until an equilibrium is reached in which the rate of elimination balances out the rate of input (Figure 3.10).

The description of clearance given above can be rewritten as

rate of elimination (μg d^{-1}) = CL (mL d^{-1}) $\times$ plasma concentration (μg mL^{-1})

i.e. the rate of elimination (in mass per unit time) is proportional to plasma concentration. When doses of a chemical with a long half-life are given every day the low plasma concentrations from the first dose will give a low rate of elimination, such that not all of the chemical would be eliminated before the next dose is given. In consequence, the next dose will give higher concentrations (due to carryover from the first dose) and, therefore, the rate of elimination will be higher. In consequence, a greater proportion of the daily dose will be eliminated on the second day. The plasma concentrations and rates of elimination will continue to increase each day until the plasma concentrations are such that the daily dose is eliminated each day (Figure 3.10), *i.e.* an

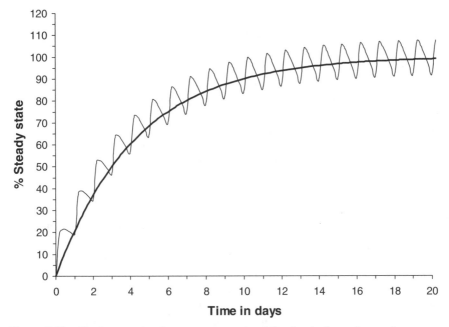

Figure 3.10   *The increase in plasma concentration following both continuous intravenous administration (thicker line) and once daily oral administration (thinner line). The chemical has a half-life of 3 days and it takes 4–5 half-lives, i.e. 12–15 days, to approach steady state (strictly speaking the true steady state is never achieved because it is an exponential increase)*

equilibrium or steady state is reached. At equilibrium the balance between input and output can be written as

$$\text{input} = \frac{\text{dose} \times F}{\text{dose interval}} = CLC_{ss} = \text{output}$$

C_{ss} is the average steady-state plasma concentration which can be calculated by

$$C_{ss} = \frac{\text{dose} \times F}{CL \times \text{dose interval}}$$

It is important to realize that clearance (CL) is the same value throughout the build-up to steady state. Once steady state has been reached,

$$CL = \frac{\text{dose} \times F}{\text{AUC for a dose interval}}$$

The equations above assume that CL is not altered by repeated exposure; the assumption is not correct if the chemical induces or inhibits its own elimination because clearance would be increased or decreased, respectively, after the period of chronic intake. The possibility that metabolism or excretion is saturated at the higher plasma concentrations during chronic intake is discussed below.

A further important toxicokinetic variable to be considered in the design and interpretation of chronic studies is the time taken to reach steady state. The extent of toxicity is usually proportional to the dose or the body load and the body load is given by the plasma concentration (at any time) multiplied by V. Because the plasma and therefore tissue concentrations increase during chronic intake until an equilibrium is reached (Figure 3.10), the amount in the body (Ab) will also increase to reach a steady state. The time taken to reach steady state is 4–5 times the elimination half-life and, therefore, the true duration of steady-state exposure in a toxicity study is the study duration minus 4–5 half-lives of the chemical. This is particularly important for chemicals that have a very long half-life; for example in rodents the steady-state body load of TCDD, which has a half-life in rats of about 1 month, will not be reached until after about 4–5 months of continuous treatment.

3.6   SATURATION KINETICS

All the parameters described above relate to first-order processes and therefore are independent of dose at low doses. However, at high doses and/or during chronic studies it is possible to overload or saturate compound–protein interactions. Under such circumstances any increase in the concentration of the compound cannot give a proportional (first-order) increase in the rate of the process. When a process is saturated the rate is at the maximum possible and is essentially independent of concentration.

In simple mathematical terms this means that the reaction changes from first to zero order. This is best described by Michaelis–Menten kinetics, *i.e.*

$$\text{rate} = \frac{V_{max} C}{K_m + C}$$

At low concentrations, C is less than the Michaelis constant K_m and, therefore, $(K_m + C)$ approximates to K_m. At such low concentrations the rate equals $(V_{max}/K_m)\,C$ and since V_{max} and K_m are constants the rate is proportional to the concentration (first-order).

At high saturating concentrations, C is greater than K_m and, therefore, $(K_m + C)$ approximates to C. At such high concentration the rate equals $(V_{max}/C \times C, i.e.\ V_{max})$ and therefore is a fixed maximum rate (zero-order).

The consequences of this for a plasma concentration–time curve are shown in Figure 3.11. Michaelis–Menten kinetics can be included in PBPK models based on *in vitro* enzyme kinetic measurements. It is important to note that the terminal elimination half-life is always determined at low concentrations and therefore is a first-order parameter, which does not show saturation kinetics. The best parameter to reflect saturation kinetics is the CL, which is based on the total AUC and includes the slower zero-order elimination phase.

A classic example of this type of data is given in the studies of Dietz *et al.* (1982) on the solvent dioxane, in which saturation of metabolism resulted in a change of clearance in rats from 13.3 mL min⁻¹ at 3 mg kg⁻¹ to 1.0 mL min⁻¹ at 1000 mg kg⁻¹. Renal tubular secretion can also be saturated, as demonstrated for cyclohexylamine in rats by Roberts and Renwick (1989), which resulted in a non-linear accumulation of the compound in the testes of rats (but not mice), which correlated with the dose–response for the testicular toxicity.

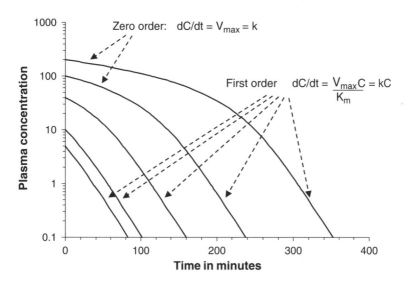

Figure 3.11 *The influence of saturation of elimination on the shape of the plasma concentration–time curve for a chemical. At low doses and low initial concentrations the decrease shows a simple exponential decrease (the example chosen represents the simplest case, i.e. a one-compartment model). At high doses high plasma levels cause saturation of the elimination process so that at very high initial concentrations the decrease is at V_{max} and is essentially independent of concentration. As the concentration deceases eventually the enzyme will no longer be saturated and the elimination will revert to a simple exponential decrease*

Important possible consequences of saturation of metabolism or excretion are that the chemical will accumulate to higher concentrations and that some normally minor alternative routes of elimination may become involved in the elimination of the chemical. Toxic effects seen at saturating doses may be of little or no relevance to lower non-saturating doses if the alternative route is a different pathway of metabolism which results in bioactivation of the chemical.

3.7 TOXICOKINETICS AND RISK ASSESSMENT

As described in the introduction, toxicokinetics is one of two aspects that link exposure to a chemical to the development of toxicity. Unlike toxicodynamics, kinetic processes can be studied ethically in humans, and this allows the potential for chemical-specific data on this aspect of inter-species differences and human variability to be taken into account is the establishment of safe human exposures, such as the acceptable daily intake (ADI). Traditionally, 10-fold uncertainty factors have been used to allow for possible species differences and human variability, and the no-observed-adverse-effect level of intake in an animal study (in mg kg^{-1} body weight) would be divided by 100 to calculate the ADI. Subdivision of each 10-fold factor into toxicokinetic and toxicodynamic aspects allows relevant chemical-specific data on toxicokinetics to replace the relevant default (see Figure 3.12). Replacement of one of the factors for kinetics or dynamics in Figure 3.12 requires an extensive database; this subdivision has been used in recent evaluations of the sweetener cyclamate, dioxins, and methylmercury.

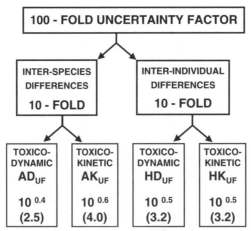

Chemical specific data can be used to replace a default uncertainty factor (UF)

A – animal to human; H – human variability; D – toxicodynamics; K - toxicokinetics

Figure 3.12 *Subdivision of the 10-fold uncertainty factors to allow for species differences and human variability in toxicokinetics or toxicodynamics (based on IPCS, 1994). The total composite factor would be the product of any chemical-specific values and the remaining default uncertainty factors that had not been replaced: for example, if the inter-species toxicokinetic UF were replaced by chemical-specific value of 6.0, then the total factor would be 2.5 × 6 × 3.2 × 3.2 = 150*

3.8 CONCLUSIONS

A common criticism of animal experiments is that they are 'not relevant' to humans. This is not true as a generalization, but there are instances where animal data are not relevant to human risk assessment owing to the nature of the target organ for toxicity or to the fate of the chemical in the body. Even when the target organ and pathways or metabolism are similar, inter-species differences in the fate of a chemical in the body complicate the interpretation of animal data in relation to human risk assessment. Information on the toxicokinetics of a chemical can provide an understanding of the extent of absorption and distribution and the pathways and rates of elimination. Such data provide a vital link between animal experiments and human safety.

BIBLIOGRAPHY

J.D. de Bethizy and J.R. Hayes, Metabolism: a determinant of toxicity, in *Principles and Methods of Toxicology*, 3rd edn, A.W. Hayes (ed), Raven Press, New York, 1994, pp. 59–100.

F.K. Dietz, W.T. Stott and J.C. Ramsey, Non-linear pharmacokinetics and their impact on toxicology: illustrated by dioxane, *Drug Metab. Rev.*, 1982, **13**, 963–981.

E.R. Garrett, Toxicokinetics, in *Toxic Substances and Human Risk*, R.G. Tardiff and J.V. Rodricks (eds), Plenum, London, 1987, pp.153–237.

J.D. Harvell and H.I. Maibach, Percutaneous absorption and inflammation in aged skin: a review, *J. Am. Acad. Dermatol.*, 1994, **31**, 1015–1021.

IPCS, *Assessing Human Health Risks of Chemicals: Derivation of Guidance Values for Health-based Exposure Limits*. Environmental Health Criteria, 170, World Health Organisation, International Programme on Chemical Safety, Geneva, 1994, 73 pp.

K. Krishnan and M.E. Andersen, Physiologically based pharmacokinetic modelling in toxicology, in *Principles and Methods of Toxicology*, 3rd edn, A.W. Hayes (ed), Raven Press, New York, 1994, pp.149–188.

J.H. Lin, Species similarities and differences in pharmacokinetics, *Drug Metab. Dispos.*, 1995, **23**, 1008–1021.

E.J. O'Flaherty, Differences in metabolism at different dose levels, in *Toxicological Risk Assessment*, Vol 1, D.B. Clayson, D. Krewski and I. Munro (eds), CRC Press, Boca Raton, FL, 1985, pp.53–91.

A.G. Renwick, Data-derived safety factors for the evaluation of food additives and environmental contaminants, *Food Addit. Contam.*, 1993, **10**, 275–305.

A.G. Renwick, Toxicokinetics, in *General and Applied Toxicology*, B. Ballantyne, T. Marrs and P. Turner (eds), 2 Vols, Macmillan, London, 1993, pp.121–151.

A.G. Renwick, Toxicokinetics – pharmacokinetics in toxicology, in *Principles and Methods of Toxicology* 3rd edn, A.W. Hayes (ed), Raven Press, New York, 1994, pp.101–147.

A. Roberts and A.G. Renwick, The pharmacokinetics and tissue concentrations of cyclohexylamine in rats and mice, *Toxicol. Appl. Pharmacol.*, 1989, **98**, 230–242.

J.A. Timbrell, Biotransformation of xenobiotics, in *General and Applied Toxicology*, B. Ballantyne, T. Marrs and P. Turner (eds), Macmillan, London, 1993, pp.89–119.

WHO, *Principles of Toxicokinetic Studies*, Environmental Health Criteria 57, World Health Organisation, Geneva, 1986.

J.R. Withey, Pharmacokinetic differences between species, in *Toxicological Risk Assessment*, Vol 1, D.B. Clayson, D. Krewski and I. Munro (eds), CRC Press, Boca Raton, FL, 1985, pp.41–52.

L. Zhang, C.M. Brett and K.M. Giacomini, Role of organic cation transporters in drug absorption and elimination, *Ann. Rev. Pharmacol. Toxicol.*, 1998, **38**, 431–460.

Chapter 4

Data Interpretation

JOHN S.L. FOWLER

4.1 INTRODUCTION

'Rubbish in – rubbish out': data interpretation in toxicology is usually a complicated issue. If you are presented with a data package, first ask 'where are the data from?' and ensure that the data come from a reliable source (see below). This is only the start. If you do not have any data, you will have to search for it yourself or commission someone else to do so. Irrespective of the case, you will have to assess the data you get and this chapter will help you in doing so.

4.1.1 The Data Package

The 'Data Package' from which you start may be in hard copy or in electronic format. Most importantly, it may be complete or incomplete and it will probably be your job to determine this.

Unless all the data have been generated recently, in response to a clearly defined question, and as part of properly managed, well-integrated projects, shortfalls will probably exist. It is important to identify as soon as possible any ambiguity or omission in the Data Package since this may lead to subsequent queries that will cost time, money or both.

4.1.2 Where Do the Data Come From?

Toxicology is an applied science and the subject covers a wide range of disciplines. As a result, useful data can arise from multiple sources. Bearing in mind the heterogeneous nature of the data, you should ask the following questions in order to assess their trustworthiness:

- Were the data generated in-house?
- Were they published in a peer-reviewed journal?
- Do you know the originating laboratory and its reputation?
- Can you check the originating laboratory?
- Do the scientists who produced the data possess adequate CVs?
- When were the data generated? Are they based on work that satisfies current standards? If not, was the standard at the time satisfactory?

- Have the data withstood independent audit?
- Do clear audit trails exist?
- Are all the data validated and compatible?
- Are they of similar (suitable) quality and reliability and can they legitimately be combined, in other words – are they of similar type?

4.1.3   Use of Data to Assess Chemical Hazard

Potential for harm exists in every chemical; however, no harm needs to happen. Even if you have so few data that you are uncertain about the toxicity, it is easy to minimise the risks. Clearly if there is no exposure, there can be no risk; therefore, at its simplest, leave the cap on the bottle! This is the principle of containment; it is always used in the early stages of evaluation of a novel chemical. You must assume the worst if you have no evidence to the contrary. If you have data, you can use it to define the exact potential for harm. In doing this, you will again have to answer some questions:

- Can the chemical quickly kill someone?
- Can the chemical cause genetic damage that may harm the children of those who are exposed to it?
- Can the chemical shorten life, make someone blind, damage the liver or kidneys, *etc*?
- Can the chemical induce or lead to the development of multiple allergies?
- How can harmful effects occur? – by swallowing it? by inhaling it? or by touching it?

Having assessed the data and decided that a chemical is hazardous, in order to reduce the chance of chemical damage, the following decisions need to be possibly taken:

- *If the chemical is not needed*
  - Do not make the chemical
  - Do not make any more chemical
  - Keep what you have in sealed containers
- *If the chemical is needed*
  - Determine how best to
    - Distribute it
    - Store it
    - Use it safely; for example can you generate it at the point of consumption, thereby minimising need for transport and stocks?
- *If the chemical is no longer needed*
  - Safely destroy remaining stocks:
    - By incineration
    - By conversion (*e.g.* use as a substrate)
    - By sequestration (in suitable safe sites, if such exist)
- *If the chemical has been spilled*
  - As far as possible, remediate the site that has become contaminated by the unwanted chemical; if necessary, change the use of the site to minimise risk to users.

4.2 RISK ANALYSIS AND RISK MANAGEMENT

An important use of toxicological data is in the process of risk analysis (see Chapter 5). The need to undertake risk analysis is central to the understanding and intelligent management of toxicological hazard and requires interpretation of a wide range of data.

4.2.1 Contents of the 'Data Package' for Risk Analysis

Before risk analysis process can begin, data must be available that pertain to:

- The intrinsic toxicological hazard possessed by the chemical
- The likely duration of contact with potential targets by the chemical
- The amount of chemical achieving interaction with the target
- The inherent susceptibility of the target to the chemical

4.2.2 Reasons Why the 'Data Package' May be Inadequate

In several predictable circumstances, 'the 'Data Package' may be (or may become) inadequate for proper risk analysis and risk management'. Two major scenarios of Data Package shortfall can be envisaged: one scientific and the other chronological/legislative.

4.2.2.1 First Scenario: Data are Inadequate for Scientific Reasons. Common situations where this might arise are as follows:

(i) When dealing with new substances.
(ii) When dealing with new applications of existing substances (leading to greater exposure).
(iii) When there are newly discovered metabolites (products of biotransformation of the known molecule).
(iv) When significant levels of impurities have been detected (*e.g.* arising from alternative pathways during synthesis of the known molecule).
(v) When there are significant amounts of degradation products (arising from aging and instability of the parent molecule).

4.2.2.2 Second Scenario: Data are Inadequate for Chronological/Legislative Reasons. Situations where this might arise are as follows:

- Danger is immediate and insufficient time is available to obtain appropriate data.
- The test laboratory is unable to supply the data; for example, the researching laboratory's project license may be too restrictive.
- The needed-test is banned
  - Certain substances (*e.g.* cosmetics) may not be tested in order to minimise animal suffering.
  - Certain test procedures (*e.g.* determination of LD_{50}) are banned.
- The needed-test is not possible because specified cell systems or animal species are not available.

4.2.3 Decision Taking Without All the Required Data

It is often the case that decisions to ensure the safe use of chemicals must be taken in the absence of some of the data that the toxicologist would like to have. Under such circumstances, it is essential to advise the responsible risk manager that the data package is not complete. The following considerations may be of assistance under such situations.

The physicochemical properties of the molecule may enable some estimates to be made of the likely exposure to sensitive systems. In this case, questions to ask include:

- Is the usual form of the molecule solid, liquid or gas?
- Can it be inhaled?
- Does it dissolve in water?
- What is the pH and osmolality of aqueous solutions?
- What is the fat solubility and what is octanol–water partition coefficient?
- Will the material be likely to interfere with sewage organisms, leach into drinking water sources or bind to topsoils or clay?

The chemical structural similarities with known molecules may also enable some useful estimates to be made, including, by use of computer-modelling, inferences about toxicodynamic and toxicokinetic properties. Relevant questions to be asked might be:

- Is the chemical structure similar to the one that may (by analogy) be activated or degraded to known products?
- Does the chemical belong to a class of chemicals with known toxic properties or does it possess reactive groups known to interact with biological systems?

4.3 DATA RETRIEVAL

Data are generated by surveys, testing, studies and experiments, and are accumulated in databases or databanks from which they must be retrieved. Many of these databases or databanks are accessible to toxicologists: many originate in the USA. Some of these will be discussed below.

It is essential that parameters of searching that admit only properly authenticated material should be used; an internet example from the medical field is as follows:

Organising medical networked information (OMNI, http://omni.ac.uk/) is a gateway to evaluated quality internet resources in health and medicine, aimed at students, researchers, academics and practitioners in the health and medical sciences. OMNI is created by a core team of information specialists and subject experts based at the University of Nottingham Greenfield Medical Library, in with key organisations throughout the UK and further afield. OMNI is one of the gateways within the BIOME service (http://biome.ac.uk/). BIOME is part of the Resource Discovery Network (RDN, http://www.rdn.ac.uk/), and is funded by the Joint Information Systems Committee (JISC).

4.3.1 The Search for Information

Typical headings of information that correspond to the data profile structure utilised by the International Register of Potentially Toxic Chemicals (IRPTC) are as follows: identifiers, properties, classification; production trade, production process, use; pathways in the environment, concentrations, environmental fate studies, effects on organisms in the environment, sampling methodology, analysis, spills, treatment of poisoning and waste management.

A few years ago, to obtain data under appropriate headings, a library had to be identified that contained or had access to the materials to be consulted, *e.g.* books, journals, abstracts and indexes. This principle still applies, but the prime need now is to locate a library resource that is connected to the internet and that has available experienced information technologists, conversant with the specific needs of toxicology searching. Some examples follow.

4.3.1.1 Current Contents Connect and Chemical Abstracts Service (CAS)

    4.3.1.1.1 Current Contents Connect® (http://connect.isihost.com/homez.html). *Current Contents Connect®* is a multidisciplinary current awareness web resource providing access to complete bibliographic information from over 8000 of the world's leading scholarly journals and more than 2000 books. Users can also search a premium collection of evaluated scholarly web sites and access evaluated, full-text web documents in three general resource types: preprints, funding information and research activities.

    4.3.1.1.2 The chemical abstracts service (CAS, http://www.cas.org/about.html). Substance identification is a special strength of CAS. It is widely known as the *CAS Registry*, the largest substance identification system in existence. When CAS processes a chemical substance, newly encountered in the literature, its molecular structure diagram, systematic chemical name, molecular formula and other identifying information are added to the Registry and it is assigned a unique CAS Registry Number. Registry now contains records for more than 24 million organic and inorganic substances and more than 47 million sequences.

Chemical abstracts provide summaries with citations from journals, patents, reports, specialist books and conference proceedings, and there is a Science Citation Index that allows identification of related papers.

4.3.1.2 Online Databanks and Databases. Access online to computer-based databases or databanks may simplify and speed up data retrieval. Databanks contain pre-selected information in summary form while databases provide access to data without pre-selection or evaluation. A very useful important databank is RTECS (the United States Registry of Toxic Effects of Chemical Substances), to be found at http://www.cdc.gov/niosh/rtecs.default.html

An example of an important database is TOXLINE. TOXLINE may be accessed through TOXNET (http://toxnet.nlm.nih.gov/), which is a portal to most of the toxicological information in the US National Library of Medicine (NLM). The TOXLINE database is the bibliographic database for toxicology of the NLM. TOXLINE provides bibliographic information covering the biochemical, pharmacological, physiological and toxicological effects of drugs and other chemicals. It contains over 3 million bibliographic citations, most with abstracts and/or indexing terms and CAS Registry

Numbers. TOXLINE references are drawn from various sources grouped into two parts, TOXLINE Core and TOXLINE Special. A standard search of TOXLINE retrieves records from both subsets but they can be searched separately.

PUBMED® (http://www.ncbi.nlm.nih.gov/gquery.fcgi) and CANCER LINE (http://hcp.cancerline.com/) are valuable free text databases. PubMed, a service of the NLM, also available through TOXNET, includes over 15 million citations for biomedical articles back to the 1950s. These citations are from MEDLINE® and from various additional life science journals. PubMed includes links to many sites providing full-text articles and other related sources.

Another useful database is NCBI (the US National Centre for Biotechnology Information; http://www.ncbi.nlm.nih.gov/). Established in 1988 as a national resource for molecular biology information, NCBI provides public databases, conducts research in computational biology, develops software tools for analysing genome data, and disseminates biomedical information – all for the better understanding of molecular processes affecting human health and disease.

One of the oldest databases is *The Merck Index*, an encyclopaedia of chemicals, drugs and biologicals with over 10,000 monographs on single substances or groups of related compounds. It is possible to access the entire content of The Merck Index (13th edition) through a website and view extensive information about:

- Names and synonyms
- Physical properties
- Preparations and patents
- Literature references
- Therapeutic uses

The Merck Index, 13th edition, is available through a website, http://library.dialog. com/bluesheets/html/bl0304.html, or as a stand-alone product.

4.4 TYPES OF DATA

Not all data arise from similar systems; for example, complex non-linear *systems* such as thermodynamics and meteorology do not behave predictably and a simple cause–effect relationship may not be assumed. Interpretation of data from such systems necessitates definition of the exact state of all the forces and matter involved, together with accurate measurement of all the interacting factors, in which case behaviour may be predictable at a certain level of probability. A pattern will emerge, but will never repeat itself exactly: this is the basis of the recently described Chaos Theory. According to this theory, health is regarded as chaotic, and in health there is an ability to respond to a large variety of adverse stimuli. On the other hand, disease is interpreted as a loss of flexibility or periodicity, and in disease there is a lessened ability to adapt and respond to external stimuli.

4.4.1 The Inductive-Hypothetico Approach

The inductive-hypothetico concept of cause-and-effect applies to simple linear systems and still forms the basis for most data interpretation in toxicology. The method of logical deduction is attributed to the Frenchman Descartes (1596–1650), who did most of his work in bed, whereas the idea of inductive reasoning and experimentation is associated with Francis Bacon (1561–1626). Bacon died of the cold he caught while he tried the experiment of stuffing a chicken with snow. The inductive approach, which is based on observation and collection of data, is familiar to all scientists, having been adopted very widely during the 19th century (*e.g.* by Darwin), and the experimentation approach is used by nearly all present-day scientists.

4.4.2 Data Arising from the Study of Chemical Toxicity

The normal plan when studying the potential toxicity of a chemical is to undertake carefully controlled experiments, preferably in the target species. If there is concern about potential human toxicity, effects are studied on surrogate species, usually rodents. Such studies may use both *in vivo* and *in vitro* methods.

Typically, when seeking to discover the biological properties of an unknown chemical, quantitative data are gathered regarding various predicted and expected effects, and additionally a search is made for other actions that were not expected. Unexpected data from the latter (screening) studies tend initially to be qualitative rather than quantitative. Interpretation of qualitative data is usually subjective, being based on experience and intuitive reasoning.

It is usually necessary, therefore, that qualitative data of interest or of concern, which have been generated from screening approaches, should subsequently be studied further and in greater depth using quantitative techniques. Objective interpretation of the quantitative data is then possible.

4.5 HANDLING QUANTITATIVE DATA USING STATISTICAL ANALYSIS

Statistical treatment of data will enable extraction of important features, for example indicators of central tendency such as the mean, mode or median, and indicators of spread about the centre, such as standard deviation and interquartile range.

4.5.1 The Null Hypothesis

In chemical testing, it is usual to set up a study-based statistical model in order to provide data suitable for statistical analysis. For example, a chemical under study (the test substance) is usually applied to a model system while comparing its actions (if any) with those arising from simultaneous application of a bland or otherwise relevant reference of standard substance (the control). Often called the 'sham-treated group', this controlled group may be exposed to the solvent vehicle used to solubilise the test substance (*e.g.* corn oil or carboxymethylcellulose depending on the solubility of the test material) and will be identical in every way with the test group, except that it will not receive the test substance. At the outset of such a study, it is assumed that there will be no difference between the 'test' and 'sham-treated' groups (the null hypothesis). Data are collected from each group to enable comparisons to be made. If subsequently data from the group are found to differ, then it is assumed that this difference arose due to prior treatment with the test substance, *i.e.* it is attributable to exposure to the chemical under investigation.

The possibility or probability that differences are indeed attributable to an effect of the chemical may be calculated using statistical techniques (see below). Application of such techniques will allow determination of the confidence that should be illustrated by applying the decision-tree approach, which is illustrated in Figure 4.1.

4.5.2   Generation of Data Relating to Chemical Safety

The data needed for risk estimation are currently provided by toxicological investigations that depend mainly on laboratory experiments. For a summary of the types of non-biological data and biological-effect data that are needed in order to make assessment of biological risk, it is useful to refer to standard texts such as Hayes (2001).

4.5.3   Presentation of Data

Since it is likely that the reader of a report relating to chemical safety is not as knowledgeable as the author, it follows that in communicating results the use of shorthand, jargon and excessive notation should be avoided. For example, since the reader is unlikely to be a statistician, particular attention should be paid to ensuring that in addition to notation, an explanation in words is provided. It is also quite likely that the reader's mother tongue might not be English, and so there is additional need for the wording used to be as simple and direct as possible and capable of only one interpretation. For example, in reporting results, when the conventional significance levels of 0.05, and 0.01 and 0.001 are utilised, they are best translated into 'statistically significant', 'highly statistically significant' and 'very highly statistically significant', respectively.

4.5.4   Expression of Results as Tables, Graphs, Figures and Statistics (Figure 4.1)

It is usually beneficial to make full use of tabulation and graphical, diagrammatic and other visually attractive methods for the presentation of data, since 'one picture is worth a thousand words'.

4.5.4.1 Choice of Statistical Test.   In terms of selecting a statistical test, the most important question is 'what is the main study hypothesis?' If there is no hypothesis, then there is no statistical test that can be applied. At the outset of a statistically based study, as stated above, it is assumed that there will be no difference between the 'test' and 'sham-treated' groups (the null hypothesis). Data are collected from each group to enable comparisons to be made. Subsequently if data from the groups are found to differ, it is assumed that this difference arose as a result of the prior treatment with the test substance, *i.e.* it is attributable to exposure to the chemical under investigation.

The investigator should then ask 'are the data independent?' This can be difficult to decide but usually results from the same individual, or from matched individuals, are not independent. For example, results from a crossover trial are not independent. The statistical analysis must reflect the design of the test. Results measured over time require special care.

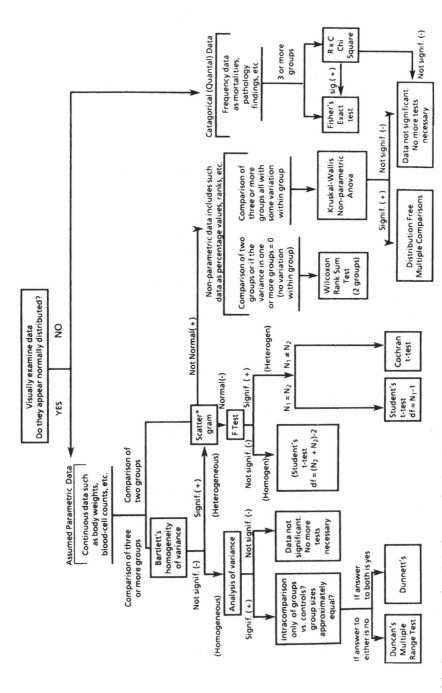

Figure 4.1 *Data types and statistical approaches in common use*

Table 4.1   *Choice of statistical test from paired or matched observation*

Variable	Test
Ordinal (ordered categories)	Wilcoxon
Quantitative (discrete or non-normal)	Wilcoxon
Quantitative (normal[a])	Paired *t*-test

[a] It is the difference between the paired observations that should be plausibly normal.

The next question is 'what types of data are being measured?' The statistical test used should be determined by the data. The choice of test for matched or paired independent data is described in Table 4.1.

It is helpful to define the *input* variables and the *outcome* variables. For example, in a toxicological trial the input variable is type of treatment – a nominal variable – and the outcome may be some clinical measure that is normally distributed. The required test is then the *t*-test (see Table 4.1). As another example, suppose we have a study in which we score histopathological changes on a five-point scale, and we wish to ascertain whether animals receiving the test article have a higher incidence of change than animals receiving only the vehicle control. The input variable is unknown versus vehicle, which is nominal. The outcome variable is the five-point ordinal scale. Each animal's outcome is independent of the others and so we have independent data.

4.6   EVALUATION OF EXPERIMENTAL DATA

4.6.1   NOEL, ADI and TLV

The no observed adverse effect level (NOEL, NOAEL) for a substance under consideration is the most commonly sought quantitative output of a laboratory experiment, since it is the 'threshold dose' below which the body is thought to be able to safely dispose of a xenophobe.

Based on estimates of exposure and knowledge of the NOEL, risk analysis may progress, for example with respect to a food substance or environmental contaminants, by calculation of an acceptable daily intake (ADI) or threshold limit value (TLV). These are not to be regarded as sacrosanct figures. An ADI provides a sufficiently large safety margin to ensure that there may be no undue concern about occasionally exceeding it, provided that the average intake over longer periods of time does not exceed it.

4.6.2   Extrapolation

The process of utilising data from laboratory surrogates in order to make estimates of risk in real-life situations is extrapolation and therefore it is to some extent uncertain. The relative imprecision of the extrapolation process is compensated for (made fail-safe) by the use of factors whose effect is to make the final estimate highly conservative: a division of the experimental 'no effect level' by factor of 10 is utilised to compensate for possible inter-individual variation (in other words, to protect the exceptionally sensitive individual) and a further division by 10 is used to compensate for extrapolation across species. Therefore, a 100-fold reduction factor (10×10) is utilised when extrapolating to humans from a NOEL obtained in a laboratory study in rodents. On some occasions, factors greater than 100 have been used, for example, when there have been uncertainties in the data that are available or doubts about the purity of the tested substance (see Chapter 5).

4.6.3 Exposure, Dose, Surface Area and Allometry

Extrapolation from surrogate to target, when possible, should always be on the basis of exposure rather than dose. If this is not possible then extrapolation on the basis of dose may be improved by taking account of relative sizes of the surrogate and the target.

Allometry, a method that has been developed for describing morphological evolution, has been utilised in attempts to improve extrapolation from surrogate to target. Allometry is the study of a relation between the size of an organism and the size of any of its parts: for example, there is an allometric relation between brain size and body size; animals with larger bodies have larger brains. On the other hand, in the case of skin area, a larger animal has relatively less surface area than does a small animal.

The body surface area-dependent dose conversion for rats and mice to humans is as follows:

- Rat (150 g) to human (60 kg): human needs 1/7 the rat dose
- Dog (8 kg) to human (60 kg): human needs 1/2 the dog dose

Various allometric and non-allometric methods have been used to predict human volume of distribution based on toxicokinetic data from surrogate species. Clearance and volume of distribution values generated by various means are then used to estimate human exposure. From both a quantitative and qualitative perspective, estimating human exposure is usually best based on the species that handles the chemical in the most similar fashion to the human. If it is not possible to determine exposure and kinetic behaviour directly in the human, then subhuman primates often provide the best estimates. However, the use of primates is usually avoided because of concern for their welfare and conservation.

4.7 ERRORS AND FAULTS IN DATA INTERPRETATION

4.7.1 ADI or TLV are not Immutable Numbers

Because their derivation depends so much upon expert judgment, ADIs or TLVs should not be regarded as immutable numbers. Uncertainties in these values may focus attention on uncertainties in the risk-analysis/risk management process. Thus, values for allowable exposure may be adjusted (often downwards for environmental contaminants, sometimes upwards for beneficial therapeutic entities) as better data are generated and as overall knowledge increases.

While these adjustments to extrapolated values are to be regarded as part of the iterative phase of safety assessment, there are other situations where errors and faults of data handling can jeopardise the data interpretation process. For example, when two groups are compared with a view to detecting a significant difference between them (using the null hypothesis) there is never a completely certain answer, only a probability. The degree of certainty is given by the P value. When the $P = 0.01$ then there is only a 1% likelihood that the difference is due to chance variation. This means that, on average, in every 100 studies demonstrating a significant difference at this level, there will be one occasion when the difference is not real but is accidental, giving a false-positive result.

4.7.2   False-Positive and False-Negative Results

False-positive results arise when test systems are oversensitive. Such results will lead to overestimates of hazard and unduly low NOAELs, which, when extrapolated, give overestimates of risk. While such errors are on the 'safe side', they may prove to be costly in other ways, for example by leading to over-investment in containment (in the case of an industrial chemical) or possibly leading to underdosage during the development of a new pharmaceutical entity.

At least as unsatisfactory an issue as 'false positives' is that of false-negative results. Clearly, an underestimate of potential for hazard can have serious consequences. Probably, the most common reason for failure to detect a difference between two groups, even when one exists, is that the experimental groups are too small. For example, the number of patients who must be enrolled in a trial investigating a new candidate drug depends on

–   The degree of variability of the expected data
–   The minimum size of the measurement that will be made
–   The P value sought, which is usually 0.01 or 0.05
–   The power of the study, usually 80 or 90% is required

4.7.3   'Eyeballing' the Data

Finally, it is essential to confirm as soon as possible that the data are trustworthy. Before spending precious time and money in detailed processing of data that are 'rubbish', establish that they seem to make sense by 'eyeballing', *i.e.* by comparing them quickly with background data, your previous experience, and any other relevant information that you may have. In this way, you may identify obvious discrepancies indicating that the data are unreliable.

4.8   CONCLUSION

The following are useful questions to ask when interpreting data.

1. Are the data valid, that, do they result from application of approved methodology, competently performed?
2. Have the data been validated, *i.e.* have they been subject to quality assurance audit (as required for all modern regulatory toxicological studies); and is there available a peer review (as there is in many modern toxicological studies that contain subjective assessments such as microscopic pathology)?
3. Has the experimental result been proved to be reproducible (at least performed in duplicate on separate occasions, the standard approach for many of the current in *vitro* genotoxicological tests)?

4. Has the chosen validation process been adequately documented: for example, have the data been generated under good laboratory practice (GLP) conditions or under a National Measurement Accreditation Service (NAMAS) scheme?
5. Do the data arise from a sufficiently large experiment (modern regulatory toxicological study designs are harmonised on an international basis according to the Guidelines for Testing of Chemicals produced by the Organisation for Economic Co-operation and Development (OECD) or to the pharmaceutical guidelines produced by the International Conference on Harmonisation (ICH))?
6. Did the experiment yield sufficient controlled data to allow within-study comparisons to be made (assuming adequate study conduct, tests following the guidelines referred to in question 5 above usually yield sufficient data to enable interpretation of the study without resorting to the use of background data)?
7. Are the data from the experiment in agreement with relevant background data? (although it should be possible to interpret the study data without recourse to other data, the availability of concurrent background data, which have been generated at the same laboratory, can be invaluable as a means of validating the study under review; such data can also be helpful by putting the study results into the proper context)?

If all of the above are satisfactory, then the final question can be asked:
Do the data arising from the test group(s) differ quantitatively and/or qualitatively from the control groups and relevant background data? If yes, then this may be attributed to an effect of the test substance.

BIBLIOGRAPHY

G.W. Bradley, *Disease Diagnosis and Decisions*, Wiley, New York, 1993.

M.F.W. Festing, P. Overend, R.G. Das, M.C. Borja and M. Berdoy, *The Design of Animal Experiments*, Royal Society of Medicine Press, London, 2002.

A.W. Hayes (ed), *Principles and Methods in Toxicology*, 4th edn, Taylor & Francis, Philadelphia, PA, 2001.

K.S. Khan, R. Kunz, J. Kleijnen and G. Antes, *Systematic Reviews to Support Evidence-Based Medicine*, Royal Society of Medicine Press, London, 2003.

R. Kiley, *The Doctor's Internet Handbook*, Royal Society of Medicine Press, London, 2000.

T.D.V. Swinscow, *Statistics at Square One*, Ninth edn, Revised by M.J. Campbell, University of Southampton, BMJ Publishing Group, London, 1997, http://bmj.bmjjounals.com/collections/statsbk/index.shtml.

N.J. Wald, *The Epidemiological Approach*, The Wolfson Institute and Royal Society of Medicine Press, London, 2004.

A. Woolley, A Guide to Practical Toxicology, Taylor & Francis, London, 2003.

Chapter 5

Risk Assessment

H. PAUL A. ILLING

5.1 INTRODUCTION

Although conceptually distinct, risk assessment and risk management are often impossible to separate in practice. Risk assessment is essentially a preliminary to setting up proper risk management procedures. Confirmation of the effectiveness of risk management procedures requires reassessment of the risks after implementation of management procedures. Principles and procedures for risk assessment will be discussed in this chapter and principles of risk management in the next, but, in practice, the two chapters need to be read together.

5.2 DEFINITIONS

A recent publication by the Organisation for Economic Co-operation and Development (OECD, 2003) gives up-to-date definitions of risk, risk assessment and risk management.

Risk. The probability of an adverse effect in an organism, system or (sub) population caused under specified circumstances by exposure to an agent.

Risk assessment. A process intended to calculate or estimate the risk to a given target organism, system or (sub) population, including the identification of attendant uncertainties, following exposure to a particular agent, taking into account the inherent characteristics of the agent of concern as well as the characteristics of the specific target system.

The risk assessment process includes four steps: hazard identification, hazard characterisation (related term: dose–response assessment), exposure assessment and risk characterisation. It is the first component of a risk analysis process.

Risk management. Decision making process involving consideration of political, social, economic and technical factors with relevant risk assessment information relating to hazard so as to develop, analyse and compare regulatory and non-regulatory options and to select and implement appropriate regulatory responses to that hazard.

Risk management comprises three elements: risk evaluation, emission and exposure control and risk monitoring.

There are many other definitions in circulation. Perhaps one of the most useful definitions of risk is that 'risk is the possibility of suffering harm from a hazard' (Cohrssen and Covello, 1989). This introduces two further concepts, the adverse effect or harm associated with the agent (the hazardous substance or process, hence the hazard), and the (likely) exposure to the hazardous substance or process. The OECD (2003) definitions of hazard and exposure are:

Hazard. The inherent property of an agent or situation having the potential to cause adverse effects when an organism, system or (sub) population is exposed to that agent.

Exposure. Concentration or amount of a particular agent that reaches a target organism, system or (sub) population in a specified frequency for a defined duration.

Not all chemical hazards are toxic hazards. Fire and explosion hazards are not usually considered toxic. *Toxicity* is defined as 'the inherent property of an agent to cause an adverse biological effect' and fire and explosion cause physical damage.

5.3 PROCESS OF RISK ASSESSMENT

The process of risk assessment consists of four steps: hazard identification, hazard characterisation (related term: dose–response assessment), exposure assessment and risk characterisation. The OECD (2003) describes these steps as:

Hazard identification. The identification of the type and nature of adverse effects that an agent has as an inherent capacity to cause in an organism, system or (sub) population.

Hazard characterisation. The qualitative and, where possible, quantitative description of the inherent properties of an agent or situation having the potential to cause adverse effects. This should, where possible, include a dose–response assessment and its attendant uncertainties. (*Dose–response assessment*: Relationship between the amount of an agent administered to, taken up or absorbed by an organism, system or (sub) population and the reaction to the agent.)

Exposure assessment. Evaluation of the exposure of an organism, system or (sub) population to an agent (and its derivatives).

Risk characterisation. The qualitative, and, where possible, quantitative determination, including attendant uncertainties, of the probability of occurrence of known and potential adverse effects of an agent in a given organism, system or (sub) population, under defined exposure conditions.

The relationship between these steps is shown diagrammatically in Figure 5.1.

5.4 HAZARD IDENTIFICATION AND CHARACTERISATION

Identifying the potential harm of a substance or process may involve any or all of the following: observation, experimental work, information retrieval and deductive work based on physicochemical parameters and structure–activity relationships. It includes identifying the type of harm (*e.g.* narcosis, skin sensitisation, cancer) as well as characterising the dose levels at which harm occurs.

5.4.1 Sources of Information

In the past much of the publicly available information was retrieved from peer reviewed articles in recognised journals. Unfortunately, this excluded many of the best studies for hazard assessment. These are studies conducted to internationally

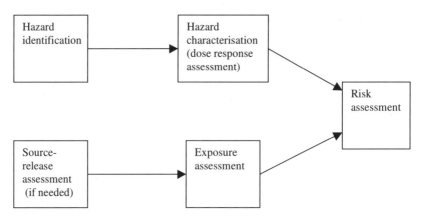

Figure 5.1 *Stages in carrying out a risk assessment*

agreed protocols (such as the OECD test guidelines and guidelines derived from them). The results of these studies are rarely of great scientific interest in terms of advancing fundamental science and often are not published. Many are regarded as confidential commercial information. Those that are available are often difficult to retrieve, as they are not entered into conventional databases, such as MEDLINE, concerned with the published literature. Frequently they are only available from the sponsor of the test. Internationally validated assessments of relevant literature are available through the International Programme on Chemical Safety (IPCS; the Environmental Health Criteria series) and, for cancer, through IARC (the International Agency for Research on Cancer) series of monographs. However, these are available for only a limited number of chemicals.

In recent years there have been major changes. The development of publicly available versions of IUCLID (International Uniform Chemical Information Database) accessible via the Internet (*e.g.* from the website of the European Chemicals Bureau) has meant that summaries of much unassessed information are now available on 'high production volume' (HPV) chemicals. This includes older studies of variable quality as well as modern studies. The ICCA (International Council of Chemical Associations) is cooperating with the OECD and its member countries in an initiative to provide risk assessments for about 1000 HPV chemicals. The documents (Screening Information Datasets (SIDS) Dossiers and SIDS Initial Assessment Reports (SIARs)) associated with this initiative are being published. These documents include careful evaluations of the evidence identifying and characterising the toxicological hazards. Thus the amount of information in the public domain is rapidly increasing.

Modern studies are usually conducted to internationally agreed guidelines, normally based on those of the OECD (1981; revised 1993 and Addenda) and the data audited in accordance with the OECD principles of good laboratory practice (OECD, 1997).

5.4.2 Types of Information

Data on human health effects may come from

- Human observations (case reports, epidemiological studies or, in suitable cases, human studies)
- Animal toxicology studies
- *In vitro* toxicology
- Structure–activity relationships.

The weight of evidence is assessed on the basis of the combined strength and coherence of inferences appropriately drawn from all of the available data. This entails rigorous examination of the quality, quantity and nature of the results of available studies. The study data can be supplemented with toxicokinetic and toxicodynamic information to improve the interpretation of the data and to confirm (or otherwise) the assumption (usually made) that behaviour in the test animal species is/are similar to that in the target species. Recently, the IPCS (1999, 2001) has provided guidance on these aspects of hazard identification.

Evidence of harm to the environment may come from

- Assessment of structure–activity relationships
- Toxicity studies in indicator species
- Studies using microcosms and mesocosms
- Field studies and field observations.

Predictive evidence for environmental effects is usually obtained in stages, with the more complex studies only being undertaken if there is a specific need.

5.4.3 Dose–Response, Dose–Effect, LD50 and the 'No Observed (Observable) Adverse Effect Level'

Some harmful effects, notably genotoxic cancers and germ cell mutations, are considered to be 'stochastic' (quantal). Either they are present or they are not present in the individual. Biological variation in the sensitivity of individuals to these effects means that the frequency of occurrence of the response within the population will depend on dose, and therefore can be represented by a dose–response relationship.

Most effects are 'non-stochastic' and the severity increases progressively with dose (continuous data). Non-stochastic health effects commonly occur through the following process:

Normal $\rightarrow$ homeostatic adjustment $\rightarrow$ compensation (input of reserve capacity with possibility of repair) $\rightarrow$ breakdown (leading to increased disability and, ultimately, death).

A similar progression can be defined for environmental effects, but with 'disorganisation' and 'disintegration' as the consequences of breakdown of an ecosystem, rather than disability and death.

If the inter-individual variation of the severity of effect with dose is small, then it is possible to determine a dose–effect curve for the whole population. However, if the variation is large, population comparisons are based on the dose required to cause a fixed level of effect in each individual in the population, and are therefore assessments of dose–response.

The relationship between dose–effect and dose–response can be shown diagrammatically (Figure 5.2).

There is a major distinction between the assessment of genotoxic carcinogens and germ cell mutagens and other types of toxic effects. It is based on the premise that simple mutagenic events may be responsible for initiating the effects leading to cancer and germ cell mutagenesis. The suggestion is that it is not possible to demonstrate experimentally whether there is a threshold. Similarly, it is thought that there is enormous variability in individual thresholds for respiratory sensitisation so, for occupational exposure purposes, it also may be treated similarly.

For other types of toxic effect, the convention is that there is a threshold, a level of exposure below which it is believed that there are no adverse effects. For these effects it is possible to define a 'no observed (observable) adverse effect level' (NOAEL), 'lowest observed (observable) adverse effect level (LOAEL) or 'benchmark dose' (BMD). These have been defined as (IPCS, 1994):

NOAEL: The greatest concentration or amount of a substance, found by experiment or observation, which caused no detectable adverse alteration of morphology, functional capacity, growth, development or life span of the target organism under defined conditions of exposure. Alterations of morphology, functional capacity, growth, development or life span of the target may be detected which are judged not to be adverse.

LOAEL: The lowest concentration or amount of a substance, found by experiment or observation, which can cause an adverse alteration of morphology, functional capacity, growth, development or life span of the target organism distinguishable from normal (control) organisms of the same species and strain under the same defined conditions of exposure.

BMD: The lowest confidence limit of the dose calculated to be associated with a given incidence (*e.g.* 5% or 10% incidence) of effect estimated from all toxicity data on that effect within that study.

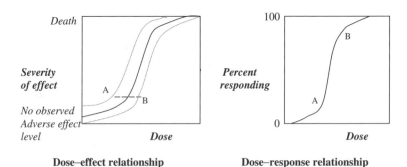

Dose–effect relationship   **Dose–response relationship**

Figure 5.2   *The relationship between dose–effect and dose–response. The dotted lines indicate the 95% confidence limits for the dose–effect relationship; taking the values for a fixed level of effect (AB) they can be transformed into the 5% and 95% dose values for the dose–response curve for the that particular level of effect*

Conventionally 'observed' relates to the data from a single study, 'observable' relates to the data from all the examined studies.

For environmental effects concentration may be more important than amount, in which case the NOAEL is replaced by the 'no observed adverse effect concentration' (NOAEC) or, often, the 'no observed effect concentration'.

Acute toxicity testing has been aimed at identifying *the LD_{50} or $LC_{50\ (time\ of\ exposure)}$*: the dose or exposure concentration and time associated with the death of 50% of the animals exposed. More recently, testing has moved from using lethality as the parameter for acute toxicity to 'evident toxicity'. For environmental purposes the EC_{50} (effective concentration for a named end point such as immobilisation) may be used in place of the LC_{50}.

5.4.4 Hazard Characterisation

The hazard characterisation may be used for one of four purposes depending on the circumstances. These are

- Determining if a chemical is persistent, bioaccumulative and 'toxic' (PBT), or very persistent and very bioaccumulative (vPvB), and therefore evaluation and management of the environmental risk that the chemical may pose should be a priority (Table 5.1).
- Classification and labelling. The hazard information is set against pre-determined criteria and the simplified hazard information is passed to the appropriate risk manager for examining management during transport (UNECE, 2003a) or through the supply chain (safety data sheets and labels; the 'globally harmonised system', UNECE, 2003b).
- Development of exposure standards ('intake' or 'uptake' standards; see Chapter 6) by setting the information against a (pre-determined) risk evaluation procedure and developing a maximum 'safe' exposure.
- Setting of the hazard characterisation against exposure information to determine a 'margin of exposure (safety)'.

When dealing with simple experiments, clear end points and definite criteria, classification (and hence labelling) is relatively simple. It may take the deterministic

Table 5.1 *EU criteria for identifying PBT and vPvB chemicals (TGD, 2003)*

Criterion	PBT criteria	vPvB criteria
P	Half life >60 d in marine water or >40 d in fresh water or half life >180 d in marine sediment or > 120 d in freshwater sediment	Half life >60 d in marine or freshwater or >180 d in marine or freshwater sediment
B	Biological concentration factor >2 000	Biological concentration factor >5 000
T	Chronic NOEC <0.01 mg/L or CMR (carcinogenic category 1 and 2, mutagenic category 1 or 2, toxic to reproduction categories 1, 2 and 3) or endocrine disrupting effects	Not applicable

(is/is not) or a more quantitative (grading as very toxic, toxic, harmful) approach. The former is hazard identification, the latter includes an element of hazard characterisation. However, when experiments are complex, end points are difficult to interpret and/or when criteria are poorly defined, interpretation becomes difficult. In these cases, a simple deterministic statement (is/is not) may be sufficient and the classification is based on 'weight of evidence'. Here, the classification criteria are based on quality and type of evidence. The classification of carcinogens, mutagens and substances toxic for reproduction are examples of this approach. The IARC classifies carcinogenic agents, mixtures or exposure circumstances as:

- Group 1 – The agent is carcinogenic to humans.
- Group 2A – The agent is probably carcinogenic to humans.
- Group 2B – The agent is possibly carcinogenic to humans.
- Group 3 – The agent is not classifiable as to its carcinogenicity to humans.
- Group 4 – The agent is probably not carcinogenic to humans.

Most schemes use the basic classification *known human carcinogen* (requiring good evidence in humans), *probable human carcinogen* (requiring good evidence in animals and/or some evidence in humans or on relevance of mechanism), and *possible human carcinogen* (usually a mixed collection of criteria). Formally, this is hazard identification. Hazard characterisation would require some measure of the potency of the chemical. However, it is likely that the more potent carcinogens will be the more easily identifiable carcinogens, and so, informally, some hazard characterisation is taking place.

5.4.5 Source–pathway–receptor

The interaction between a chemical and a receptor is the critical interaction for a toxic effect to take place. The chemical has to get to that receptor for the interaction to take place. It has to be released from a source, pass through one or several media, and, in many cases be taken into the target organism/system to interact with the receptor before harm can arise from a hazard. The concept is illustrated in Figure 5.3.

Although the concept may be relatively simple, in practice, the complexity of these systems can be gleaned from Figure 5.4, in which the target organism is the human. Similarly complex chains can be developed for other targets.

5.4.6 Measurement and Modelling

Ideally, exposure assessment should include direct measurements of chemical in appropriate media. For humans, uptake is measured using samples of blood, urine or exhaled air. These are the nearest sampling sites to the receptor at the 'site of action' and are surrogates for sampling at that receptor. Intake (exposure levels or daily intakes in food/drinking water) is measured in ambient medium (air, water, soil, total diet) and is related to uptake using simple models. Relating input to uptake require more complex models in order to link emission to air, discharge to

Source	Input	Pathway	Intake	Uptake	Receptor
		Environment			Biosphere
Point	Emission (to air)	Environmental medium (air, water, soil, ecosystem)	Air (indoor, outdoor, workplace)	into organism/ ecosystem	Species
	Discharge (to water, soil)		Soil		
Diffuse	Residue in individual foodstuff	Food and drinking water (food basket/total diet)	Drinking water/ Food		Organism (e.g. humans)

Figure 5.3 *The passage of a chemical from source to receptor (idealised)*

water, or presence of a residue in a foodstuff to the receptor in the target organism. These latter models include scenarios describing dispersion from the source, transport through the medium (air, water, soil or food) and transfer between media (*e.g.* fugacity models) as well as models of intake. These models can be classified as (IPCS, 1999)

- Simple dilution models where a measured concentration in an effluent is divided by a dilution factor or the chemical release rate is divided by the bulk flow rate of the medium.
- Equilibrium models, which predict the distribution of a chemical in the environment based on partitioning ratios or fugacity (the escaping tendency of a chemical from one environmental phase to another).
- Dispersion models, which predict reductions in concentrations from point sources, based on assumed mathematical functions or dispersion properties of the chemical.
- Transport models which predict concentration changes over distances that can represent dispersion, biochemical degradation and absorption.

The reliability of modelled estimates of chemical concentration in the general environment or the food supply depends on

- how well the assumptions match reality (*i.e.* how realistic assumptions such as steady state conditions and homogenous media properties are);
- whether the model performance has been demonstrated under conditions similar to those of concern; and
- the quantity and quality of input data.

Modelling which uses input values based on default assumptions is generally precautionary. It is useful for screening purposes as it highlights areas where specific additional data are required to estimate exposure more accurately. Further information is contained in IPCS (1999), TGD (2003) and Illing (2001).

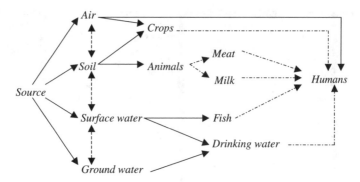

Figure 5.4 *Schematic representation of human exposure routes via the environment. Environmental routes ——▶; transfer between environmental compartments ◀---▶; intake through food chain ------▶*

There are two main circumstances where exposure information is required, accidents (uncovenanted releases) and occasions, where low-level exposure is expected (covenanted or anticipated exposure).

5.4.7   Human Exposure

Possible sources of exposure for humans are given in Figure 5.5. Generally, deliberate exposure is the administration of known amounts of chemical for a specified reason to the target organism, and only needs further examination if there is a potential for collateral damage to other non-target organisms. The example is a plant protection product for which the collateral damage may be to beneficial insects, workers spraying the pesticides or handling crops post-spraying, consumers or non-target species eating the crop. The EU has developed a system, called EUSES, for examining exposure scenarios for anticipatable covenanted exposure.

5.4.7.1   Exposure and Food.   Exposure to chemicals through the food supply is an important consideration when considering human exposure. Some chemicals are deliberately applied to food crops and animals and may leave residues in or on edible material; others are there incidentally, through their presence in air or soil. Others, again, may leach from packaging. When it comes to human exposure all are contaminants. Residue levels (*e.g.* for pesticides, for veterinary products, for other contaminants) in individual foods need to be related to total intake in order to characterise the risks associated with that residue.

Methods used for estimating exposure to residues in food include consumption data and residues data (see Illing, 2001 for a fuller explanation). The consumption data can include model diets, regional diets, national diets or household/individual diets. The exposure assessment may be based on 'average' consumers or specific groups of consumers ('critical groups') and may be based on 'high-level' exposure (the exposure is based on the upper end of the distribution of data from a representative sample of consumers) and/or 'best estimate' (the median value). Further the data inserted into the model may also be 'best estimate', often used for dealing with agricultural pesticides

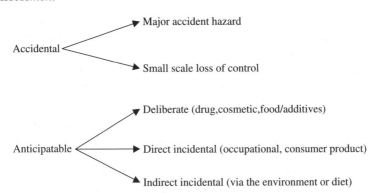

Figure 5.5 *Potential exposure scenarios for consideration in human risk assessment*

residues when the exposure data are obtained from field trials, or a 'worst case', often used for food additives and contaminants.

It may also be necessary to calculate bioconcentration or biotransfer factors for uptake and retention of contaminants in individual foods. For this purpose foods may be grouped into fish, leaf crops, root crops, meat and dairy products. For deliberately administered chemicals this includes the possibility of calculating withdrawal periods during which direct application of chemical to the food source is not permitted.

5.4.7.2 Consumer Exposure. The assessment of consumer exposure involves an initial screening, dealing with the question whether or not the substance is actually used in products and a quantitative exposure assessment for those circumstances where actual exposure occurs.

The quantitative assessment involves setting up consumer exposure scenarios. The scenario describes the use of the substance, the pathways (release distribution and elimination) in order to describe the contact amount/concentration and the behaviour of the relevant exposed population. Reasonable worst case is used. The quantitative exposure assessment needs to deal with contact parameters, concentration parameters and behavioural data. Contact parameters include intended use, frequency, duration (per event) and site of use, physical form (aerosol, liquid, *etc.*), amount of product used per event and contact surface (skin, *etc.*) (if appropriate). Concentration parameters may include weight fraction of substance in product, concentration of substance in the product as used (after dilution/evaporation, *etc.*, if different), amount of substance in an article, amount of residual monomer, emission rate constant (if available). Behaviour data are used to identify the time spent indoors (in different rooms or vehicles) and outdoors (garden, playground, street).

Separate algorithms are used to calculate exposure by inhalation, ingestion and skin contact. Where appropriate consumer exposure is aggregated from several scenarios. Although measured data are preferable, it is rarely available and much of the data used are default information.

5.4.7.3 Occupational Exposure. Workers exposed to chemicals in the workplace are exposed principally by inhalation and dermal contact. Thus models used to estimate exposure are therefore identifying amounts of chemical in the ambient medium (and hence making assumptions concerning intake and uptake) or in biological media (biological and biological effects monitoring, making assumptions related to uptake). Occupational exposure measurements are often available from occupational hygiene surveys. When monitoring data are incomplete, exposure information can be obtained from models. The EASE (Estimation and Assessment of Substance Exposures) model has been incorporated into the overall EUSES system.

5.4.7.4 Modelling for Major Industrial Accident Hazards.   Major industrial accident hazards exposure modelling is a special (and, in some ways, simplified) case of environmental exposure modelling. However, it does exemplify the complexity of modelling. For these hazards, a model has to be established that predicts likely airborne exposures following a single, short-term exposure. Exposure through other environmental compartments may also be important, but is dealt with descriptively or by adapting models used for anticipatable (covenanted) exposure.

The important factors for the model are

- a source term, identifying the frequency and size of possible releases
- dispersion information for different weather conditions
- a knowledge of the relative frequencies of the different weather conditions
- a criterion (developed from the hazard characterisation) against which to set the exposure information.

Major accident hazards are unusual in that there is a need to deal with frequency and size of relatively short-term unintended discharges from a point source, thus steady state is rarely an appropriate assumption.

The source terms for inputting into the dispersion model may be obtained by 'fault tree' analysis, identifying the frequency of events (failure rates) by 'event tree analysis', leading to an estimate of failure rates or by engineering judgement based on historical event frequency. The sizes and duration of the postulated/actual releases are also calculated and combined with the failure information to give the overall source term.

The dispersion of the toxic cloud released is modelled using knowledge of the buoyancy of the cloud and weather conditions at the release site. The model yields a set of 'isopleths' (concentration–time contours for a particular set of weather conditions) for a particular size and duration of source term. These isopleths are combined to yield isopleth envelopes for particular type of weather conditions and then concentration–time–frequency relationships based on the frequencies with which the different weather conditions occur. The physical geography of the site and its surroundings also need to be taken into consideration. These relationships are turned into risk contours by insertion of information from the hazard characterisation (the selected level of toxicity) (Figure 5.6).

5.4.8   Environmental Exposure

Most exposure assessment involves a screening step and, if required, a quantitative exposure assessment. The screening step is used to identify whether there is expected exposure and, if so, whether it is negligible. If actual or potential exposure has been identified, a quantitative risk assessment may be required. Initial exposure assessments may be conducted on 'worst-case' assumptions, using default values if actual measurements are not available.

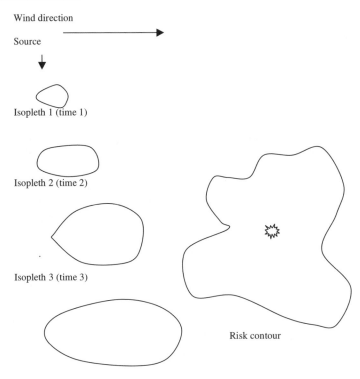

Wind direction

Source

Isopleth 1 (time 1)

Isopleth 2 (time 2)

Isopleth 3 (time 3)

Risk contour

Isopleth envelope for one set of
Weather conditions

Development of isopleth envelope. (The information on weather is three dimensional and includes vertical as well as horizontal directions. Plumes from a stack can rise or go to ground. Level, dependant on weather conditions. The isopleth envelope is two-dimensional and refers to near ground conditions. The isopleth required is usually selected on the basis that it represents a concentration time relationship for toxic effects of relevance to the land use or emergency planner.

Incorporation of meteorological information on frequencies and types of weather conditions, physical geographical information and a selected level of toxicity (in terms of concentration and time of exposure) yields a risk contour for that level of toxicity. Several risk contours may be developed for different levels of toxicity, e.g. discomfort, disability and death/permanent incapacity when dealing with emergency planning, and different sizes of release.

Figure 5.6 *Development of risk contours from dispersion modelling*

A full environmental exposure assessment consists of four steps:

- Identification of the target compartments (air, surface or ground water, soil, sediment, biota)
- Estimation of emissions or releases

- Estimation of the environmental behaviour
- Estimation of environmental concentrations.

It may refer to local, regional or continental exposures. A local environmental exposure assessment usually relates to one or several point sources. A regional or a continental environmental exposure assessment is usually more relevant when dealing with diffuse sources.

Identification of the target compartments is undertaken for all stages of the life cycle of the chemical, manufacture, formulation, industrial, professional and consumer use and end of product life disposal or recycling. The emissions and releases to air, soil and wastewater are estimated using emission scenarios and the physico-chemical properties of the substance. The interaction between the chemical and sewage treatment works is of particular importance when dealing with discharges to the wastewater. Chemicals can be distributed within an environmental compartment (intra-media transport) or be transported from one environmental medium to another (inter-media transport). The main transport mechanisms are

- *Advection*: transport of a substance from one place to another as a result of the flow of the medium in which it occurs (*e.g.* transport with infiltrating rain from soil to ground water).
- *Dispersion*: transport from one place to another as a result of differences between concentration gradiFents (*e.g.* diffusion from surface water to sediment).

Dispersion also occurs between phases (*e.g.* air/water or solids/water or water/biota) if they are not at equilibrium. Partition coefficients can be used to estimate these distributions.

Degradation (biodegradation, hydrolysis and photodegradation) will affect the exposure concentrations by reducing exposure. This is particularly important, as it is the intended effect when wastewater is treated in a sewage treatment plant.

Another parameter is of particular concern when dealing with biota. This is the biological concentration factor. Concentration of chemicals in organisms, and hence in food chains can lead to so-called secondary poisoning in, for example, the top predator.

The last step in this process is the estimation of the concentration of chemical in the receiving compartment – soil, air, surface water and ground water. Details of the processes used by the EU can be found in the Technical Guidance Document (2003).

The information from an environmental exposure assessment is usually intended for incorporation into an environmental risk assessment using effects information from a variety of indicator species. If human health is the critical end point the exposure data thus generated could also be used for human health risk assessment.

5.5   RISK CHARACTERISATION

5.5.1   Humans

The risk characterisation is the last stage of the risk assessment. It is 'a summary, integration and evaluation of the major scientific evidence, reasoning and conclusions of a risk assessment. It is a concise description of the estimates of potential risk and the strengths and weaknesses of those estimates'. This is the stage when the hazard characterisation and exposure assessment are combined. It is therefore purely scientific and technical.

In practice a risk characterisation is rarely conducted independent of the risk evaluation (the first stage of risk management and therefore defined in the next chapter). The usual procedure described for guideline (standard) setting (the 'uncertainty factors' approach) inextricably mixes risk characterisation and risk evaluation, and hence mixes the risk assessment with risk management. For standard setting purposes a risk management system pre-determines the definitions of a particular maximum exposure, and the management actions that flow from the exposure statement. Societal, political and other factors are mixed with technical factors concerning the quality and type of data and incorporated into the (usually pre-determined) risk criterion against which the hazard characterisation is set in order to define the maximum exposure level (amount) deemed appropriate for the particular exposure circumstances. The process employs 'uncertainty factors' and is described in the next chapter.

A similarly mixed approach is used in the US and some European countries for risk assessment for carcinogens. However, the assumption is made that, for genotoxic carcinogens, it is theoretically possible that 'one hit' in DNA may result in a malignant tumour. Although in practice this is unlikely, it is not possible to identify a threshold level below which no effect would be expected. Thus the process assumes that for genotoxic carcinogens there is no threshold dose below which carcinogenic effects will not occur. It therefore employs dose–response information and uses models to extrapolate from the frequencies seen experimentally to that identified as the appropriate risk criterion. The UK prefers to identify that a chemical is a carcinogen and then move immediately to risk management.

The approach for airborne major accident hazard risks also mixes risk characterisation and risk evaluation. The intent is to derive concentration–time relationships associated with a selected level (or several selected levels) of toxicity to set with the exposure information. The toxicity in question is usually acute toxicity as exposure is likely to be for a relatively limited time. The levels of toxicity selected need to take into consideration societal judgements and political decisions concerning the acceptability of the risks. Other major accident risks tend to be evaluated in a similar manner to the corresponding covenanted environmental exposure.

The most commonly used process that constitutes a pure risk characterisation is the 'margin of exposure' approach.

Margin of exposure: It is the ratio of the NOAEL for the critical effect to the theoretical, predicted or estimated exposure dose or concentration.

For some experts the 'margin of safety' is an alternative term and has the same meaning as the margin of exposure (OECD, 2003). For others, the margin of safety means the margin between the reference dose (see Chapter 6) and the actual exposure or concentration. Thus, if the term 'margin of safety' is employed it is important to know how it has been defined.

With the 'margin of exposure' approach, social, political and other factors affecting the risk evaluation are separated from the scientific and technical process of risk assessment.

For both development of exposure standards and identification of margins of safety, either the dose–response relationship or one of the LD_{50}, $LC_{50 \text{ (time of exposure)}}$, EC_{50}, NOAEL or C, LOAEL or BMD is the critical end point for the hazard characterisation. The dose–response is used for assessment of risks from major hazards and for some carcinogenicity assessments; otherwise the LD_{50}, NOAEL, *etc.*, are used.

5.5.2 Environmental Risk

For environmental risks the risk characterisation process is also mixed with risk evaluation. There are two levels at which it can be conducted:

- a qualitative risk assessment; and
- a quantitative assessment using the predicted environmental concentration and predicted no effect concentration (PNEC). This should also include an assessment of risks due to accumulation in the food chain and secondary poisoning.

These assessments may be for inland aquatic ecosystems, terrestrial ecosystems or top predators, for microorganisms in sewage works or for the atmosphere. The characterisations for aquatic ecosystems and for top predators may also be undertaken for marine environments.

The quantitative risk assessment is derived from comparing the PNEC and the predicted exposure concentration (PEC) under appropriate circumstances. The PNEC is derived from LC_{50}, EC_{50} or NOAEC data using 'assessment' (uncertainty) factors. It is therefore equivalent to the 'reference dose' for human health risk evaluation and the resultant ratio (sometimes called the 'risk characterisation ratio') is, in effect, a determination of the OECD's 'margin of safety', second definition – the difference between reference dose and predicted or actual exposure.

Initially environmental risk assessment is a relatively technical discussion concerned with the need to acquire further data. If, eventually, refining and adding to the data available is not possible, it becomes a criterion against which the need to activate risk reduction measures is judged. At this point the ratio constitutes a pre-determined criterion for risk evaluation and the process is a mixture of characterisation and risk evaluation.

5.6 CONCLUSIONS

This chapter has concentrated on risk assessment. This is a scientific and technical exercise aimed at gathering information on effect and exposure and combining that information into a risk characterisation, ideally one set out in a form suitable for use by risk managers. In practice, the process of risk assessment is not separated from risk management. The first stage of risk management, risk evaluation, involves societal and political considerations and is set within a risk management system. As the risk characterisation and risk evaluation are often conducted simultaneously by the technical expert(s) using pre-determined evaluation criteria, the division between this chapter and the next is arbitrary.

BIBLIOGRAPHY

J.J. Cohrssen and V. Covello, *Risk Analysis: A Guide to Principles and Methods for Analysing Health and Environmental Risks*. National Technical Information Service, Springfield, VA, 1989.

EUSES, European Union System for the Evaluation of Substances (available online through the European Chemical Bureau website).

P. Illing, *Toxicity and Risk: Context, Principles and Practice*, Taylor and Francis, London, 2001.

IPCS (International Programme on Chemical Safety), *Assessing Human Health Risks of Chemicals: Derivation of Guidance Values for Health Based Exposure Limits*. Environmental Health Criteria 170, UNEP-ILO-WHO, World Health Organisation, Geneva, 1994 (available online through the IPCS-INCHEM website).

IPCS (International Programme on Chemical Safety), *Principles for the Assessment of Risks to Human Health from Exposure to Chemicals*, Environmental Health Criteria 210, UNEP-ILO-WHO-IOMC, World Health Organisation, Geneva, 1999 (available online through the IPCS-INCHEM website).

IPCS (International Programme on Chemical Safety), *Principles for Evaluating Health Risks to Reproduction Associated with Exposure to Chemicals*, Environmental Health Criteria 225, UNEP-ILO-WHO-IOMC, World Health Organisation, Geneva, 2001 (available online through the IPCS-INCHEM website).

OECD, *OECD Guidelines for Testing of Chemical*, two volumes and addenda, Organisation for Economic Co-operation and Development, Paris (revised 1993, with addenda).

OECD, OECD Principles of Good Laboratory Practice (as revised in 1997), *OECD Environmental Health and Safety Publications Series on Principles of Good Laboratory Practice and Compliance Monitoring*, No 1. ENV/MC/CHEM (98)17. Organisation for Economic Co-operation and Development, Paris, 1998 (available online through the OECD website).

OECD, *Descriptions of Selected Key Generic Terms Used in Chemical Hazard/Risk Assessment. Joint Project with IPCS on the Harmonisation of Hazard/Risk Assessment Terminology*, OECD Environment, Health as Safety Publications Series on Testing and Assessment, No 44. ENV/JM/MONO(2003)15, Organisation for Economic Co-operation and Development, Paris, 2003 (available online through the OECD website).

TGD, *Technical Guidance Document in Support of Commission Directive 93/67/EEC on Risk Assessment for Notified New Substances, Commission Regulation (EC) No 1488/94 on Risk Assessment for Existing Substances and Directive 98/8/EC of the European Parliament and Council Concerned with the Placing of Biocidal Products on the Market*, 2nd edn, EUR 20418 EN/1, European Communities, 2003 (available online through the European Chemicals Bureau website).

UNECE (UN Economic Commission for Europe), *UN Recommendations on the Transport of Dangerous Goods. Model Regulations*, 13th edn, United Nations, New York and Geneva, 2003a (available online through the UNECE website).

UNECE (United Nations Economic Commission for Europe), *Globally Harmonised System for the Classification of Chemicals (GHS)*, ST/SG/AC.10/30, United Nations, New York and Geneva, 2003b (available online through the UNECE website).

Chapter 6

Risk Management

H. PAUL A. ILLING

6.1 INTRODUCTION

The previous chapter was concerned with the technical and scientific process of risk assessment. This chapter is concerned with risk management, a process that involves societal and political judgements as well as technical considerations. The two are not easily separated. *Risk management* (OECD, 2003) consists of three elements:

- risk evaluation;
- emission and exposure control and
- risk monitoring.

Risk evaluation (OECD, 2003) is defined as:

> Establishment of a qualitative or quantitative relationship between risks and benefits of exposure to an agent, involving the complex process of determining the significance of the identified hazards and estimated risks to the system concerned or affected by the exposure, as well as the benefits bought about by the agent.

The process requires societal and political inputs, both in terms of the perceptual and legal frameworks within which decisions are taken and in terms of the criteria that lie behind the decisions taken. This chapter is concerned with matters that transcend scientific and technical knowledge.

Often, the last stage of risk assessment (risk characterisation) is inter-mingled with the first stage of risk management (risk evaluation). Where this is the case the combined process is considered in this chapter rather than Chapter 5.

In practice, once the risks have been evaluated, management is:

- ensuring adequate control of the risk (emission and exposure control, and risk monitoring)
- ensuring that the consequences of any residual risk that may remain when the risks are properly controlled can be dealt with (emergency planning).

6.2 THE RISK EVALUATION AND MANAGEMENT PROCESS

At this point it is necessary to examine concepts behind potential governmental (and supranational) approaches to risk evaluation and management. Specific legislation will not be dealt with here; only general principles will be discussed.

The starting point for consideration is national government. This can elect to take decisions directly or to set up means of delegating these decisions, either upwards to international bodies, such as the European Union (EU) and various United Nations Organisations, or downwards, to local government, companies or individuals. International companies also can act across national boundaries by imposing their own standards on local subsidiaries, but these have to be set in a manner acceptable to the countries in which they are operating.

Legislation is passed by governments (or by supranational organisations) which set up a decision-making structure and a general framework for risk management. The legislation may establish an organisation that considers the detailed technicalities of the problem and comes to a decision on whether to allow certain risk to be taken, and what detailed risk management procedures should be undertaken. Alternatively, legislation may leave legal liability associated with the consequences arising from the use of certain chemicals with the producer, user or consumer. Under these circumstances the producer or user performs the risk evaluation. Legislation may cover safety in the workplace, safety of the product or environmental protection. Until recently, legislation followed governmental departmental boundaries. Now a more holistic approach is being adopted and thus the problems of conflicting legislative requirements are being resolved.

Decisions on risk evaluation and management may be taken by consensus or by confrontation. In consensus-based procedures, technical aspects of the evaluation are often left to expert committees, consisting of people appointed for their technical and experience, or, for less contentious decisions, to single experts or officials. The wider societal input to decision taking can be made by consensus through representatives of the relevant interest groups (*e.g.* trade unions and industry leaders in the case of workplace safety), acting on behalf of society as a whole.

In a confrontational approach, all views on a problem may be presented using a judicial or quasi-judicial process. An appointed judge or arbitrator assesses the views expressed and either delivers an opinion to be promulgated by the appointed authority or, if the authority has been fully delegated, the decision. The latter approach is frequently used for decisions concerning the need for a particular industrial plant that may constitute a major hazard. The judge (arbitrator) would consider such problems as where to site the plant, what further developments to allow in the vicinity of the plant, *etc.* Often, confrontational approaches are employed when consensus is not achieved and one party (usually the intending supplier) is sufficiently aggrieved by a decision to be prepared to challenge it in the courts.

6.3 RISK CONSIDERATIONS

The basis for a regulatory process involving a control strategy taking into account risk and benefit considerations was originally set out in a report of a study group set up by the UK's Royal Society in 1983. The group suggested:

- An upper limit that should not be exceeded for any individual
- Further control, so far as is reasonably practicable, making allowance, if possible, for aversions to the higher levels of risk detriment
- A cut-off in the deployment of resources below some level of exposure or detriment judged to be trivial.

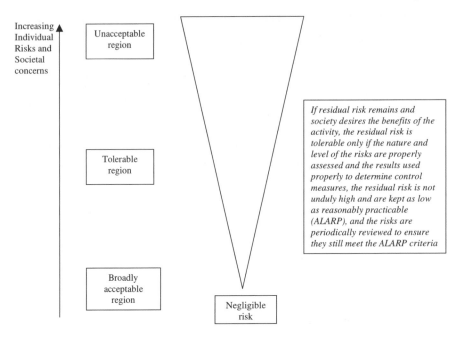

Increasing Individual Risks and Societal concerns

Unacceptable region

Tolerable region

Broadly acceptable region

Negligible risk

If residual risk remains and society desires the benefits of the activity, the residual risk is tolerable only if the nature and level of the risks are properly assessed and the results used properly to determine control measures, the residual risk is not unduly high and are kept as low as reasonably practicable (ALARP), and the risks are periodically reviewed to ensure they still meet the ALARP criteria

Figure 6.1 *Outline of the Royal Society approach to risk management/'tolerability of risk' framework (based on Health and Safety Executive, 2001)*

This approach can be set out diagrammatically (Figure 6.1).

The most recent restatement of this approach (known as the 'tolerability of risk' approach) is in a document, 'Reducing Risks, Protecting People' published by the UK Health and Safety Commission in 2001. It calls the three levels of risk defined by the Royal Society as 'unacceptable', 'tolerable' and 'broadly acceptable' risk.

'Reducing risks, protecting people' also identifies that there are three 'pure' criteria used by regulators in the health, safety and environment fields. These are:

- An equity-based criterion, which starts with the premise that all individuals have unconditional rights to certain levels of protection. In practice, this often converts into fixing a limit to represent the maximum level of risk above which no individual can be exposed. If the risk estimate derived from the risk assessment is above this limit and further control measures cannot be introduced to reduce the risk, the risk is held to be unacceptable whatever the benefits.
- A utility-based criterion that applies to the comparison between incremental benefits of the measures to prevent the risk and the cost of those measures to prevent the injury or detriment. This balance may be deliberately skewed towards benefits by ensuring there is gross disproportion between the costs and the benefits.
- A technology-based criterion that essentially reflects the idea that a satisfactory level of risk prevention is attained when 'state-of-the-art' control measures (technological, managerial, organisational) are employed to control risks whatever the circumstances.

These 'pure' criteria are not mutually exclusive and the 'tolerability of risk' approach employs all three criteria in combination.

The actual level of risk considered 'acceptable', 'tolerable' or 'unacceptable' must be a societal judgement taken through a political process, and often depends on the circumstances surrounding the exposure (see, *e.g.* IPCS, 1999a; Illing, 1999, 2001). A society's perception of the risks is a crucial factor to consider when criteria for 'acceptable', 'tolerable' and/or 'unacceptable' risk are being drawn up and when risk management procedures are being put into place (IPCS, 1999a; Illing, 1999). Risk perception is based on intuitive judgements. Different people perceive risks differently. Some of the factors affecting such judgements are discussed in Slovic (1997) and summarised in Illing (2001).

It is also necessary to distinguish between individual and societal (or population) risk when drawing conclusions. Individual risk is independent of population density, while societal (or population) risk depends on the population at risk and therefore considers the population contained in a defined geographic area. Individual risk is the usual form for dealing with small-scale accidental and anticipatable risks. Both individual and societal risks are considered when evaluating major accident risks. Population risk is usually the more important form of risk when considering risks for environmental effects.

When dealing with quantified risk assessment these definitions need to be linked to some form of numerical value for a defined effect. However, most toxicological risk assessments are qualitative or semi-quantitative, and often the criteria used are couched in much more general language.

Even when risks are properly managed there will be a residual risk. That residual risk is very small when the risk is deemed 'acceptable', and will be greater when the risk is in the 'tolerable' region. In addition, there will be residual risk due to inadequate risk management measures or non-compliance with risk management measures. 'Residual risk' therefore covers risks not known or not examined at the time of the evaluation, as well as risks arising from failure to control adequately (incidents and accidents, including major accidents).

6.4 CRITERIA FOR RISK EVALUATIONS: HUMAN HEALTH

6.4.1 Equity and Risk–Benefit

There are two extreme systems that can be applied when setting criteria for guideline values for chemicals.

1. A pure equity-based system. It starts with the premise that all individuals have unconditional rights to certain levels of protection. In a pure equity-based system exceeding the criterion is unacceptable, and exposure will need to be reduced until it is below that guideline value. In practice, this leads to precautionary decisions taken on 'worst-case' scenarios that may have little resemblance to reality.
2. A pure 'risk–benefit'-based system based on utility and technology-derived criteria. The utility criterion tends to ignore ethical considerations. A pure technology-based criterion ignores the balance between cost and benefit, and requires state-of-the-art technologies be transferred from one field to another, for example requiring the technology of 'clean rooms' (used in electronics and medicine) to be employed for manufacturing wood furniture. A pure risk–benefit-based system is therefore not practicable.

A possible alternative to the extreme systems described above is a mixed criteria system. In this system there are two equity-based criteria. The first is that for a 'broadly acceptable' level of risk. The 'broadly acceptable' risk is where the risk is considered insignificant and, if control is required to achieve it, is adequately controlled. The second is that the risk is 'unacceptable' under all circumstances. If the risk is 'unacceptable' the only appropriate risk management regime is to eliminate exposure. This is normally by banning the use of the chemical. If dealing with either a natural or an industrial point source of a chemical (*e.g.* a volcano or a large storage tank), then control can be by banning access to an area surrounding the source (creating an exclusion zone). Between these two criteria lies the 'tolerable' region of risk.

The 'tolerable' region represents the area where society indicates that it is prepared to tolerate the risk in order to secure benefits in the expectation that (Health and Safety Executive, 2001):

- the nature and levels of risks are properly assessed and the results used properly to determine control measures;
- the residual risks are not unduly high and are kept as low as reasonably practicable (ALARP) and
- the risks (in effect the guideline values) are periodically reviewed to ensure they still meet the ALARP criteria.

6.4.2   Conventional Toxicity (Non-Stochastic Effects)

Generally, when evaluating anticipatable risk, the aim is to achieve low levels of risk. In the first instance, health-based or scientifically based approaches are aimed at guidelines associated with a combination of frequency and severity of ill-health effect that is considered 'broadly acceptable' (sometimes worded 'without significant risk to health'). If a guideline based on the 'broadly acceptable' risk level is not attainable, then it may be necessary to bring other factors, such as practicability, into play and to set a guideline value within the 'tolerable' band of risk, together with appropriate risk management measures. The latter guideline may be set in a risk management regime indicating that the guideline must not be exceeded and exposure should be minimised as far as possible, or ALARP below the guideline value. Alternatively, it may be a temporary guideline with a given a time limit for its achievement, a target on the way to achieving a 'broadly acceptable' level of risk.

Guidelines may be for 'input', 'intake' or 'uptake'. Because of the assumptions that have to be made in the models used for moving from an 'input' guideline to an 'intake' or 'uptake' guideline (or vice versa), 'input' guidelines tend to be operational, rather than purely health-based.

The usual method for extrapolating from conventional toxic effects (non-stochastic effects) to an 'intake' or 'uptake' guideline is to adjust the 'no observed adverse effect level' (NOAEL), *etc.* (see Chapter 5) for the critical effect in the pivotal study, using uncertainty factors, to yield a 'reference dose' (RfD) (also called the 'tolerable intake' in IPCS, 1994, 1999a). The RfD (OECD, 2003) is defined as:

An estimate of the daily exposure dose that is likely to be without deleterious effect even if continued exposure occurs over a lifetime.

The 'uncertainty factors' approach can be described by the equation

$$RfD = \frac{NOAEL \ (or \ LOAEL \ or \ BMD)}{uncertainty \ and \ modifying \ factors}$$

(LOAEL is lowest observed adverse effect level and BMD is Benchmark dose). The critical effect is that which, under the envisaged exposure conditions, yields the lowest maximum exposure when the NOAEL (*etc.*) and uncertainty factors are combined. The pivotal study is the study from which the critical effect NOAEL is derived (Figure 6.2).

The uncertainty and modifying factors cover:

- Nature of toxicity – a factor of 10 is incorporated when the critical effect is an irreversible phenomenon, such as a teratogenic effect or a non-genotoxic carcinogenic effect.
- Adequacy of database/quality of data – 1, 3, 5 or 10 are preferred, but up to 100.
- Extrapolation from LOAEL to NOAEL – usually a factor of 3, 5 or 10.
- Inter-species extrapolation – usually 10 (4 for toxicokinetics and 2.5 for toxicodynamics).
- Inter-individual variation in humans – usually 10 (3.2 for toxicokinetics and 3.2 for toxicodynamics).
- Duration of study (*e.g.* 28 day to 90 day or to 2-year repeated dosing).
- Route to route extrapolation (usually from parenteral or oral to inhalation). (IPCS, 1994, 1999a, 2001; Illing, 2001).

If evidence is available identifying a more accurate value for use as the appropriate factor that value should be used in place of the default values above. The most common overall default value for a good quality data set derived from animal studies continues to be 100, increased to 1000 when teratogenic effects or non-genotoxic carcinogenic effects are being examined.

The reference dose can be applied directly to an uptake guideline. The guideline for a specific intake may be obtained by applying the RfD directly on the basis that it is the primary route of exposure, or by allocating between intake guidelines according to the expected proportion taken in by that route (IPCS, 1994).

In essence, the uncertainty factors approach is using a fixed criterion of acceptability and applying it to the NOAEL, *etc.* to derive a maximum level of exposure considered 'broadly acceptable'. It therefore mixes risk assessment and risk evaluation. The 'margin of exposure' approach (see previous chapter) identifies the actual difference between the exposure and the effects level (the risk assessment) and separates it from the risk evaluation, the decision concerning its adequacy.

For both approaches the risk evaluation needs to take into account how risks from different exposure scenarios are perceived. Environmental air quality guidelines and, to a slightly lesser extent, food and drinking water guidelines for unintended contaminants, are perceived as involuntary exposure, with lack of personal control over exposure and lack of visible benefit. The initial 'health-based' guideline is therefore an equity-based guideline using substantial uncertainty factors/margins of exposure. This essentially equity-based approach is also used for food additives and covenanted contaminants (such as pesticides (plant protection products and biocides)

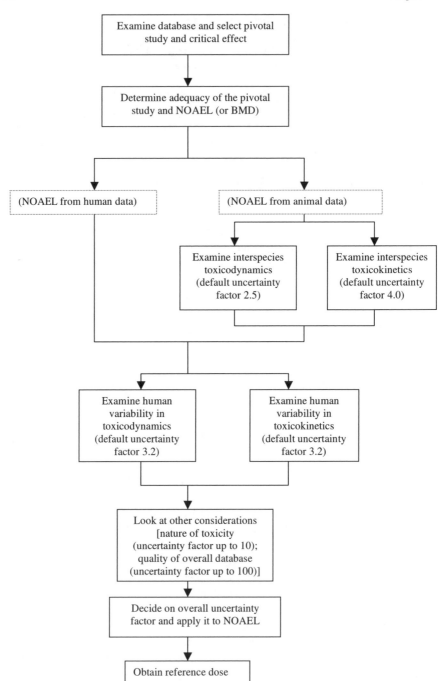

Figure 6.2   *Procedures for the derivation of uncertainty factors (from Illing, 2001; based on IPCS, 1994)*

and veterinary product residues in food). Furthermore, when 'gate keeping', setting guidelines at the same time as giving permission to use a chemical, is undertaken a precautionary approach is possible. There is a particular incentive to keep uncertainty factors (or margins of exposure) high to prevent the public outcry if ill health is perceived as being associated with an 'unsafe' chemical.

Uncertainty factors (or margins of exposure) may be much diminished when the guideline has to be incorporated into a mixed criteria system. Occupational exposures are, at least to some extent, perceived as voluntary, involving some personal control over outcome through choice of workplace and control options, and there is a benefit to society that should be fed back to the individual. Exposure to human medicines is, by comparison, largely voluntary and highly controlled. There is also a clear benefit arising from the exposure (the therapeutic effect). Thus, for the medicines the 'tolerability of risk' approach is most appropriate, and if the benefit is sufficient and no better approach is available, margins between effective dose and toxic dose may be small.

There are circumstances when implementation of a 'broadly acceptable' health-based risk criterion may be impossible. For example, current exposure may exceed that criterion. Under those circumstances it may be necessary to see if the risks associated with the attainable level of exposure are 'unacceptable in any circumstance', in which case manufacture and use of the chemical is likely to be prohibited. In pure equity-based systems the likelihood is that exceeding the 'broadly acceptable' risk criterion will be unacceptable under any circumstance. In the 'tolerability of risk' framework, there will be a range of exposures that are 'tolerable', provided that the risk is properly managed. If an exposure arises from natural sources it may not be possible to ban the presence of the chemical. If exposure arises from a point source it may be possible to base management on equity and delineate exclusion zones, but if the chemical is of diffuse natural origin there may be no alternative to management based on the 'tolerability' model.

6.4.3 Stochastic Effects

A variety of risk evaluation processes are used for the evaluation of carcinogens. Extrapolation from human epidemiological data is sometimes possible, in which case the extrapolation is likely to be relatively limited. More often extrapolation from animal studies is required. In the US and some European countries this is undertaken using information from the dose–response curve. The information is extrapolated to extremely low-risk levels (often the level of one excess cancer in a population of 1 million) using mathematical models. The actual criterion is a societal decision. Traditionally the preferred model was the 'linearised multistage model' allowing for a 95% confidence boundary; currently linear extrapolation is the preferred default model (IPCS, 1999a). UK regulators (see IGHRC, 2002 for a recent re-statement of the UK position) do not favour this approach. They claim:

- The methods are not validated
- They are often based on incomplete or inappropriate data
- They are derived more from mathematical assumptions than from knowledge of biological mechanisms
- They demonstrate a disturbingly wide variation in risk estimates, depending on the models used.

The claim is that the models may give an impression of precision that cannot be justified in the light of the approximations and assumptions on which they were based.

Once a genotoxic carcinogen has been identified, the UK moves straight from hazard characterisation to risk management. This implies that no 'broadly acceptable' risk can be derived for genotoxic carcinogens. The risk management policy adopted is to eliminate exposure or, where this is not possible, to reduce exposures so that they are as low as is reasonably practicable. Where elimination of exposure is possible, the assumption is made that no exposure is acceptable. This is a pure equity-based risk evaluation. Where elimination is not possible the risk management approach is that associated with 'tolerable' risk.

Germ cell mutagens (commonly) and occupational asthmagens (in the UK) are treated similar to carcinogens.

6.4.4 Major Accident Hazards

For major accident hazards the aim of the risk evaluation is to identify appropriate risk contours delineating geographical areas that correspond to particular zones around the source. This involves developing toxicity information in a suitable form for integrating with the exposure information. The information required is a concentration–time relationship for a 'selected level of toxicity' for use in developing the risk contour. The choice of 'selected level of toxicity' includes choices concerning frequency and type of effect; thus it involves societal and political judgements concerning the acceptability of the risk. The information presented here is a summary of the information collated in Illing (2001).

6.4.4.1 Land Use Planning.   For land use planning, surrounding a major accident hazard site the UK uses a single 'selected level of toxicity' – the *'dangerous dose'*. At this dose (Fairhurst and Turner, 1993):

- Severe distress will be caused to almost everyone
- A substantial fraction will require medical attention
- Some people are seriously injured and require prolonged treatment
- Any highly susceptible person may be killed

In practice, it is a combination of concentration and time that is required for inhalation of the chemical and thus the LCT_{50} (concentration–time relationship resulting in 50% mortality) or ECT_{50} (concentration–time relationship resulting in 50% responders to a given level of effect) are the important statements. In the early years of the 20th century, when examining lethality, the relationship

$$Ct = \text{constant}$$

was obtained, where C is concentration and t the time and it subsequently became known as the Haber rule. More recent observations suggest that the general relationship is

$$C^n T = \text{constant}$$

Although a wide range of values for n has been cited, usually the value appears to be close to 1 or to 2. In the absence of further information, a default value of 1 is commonly used. Normally the 50% values are transformed to another set of values representing a low percentage (1–5%) of deaths or effects (the 'dangerous dose'). Probit analysis, using 'best estimate' data for a single strain and species, may be used to carry out this transformation if direct observation is difficult. As a default, an arbitrary ratio can be used to compare the 'dangerous dose' and the LD_{50}/LCT_{50}:

$$\text{dangerous dose} = LD_{50} \text{ (or } LCT_{50}) /2.57$$

(Franks *et al.*, 1996). If a concentration–time relationship has been derived from a modern study based on the 'fixed dose' procedure and 'evident toxicity', the set of values representing a low percentage of deaths would be obtained directly. Toxicological judgement is applied to allow for heterogeneity in the human population.

6.4.4.2 Emergency Planning. Emergency planning is concerned with mitigating the consequences after an event has occurred. It involves both 'on-site' and 'off-site' planning and may cover immediate 'emergency' shutdown, responses of emergency services, the medical management of immediate and longer term health effects, and the management of food and drinking water supplies and potential environmental effects (IPCS, 1999b). The planning is concerned with a much wider range of biological effects. Generally, for immediate effects on human health, these include delineation of zones around the source in which different 'selected levels of toxicity', namely 'death/permanent incapacity', 'severe health effects' (disability, requiring rescue and hospital treatment), or 'mild health effects' (discomfort or distress, detection or nuisance) are likely to be present. A number of exposure limits, such as ECETOC's 'emergency exposure index' (ECETOC, 1991), the US National Advisory Committee's 'acute exposure guideline value' and the American Industrial Hygiene Association's 'emergency response planning guideline' (IPCS, 1999b), have been developed using toxicological information. They are intended for use by rescue personnel in emergency situations.

6.5 CRITERIA FOR RISK EVALUATION: ENVIRONMENT

The EU has developed a system for risk evaluation for environmental effects (TGD, 2003). This risk evaluation includes a number of goals, such as protecting:

- The aquatic ecosystem (including sediments and marine ecosystems)
- The terrestrial ecosystem
- Top predators
- Microorganisms in sewage treatment works
- The atmosphere

The risk evaluation is based on the 'risk characterisation ratio' (RCR), which depends on the 'predicted exposure concentration (PEC)' and the 'predicted no effect concentration (PNEC)'. If the RCR (PEC/PNEC) is greater than 1 there is little concern for that particular end point, if it is higher than 1 then it is likely that further information will be required. A schematic representation of the process is given in Figure 6.3.

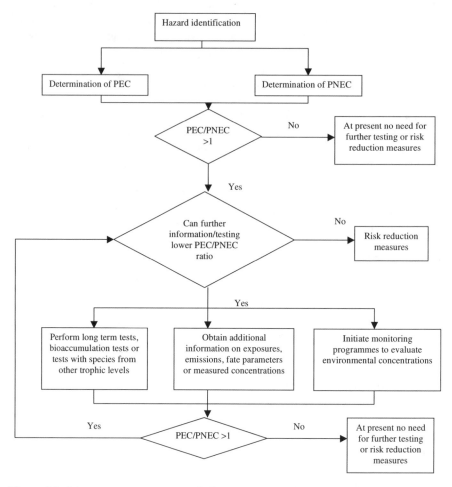

Figure 6.3 *Schematic representation of the environmental risk evaluation process (from TGD, 2002, with minor modification)*

As uncertainty factors have been included in the determination of the PNEC, the RCR incorporated societal judgements on concern ('acceptability'), and hence mixes risk characterisation and risk evaluation. In addition, uncertainty factors cover:

- The variation within a given laboratory and between different laboratories performing the assays
- Intra-species variations due to variations in the physiological state of individuals of the same species
- Extrapolation of acute or short-term toxicity towards chronic or long-term toxicity, respectively
- Extrapolation of laboratory results to the field

Uncertainty factors generally vary from 0 to 1000, depending on the nature and quality of the evidence, with larger safety factors being associated with more limited evidence (Table 6.1). Those for sediment vary from 10 to 100.

Effects on microorganisms in sewage plants are also considered using assessment factors. Additional factors for bioconcentration and biomagnification are included

Table 6.1 *Basis for uncertainty factors for inland water, soil and saltwater*

Assessment factor	$PNEC_{aquatic}$ (freshwater)	$PNEC_{soil}$	$PNEC_{water}$(saltwater)
10,000			Lowest short-term $L(E)C_{50}$ from freshwater or saltwater representatives of 3 taxonomic groups of three trophic levels
1000	≥ 1 short-term $L(E)C_{50}$ from each of three trophic levels	≥ 1 short-term $L(E)C_{50}$	Lowest short-term $L(E)C_{50}$ from freshwater or saltwater representatives of 3 trophic levels + 2 additional marine taxonomic groups *or* 1 long-term NOEC
500			2 long-term NOECs from freshwater or saltwater species at 2 trophic levels
100	1 long-term NOEC	1 long-term NOEC	Lowest long-term NOEC from 3 freshwater or saltwater species representing 3 trophic levels
50	2 long-term NOECs for different trophic levels	Long-term NOECs for 2 trophic levels	2 long-term NOECs from freshwater or saltwater species + 1 long-term NOEC from an additional marine taxonomic group
10	Long-term NOECs from ≥ 3 species representing 3 trophic levels	Long-term NOECs from ≥ 3 species representing 3 trophic levels	Lowest long-term NOEC from 3 freshwater or saltwater species representing 3 trophic levels + 2 long-term NOECs from additional marine taxonomic groups
5–1	SSD method	SSD method	
Case-by-case	Field data or model ecosystems	Field data or model ecosystems	

Note: SSD method, species sensitivity distribution method. Tests are for appropriate species for that medium unless otherwise stated.

when looking for possible effects in top predators (secondary poisoning). More qualitative assessments are used for assessing atmospheric risks.

At present there are no adequate criteria for evaluating some types of environmental concern. For example, there are no criteria for evaluating chemicals with respect to their potential for acting as environmental oestrogens.

6.6   TOLERABLE RISK

The concept of tolerable risk carries with it the view that there is a benefit associated with, and outweighing, the risk. Decisions on managing these risks depend on the practicality of possible risk management procedures. Phrases such as 'as low as is reasonably achievable (practicable) [ALARA(P)], 'best available technique' and 'best practicable environmental option' are intended to cover these requirements. This can lead to multiple standards, one setting the minimum that is considered achievable by everyone and others indicating a timetable for achieving an equity-based 'broadly acceptable' risk, or to a standard that carries with it a continuing duty to improve. If the risk is deemed 'intolerable', then the risk management is aimed at eliminating use of or exposure to that chemical and/or remediation of the damage caused by that chemical.

6.7   CRITERIA FOR RISK EVALUATION: FURTHER COMMENTS

Deisions on whether a risk is broadly acceptable, tolerable or unacceptable, and about appropriate risk management, are essentially societal decisions. They depend on public perception and opinion. Consequently, individuals and groups of individuals within that society may not agree with the decision arrived at and may seek to change public perception.

The approach outlined above allows for evidence being insufficient by applying the 'precautionary principle'. Additional uncertainty factors are incorporated in order to allow for uncertainties in the information. This may take the form of adopting a lower exposure limit as 'broadly acceptable', or to treat a risk as unacceptable. If further evidence can be gathered, then reduction in the uncertainty concerning the risks properly leads to a re-evaluation of how they should be managed.

6.8   RISK MANAGEMENT

Risk management is concerned with the consequences of risk evaluation. When dealing with broadly acceptable risks or unacceptable risks, no further management action is needed, except that, where there is a ban, it will need to be enforced. However, decisions may be much harder when dealing with tolerable risk. Ideally, management is required to reduce tolerable risks to the point where the risk is broadly acceptable. This may not be possible, but at least risks should be reduced as far as is reasonably practicable.

6.8.1   Anticipated Exposure and Minor Accidents

For anticipated exposure and for minor, foreseeable accidental exposure, the approach to controlling risks associated with the workplace may be summarised as a combination of prevention (or minimisation) of exposure and dealing with the consequences (ill health and or pollution). Prevention and minimisation can be by:

- *Substitution*: use another, less risky process or chemical (but beware the dangers of different types of risk; how do you trade fire versus acute ill-health versus cancers versus environmental damage, and how do you trade the known and the unknown risk).
- *Engineering control*: change the process to prevent/reduce exposure and/or minimise emissions, use (or improve) process containment, and, if process improvement is inappropriate, improve workplace ventilation and/or secondary containment (collect the chemical before it is vented/discharged from the workplace, for example using suitable filters, or dilute it to 'safe' concentrations) to prevent/minimise workplace ill-health and environmental pollution.
- *Personal protective equipment*: includes respiratory protective equipment, clothing, gloves and footwear to prevent exposure.

As a general principle substitution is preferred to engineering control, which is preferred to protective equipment. Proper maintenance of process and equipment is essential to ensure that they function correctly. Measurements of concentrations of chemicals in the workplace and the surrounding environment and health and pollution monitoring may be needed to ensure that controls are adequate. All of these approaches need to be adopted within a management attitude that supports safety (and pollution) awareness.

6.8.2 Product Safety

Product safety is concerned with whether the use to which the product containing the chemical will be put is safe and non-polluting. One way of achieving this is for the organisation marketing the product to carry out its own risk assessments, based on the likely behaviour of the recipient. If society considers that a matter is sufficiently important, then the decision may be taken by government or internationally. They will base their decision on what they consider to be a broadly acceptable risk level for the population as a whole. Where there is consensus on what constitutes a broadly acceptable risk the use of internationally derived values is useful. For example, the Joint Evaluation Committee on Food Additives (of the World Health Organisation and Food and Agriculture Organisation) collects and evaluates data for chemicals in food and identifies the maximum levels (acceptable daily intake if intentionally added, tolerable daily intake if present as contaminants: beware terminology) constituting a broadly acceptable risk.

There are circumstances when tolerable risk is a relevant consideration when dealing with product safety. This is usually when some form of risk–benefit evaluation is undertaken. Two examples illustrate this. Firstly, government may permit the use of a drug with a high incidence of side effects on seriously ill patients. This can usually be done after careful judgement by the specialist charged with the patient's care and with the patient's consent, when the overall benefit to the individual is likely to outweigh the risk. Secondly, a food or drinking water additive may be intentionally introduced to prevent food spoilage or dental decay. The latter example is concerned with a societal judgement concerning individual risks. It can result in controversy if a significant number of the individuals affected do not perceive the risks in the same way as the decision taking body.

6.8.3   Enforcement

Enforcement of the outcomes of risk management decisions may be through disciplinary committees of relevant professions or groups or by statutory or insurance-based inspection. The latter may lead to sanctions such as fines following a court appearance or increased insurance premiums. Where failures do occur, or new risks emerge, they should be investigated, lessons learnt, and, if appropriate, retribution sought. Statutory bodies may instigate this process or it may be the consequence of individuals seeking redress through a judicial process.

6.8.4   Major Accident Hazards

The risk evaluation is intimately inter-mingled with the risk management. The management is in the form of decisions concerning the siting of the industrial plant and of new developments in the vicinity of the plant, in dealing with process design and in dealing with residual risk through adequate emergency planning. All of these can be addressed when evaluating the risks and developing safety cases for this type of hazard.

6.9   CONCLUSIONS

Risk assessment and risk management are fast developing fields. In the 10 years since the first edition of this book there has been a massive increase in the interest shown in this field and a considerable clarification of our knowledge and procedures. Risk assessment and risk management are iterative procedures. The risk assessment has to be revisited when there is new information on risks or a new interpretation of a study. The risk evaluation has to be reviewed when this happens or when there are new insights into how society perceives risk. Many decisions concerning the risks associated with chemicals exposure become internationalised, and there is a danger that they become remote from the people affected. Hence, unless carefully taken, these decisions are unacceptable to the affected population. Further the whole process needs revisiting periodically to ensure that management and enforcement procedures are still appropriate, *in situ* and effective. Risk assessors and risk managers are never without work!

BIBLIOGRAPHY

ECETOC. Emergency Exposure Indices for Industrial Chemicals, Technical Report No. 43. European Chemical Industry Ecology and Toxicology Centre, Brussels, 1991.

S. Fairhurst and R. Turner, Toxicological assessments in relation to major hazards, *J. Hazard. Mater.*, 1993, **33**, 215–227.

A.P. Franks, P.J. Harpur and M. Bilo, The relationship between risk of death and risk of dangerous dose for toxic substances, *J. Hazard. Mater.*, 1996, **51**, 11–14.

Health and Safety Executive, Reducing risks, protecting people, *HSE's Decision-Making Process*, HSE Books, Sudbury, Suffolk, 2001.

IGHRC (Interdepartmental Group on Health Risks from Chemicals), Assessment of chemical carcinogens. Background to general principles of a weight of evidence approach. CR8. Institute for Environment and Health, University of Leicester, 2002.

H.P.A. Illing, Are societal judgements being incorporated into the 3 uncertainty factors used in toxicological risk assessment? *Regul. Toxicol. Pharmacol.*, 1999, **29**, 300–308.

P. Illing, *Toxicity and Risk: Context, Principles and Practice*, Taylor and Francis, London, 2001.

IPCS (International Programme on Chemical Safety), 'Assessing human health risks of chemicals: Derivation of guidance values for health based exposure limits', *Environmental Health Critera 170*, UNEP-ILO-WHO, World Health Organisation, Geneva, 1994 (available online through the IPCS-INCHEM website).

IPCS (International Programme on Chemical Safety), 'Principles for the assessment of risks to human health from exposure to chemicals', *Environmental Heath Criteria 210*, UNEP-ILO-WHO-IOMC, World Health Organisation, Geneva, 1999a (available online through the IPCS-INCHEM website).

IPCS (International Programme on Chemical Safety), Public Health and Chemical Incidents. Guidance for National and Regional Policy Makers in the Public/Environmental Health Roles. International Clearing House for Major Chemical Incidents. University of Wales Institute, Cardiff, 1999b.

IPCS (International Programme on Chemical Safety), Principles for evaluating health risks to reproduction associated with exposure to chemicals, *Environmental Health Criteria 225*, UNEP-ILO-WHO-IOMC, World Health Organisation, Geneva, 2001 (available online through the IPCS-INCHEM website).

OECD, Descriptions of selected key generic terms used in chemical hazard/risk assessment. Joint project with IPCS on the harmonisation of hazard/risk assessment terminology. OECD Environment, Health as Safety Publications Series on Testing and Assessment No 44. ENV/JM/MONO(2003)15. Organisation for Economic Co-operation and Development, Paris, 2003 (available online through the OECD website).

Royal Society Study Group, *Risk Assessment*, Royal Society, London, 1983.

P. Slovic, Trust, emotion, sex, politics and science: Surveying the risk assessment battlefield, in M. Bazerman, D. Messik, A. Tenbrunsel and K. Wade-Benzoni (eds), *Environment, Ethics and Behaviour*, New Lexington Press, San Francisco, 1997 (reprinted in Risk Anal. 1999, **19**, 689–702).

TGD, *Technical Guidance Document in Support of Commission Directive 93/67/EEC on Risk Assessment for Notified New Substances, Commission Regulation (EC) No 1488/94 on Risk Assessment for Existing Substances and Directive 98/8/EC of the European Parliament and Council Concerning the Placing of Biocidal Products on the Market*, 2nd edn, EUR 20418 EN/1, European Communities, 2003 (available online through the European Chemicals Bureau website).

Chapter 7

Exposure and Monitoring

DOUGLAS M. TEMPLETON

7.1 INTRODUCTION

Exposure and monitoring are terms used together in occupational medicine to refer to the various means of assessing exposure to harmful (or potentially harmful) substances in the workplace, their intake and accumulation, the health risks that such exposures entail, and how we do or should keep track of these exposures. Exposure in the environment of the workplace is referred to as ambient exposure. Ambient exposure to a substance that is absorbed, generally through the lungs or mucous membranes of the upper airways, across the skin, or in the digestive system, gives rise to an absorbed dose that can be defined as the amount of a substance taken into the body. The ultimate goal of monitoring is only realized when an endpoint is identified that can be measured accurately and can be shown to link exposure to the substance with predictable adverse effects at a given absorbed dose. Medical assessment of the adverse effects themselves (*e.g.* the development of lung disease or malignancy) is beyond the scope of monitoring and is part of a programme of health surveillance. However, when there is the potential for harmful exposures, monitoring programmes and health surveillance go hand in hand. Many of these same principles apply in epidemiological studies that assess environmental exposures of populations.

The distinction between health surveillance and monitoring may be blurred; the monitoring programme may consist in part of additional investigations that are not part of a routine annual medical examination, and which may or may not give a positive result because of exposure to the substance of concern. For example, workers in the hard metal industries may, under some circumstances, have an annual medical examination that includes evaluation of selected pulmonary function tests and blood samples taken for serum cobalt measurement. A small number of these workers will develop interstitial pulmonary disease but the link, if any, with cobalt is not well understood. In this example, pulmonary function tests can be considered part of health surveillance in relation to all aspects of employment, while cobalt measurement is part of a monitoring programme, specifically documenting cobalt exposure.

This chapter considers some basic definitions and aspects of monitoring, the criteria for development of a successful monitoring programme, the expanding role of biomarkers in monitoring, and some ethical considerations that arise when monitoring is implemented. Examples of monitoring of exposures to both organic and inorganic substances are used to illustrate these principles.

7.2 GENERAL PRINCIPLES

It is usual to distinguish between ambient monitoring and biological monitoring of exposure. In the former, the workplace is monitored, for example by air sampling for the presence of dusts, swipe tests for radiation contamination, *etc.* From the results, inferences about exposure of the whole work force may be drawn. Biological monitoring of exposure, however, is intended to assess the absorbed dose of the individual worker. In some cases, ambient monitoring gives a good indication of individual exposure, while in others even personal samples do not give a very good indication of the absorbed dose. For example, several independent studies from different countries conclude that exposure to soluble compounds of cobalt at 0.05 mg m^{-3} in the workplace air consistently produces a concentration of cobalt in workers' urine of about 30 µg L^{-1} at the end of a shift. The major route of exposure in these studies is inhalation. In contrast, exposure to cadmium results in quite different absorbed doses in individual workers. Here, ingestion is an important additional route of exposure and individual differences in hygiene (*e.g.* washing the hands before eating) and hand-to-mouth activity (including handling cigarettes) determine quite different internal exposure levels, even when personal sampling devices indicate similar ambient exposures.

A third stage of monitoring is monitoring of effects. Here, biological effects are sought that arise from exposure but that are themselves reversible and/or without serious consequences for health. It must be stressed that this is quite distinct from diagnosis of occupational diseases. The latter represent a failure of preventive programmes, a major component of which may be monitoring. However, as noted above, this distinction, although clear in principle, is not always so clear in practise. Effects thought harmless at one time may later be found detrimental. New toxic effects are always being elucidated; yesterday's inconsequential exposures may be today's health problems. Individual variation in sensitivity is another reason to be cautious in interpreting effects. The very concept of 'biological effects of no consequence' may be controversial.

Temporal issues as they relate both to exposures and effects are also important in monitoring. Exposures may be short or long term, and monitoring has a role in both. Short-term exposures during a working week, a single shift, or even for a very brief time during a shift may all have special consequences. At the other extreme, cumulative exposure over a lifetime may be more significant. High short-term exposures, when they give rise to adverse effects, tend to produce acute toxicity. Cumulative exposure in the long term is more often associated with chronic disease. Although this generalization may seem obvious, it is by no means always true. A single brief exposure to a high dose of radiation may produce a chronic leukemia with onset many years later, whereas exposure to cadmium may present acutely as sudden renal failure after many years of accumulation.

The above comments notwithstanding, closer correlations between exposure and consequence are more usual. This principle is well illustrated by exposures to different compounds of nickel. At one extreme, nickel carbonyl is exceptionally toxic; inhalation of this volatile form of nickel at certain concentrations in air can cause death rapidly, and in cases of accidental exposures urinary nickel is monitored hour by hour. Survival of acute nickel carbonyl poisoning, however, is without known long-term effects and subtle on-going exposures are apparently without consequence. At the other extreme, inhalation of insoluble nickel compounds in dusts has no immediate consequences and is fairly difficult to detect by biological monitoring. It does, however, result in deposition of nickel in the upper airways that may be detected on nasal mucosal biopsy even years after retirement from nickel-related occupations, and its presence poses an increased risk of eventually developing cancer in the upper airways. The majority of workers exposed occupationally to nickel fall between these extremes. They are exposed to a mixture of soluble and insoluble forms of the metal, are monitored for during-shift exposure by urinary nickel measurement, and may be (though unproven) at slightly increased risk of developing dermatitis and/or upper-airway cancer.

Several terms are used to describe ambient and internal exposures; among the most commonly used for setting permissible exposure levels are those of the American Conference of Governmental Industrial Hygienists (ACGIH) (2004) and the Deutsche Forschungsgemeinschaft (DFG) (2004). ACGIH publishes threshold limit values (TLVs) for a number of substances that are guidelines to upper ambient exposures. Because a particular level of exposure may be tolerable in the short term but not so for on-going exposure, three different TLVs are used. A time-weighted average (TWA) TLV is the TWA exposure during an 8-h day and 40-h working week to which nearly all workers may be exposed indefinitely without adverse effects. A short-term exposure limit (STEL) is a 15-min TWA exposure that should not be exceeded at any time even if the 8-h exposure is within the TLV TWA. Exposures at the STEL should not be for more than 15-min, not more than four times per day, and not more than once in a 60-min period. A ceiling value (TLV-C) is one that should never be exceeded for any length of time.

On the other hand, DFG defines a single maximum concentration value in the workplace, Maximale Arbeitsplazkonzentration (MAK), which it considers as 'the maximum permissible concentration of a chemical compound present in the air which, according to the current knowledge, does not impair the health of the employee or cause undue annoyance'. For some substances there is concern that no level of exposure is free from adverse health consequences. For example, if a substance is carcinogenic, it may have no threshold value for increased risk yet exposure in the workplace may be unavoidable. Under these circumstances, DFG uses a technical exposure limit, Technische Richtkonzentration (TRK), based on attainability with current technologies to serve as a guideline for necessary protective procedures. Returning to the above example of nickel, considered by DFG as carcinogenic and therefore presumed to be without any risk-free level of exposure, the inevitable use of the metal in industrialized society dictates a realistic approach. DFG believes that current technology can reasonably assure exposures below 0.5 mg m^{-3} while maintaining essential industries, and sets this as the TRK.

MAK, TRK, and TLVs describe ambient exposures. Internal exposure is addressed by ACGIH with a biological exposure index (BEI), which represents the warning level of a particular indicator in an appropriate biological sample. The comparable (though not strictly equivalent) term used by the DFG is the biological tolerance value, Biologische Arbeitsstoff Toleranz (BAT), which it defines as, 'the maximum permissible quantity of a chemical compound (or) its metabolites, or (the magnitude of) any deviation from the norm of biological parameters induced by these substances in exposed humans'. Examples of these different quantities should make this definition clearer. The BAT for aluminum is simply reported as 200 µg L^{-1} in urine; the element itself is measured. In contrast, carbon disulfide is extensively metabolized and one of its metabolites (2-thiothiazolidine-4-carboxylic acid (TCCA)) is measured in urine. The BAT for carbon disulfide is accordingly give as 8 mg L^{-1} of TCCA in urine. In the case of acetylcholine esterase inhibitors, neither the chemical nor its metabolites is measured. Rather, the BAT is reported as a reduction in erythrocyte acetylcholine esterase activity to 70% of the reference value. In each of these above examples, the sample is taken at the end of the working shift.

There is a basic difference between BAT and BEI values, and neither is without problems. The BAT is health-based and assumes sufficient information to decide on levels with no adverse effects. Such information is frequently unavailable. The BEI, however, is the level of the parameter expected in an average worker exposed at the TLV, and so assumes inhalational exposure and absorption. Individual variation means that some workers will have levels above the BEI, and the importance of these absorbed doses is generally poorly understood.

Other jurisdictions may use their own terminology. Many define biological action levels for substances that conceptually approximate the BEI and BAT.

7.3 CRITERIA FOR A MONITORING PROGRAMME

In order to monitor exposure in a specific instance, or to assess the worth of a monitoring programme, a number of questions should be considered.

7.3.1 Is Biological Monitoring a Useful Supplement to Ambient Monitoring?

If ambient monitoring shows acceptably low levels of contamination, there may be little to add by biological monitoring. However, under some circumstances, absorbed doses may be unacceptable in certain individuals due to additional or unpredictable routes of exposure (*e.g.* oral) and this will only be detected by individual biological monitoring. The goals of the programme need to be defined. Is biological monitoring to be used as an adjunct to ambient monitoring only to assess exposure, or is there also sufficient information to relate absorbed dose to risk?

7.3.2 Is there Sufficient Information on the Handling of the Substance by the Body to Justify Biological Monitoring?

After the goals of the programme are defined, it must be decided what sample will be collected to meet them. Information on absorption, distribution, excretion, toxicokinetics, toxicodynamics, and metabolism must all be taken into account to decide on an appropriate sample to collect. It must be established that there is an analyte in the specimen (typically urine, blood, or serum) that correlates with exposure or, if desired, risk. The rate of appearance and disappearance of the analyte in the specimen must also be taken into account in order to decide the timing of the collection. For instance, there is no point in collecting an end-of-shift urine sample for a metabolite that only appears in the urine at a later time, or in delaying until the next day blood sampling for a substance that is cleared from the circulation with a half-life of minutes. Establishing correlations between exposures and the handling of substances or metabolites requires a good deal of scientific effort, and is frequently a major impediment to a new monitoring programme.

7.3.3 Is there a Reliable Analytical Method for Measuring the Chosen Parameter?

An appropriate analytical method must be chosen to measure the intended analyte at concentrations that can be expected to be significant for the monitoring programme. It is also desirable to be able to measure background levels of the analyte (if present) in a reference population not exposed occupationally to the substance, so that the spectrum of exposures can be appreciated. Acceptable protocols frequently involve relatively sophisticated technology such as gas–liquid chromatography, gas chromatography–mass spectrometry, or Zeeman-corrected electrothermal atomic absorption. Although generally available in the developed world, adequate monitoring may still necessitate the participation of an experienced analytical laboratory to ensure proper application of the techniques. When available for a given analyte, interlaboratory quality assurance programmes are indispensable for ensuring reliable results. The development of new biomarkers will rely increasingly on biotechnologies for which quality assurance programmes are generally unavailable. For example, the implementation of lymphocyte proliferation tests for metal sensitization is very difficult to standardize between different laboratories.

7.3.4 Is the Measurement Interpretable?

Even if a correlation between measurement of a substance and exposure or risk has been established in a study population, sources of variation in a particular workforce must be appreciated. Causes of individual variation and non-occupational exposure must be considered. Sometimes these will be unimportant, as for example, when the only source of a unique metabolite of a pesticide residue would be occupational exposure to that pesticide. In other cases, for instance with many trace metals, other environmental and non-occupational exposures may contribute significantly to the measured value.

7.3.5 Are the Consequences of the Measurement Foreseeable?

Before monitoring is begun, it should be decided how the results will be used. Can it be determined when an exposure is inappropriate? For many substances there are published guidelines or legislated values, which are not to be exceeded. If there are not, a decision must be made on whether to monitor, and if so what to set as an action level based on the best knowledge available at the time. A plan should be in place for correcting operational practises if predetermined levels are exceeded.

The reader will rightly conclude that successful monitoring is not always possible, not always necessary, and, when achieved, the culmination of substantial scientific effort. Nevertheless, there is a moral and legal obligation to develop and implement sound programmes for monitoring, when people are exposed to harmful substances in the workplace.

7.4 BIOMARKERS AND SENSITIVITY SCREENING

Monitoring exposure to a chemical element by measuring that element in a body compartment, or exposure to an organic compound by measuring an oxidized or conjugated metabolite are conceptually straightforward. However, to get a direct indication of biological effects, or to facilitate monitoring in situations when concentrations of a substance do not correlate with exposures, there is great interest in developing the use of biomarkers. A biomarker may be defined as an indicator signalling an event or condition in a biological system or sample and giving a measure of exposure, effect, or susceptibility. A few examples of each are described here.

A biomarker is particularly useful if it can help identify exposures before clinically apparent disease occurs. A measure of the validation of a biomarker of early effect is the strength of correlation between the biomarker and eventual development of disease. This can be addressed through relative risk (the ratio of the risk of disease among the exposed to the risk among the unexposed) and the attributable proportion (the proportion of exposure-related disease which is mediated through the marker). Several criteria should be met in defining a useful biomarker. As with any monitoring analytical technique, accuracy, reproducibility, sensitivity, and specificity are important considerations. Access to data from multiple studies is important for the development and validation of a biomarker. The biomarker of effect should also have biological plausibility and mechanistic validation in experimental models.

7.4.1 Biomarkers of Exposure to Non-Carcinogens

Oxidative damage and oxidative stress are caused by exposure to a wide range of substances, including organic oxidants, Fenton-active catalytic metals, and ionizing radiation. Damage occurs to proteins, lipids, and nucleic acids. Common measurements include protein carbonyls and 8-hydroxy-2'-deoxyguanosine. Among oxidized lipid products, the degradation product 4-hydroxy-2-nonenal has been well studied. Activation of signal transduction pathways dependent on reactive oxygen species, such as activation of the nuclear transcription factor NF-kB, is an emerging area of study.

Because haemoglobin is plentiful and has a relatively long half-life, it has proven useful in several situations. Carbon monoxide and dichloromethane form carboxyhaemoglobin. Present at about 1% in blood from reference individuals, a value of 3.5% carboxyhaemoglobin is generally taken as the limit of safe exposure to these agents. Some aromatic amines, *e.g.* aniline, generate metabolites that can oxidize Fe(II) in haemoglobin to Fe(III), forming methaemoglobin. Present at less than 2% in the blood of reference individuals, methaemoglobin should not exceed 5%. Certain alkylating agents and pesticides form adducts with amino acids in haemoglobin that can be measured specifically. For example, *N*-aryl pesticides form aromatic amine metabolites that in turn form sulfinic acid amide adducts with cysteine-93 of β-haemoglobin.

The inhibition of enzyme, cell, or tissue function are all possible indicators of adverse exposures. As mentioned above, decreased erythrocyte acetylcholine esterase activity is a well-established means of monitoring exposure to organophosphorus pesticides. Suppression of natural killer cell activity and cytotoxic T lymphocyte function in chronic organophosphorus exposure are also being studied. A control enzyme in haem synthesis, δ-aminolevulinic acid dehydratase, is extremely sensitive to lead. Therefore, decreased activity of this enzyme in blood and increased excretion of the natural substrate δ-aminolevulinic acid in urine have both been used to monitor lead exposure. Lead also interferes with later steps in haem synthesis, and the accumulation of free protoporphyrins or zinc porphyrins in erythrocytes has been used as indicators of lead exposure, although it should be noted that direct measurement of lead in blood by atomic absorption or voltametric methods is more sensitive than any of these approaches. Cadmium causes damage to the renal tubules, and increased urinary excretion of β_2-microglobulin and retinol-binding protein, early indicators of renal tubular dysfunction, may be good indicators of cadmium exposure (see Chapter 16). New markers of toxic processes are continually sought. For example, peripheral benzodiazepine receptor is a specific marker for glial cells that is showing promise as a marker of neurotoxicity, following chemical exposure.

7.4.2 Biomarkers of Exposure to Carcinogens

One of the most important uses of biomarkers is in monitoring for exposure to carcinogenic or mutagenic substances. Such substances may be present themselves in very small quantities, and cumulative genetic damage may be more relevant to assessing the cumulative effects of many structurally distinct compounds. Some years ago, the International Agency for Research on Cancer presented six classes of biological tests for assessing exposure to carcinogens. None is yet rigorously established for evaluating risk of exposures in humans. The six classes are:

- *Mutagenicity of urine.* Inconsistent results have been obtained in many instances. Smoking is an important confounding factor as the urine of smokers is mutagenic in the assays.
- *Thioethers in urine.* Many electrophilic genotoxins form glutathione conjugates that are converted into thioethers and excreted in urine.
- *Alkyl adducts of DNA or proteins.* Many carcinogenic substances are alkylating agents. Alkylated nucleic acids are excised from DNA and excreted in urine, while haemoglobin is a useful indicator of cumulative protein modification. Examples include the measurement of 3-methyladenine or 7-methylguanine in urine after exposure to methylating agents, or *N*-3-hydroxyethylhistidine in haemoglobin resulting from ethylene oxide exposure.
- *Chromosomal aberrations.* DNA can be obtained from peripheral lymphocytes or cells exfoliated from the mouth or bladder. Damage studied includes sister chromatid exchange, single-strand breaks, and point mutations affecting the expression or activity of the enzyme hypoxanthine–guanine-phosphoribosyl transferase.

- *Sperm analysis.* Functional, morphological, and chromosomal studies may all be useful.
- *Oncogene studies.* This exciting field is still emerging as a monitoring tool. An example is the induction of an activating mutation in *ras* by exposure to polycyclic aromatic hydrocarbons.

Several laboratory methods are currently used for assessing DNA damage and repair. The comet assay can detect single- or double-strand breaks in DNA in single cells, with lymphocytes commonly used in epidemiological studies. The host cell reactivation assay compares the ability of lymphocytes to repair a damaged, transfected plasmid. The mutagen sensitivity assay compares the ability of blood cells from different individuals to repair DNA strand breaks induced in culture by several agents that test different aspects of DNA repair. Repair is also detected by the incorporation of [^{3}H]thymidine in the so-called unscheduled DNA synthesis assay.

7.4.3 Biomarkers of Susceptibility

Biomarkers also have a role in assessing individual differences in susceptibilities to chemicals in the workplace. There is evidence that people with glucose-6-phosphate dehydrogenase deficiency (coded by the *G6PD* gene) and therefore with more fragile red cells may be at increased risk of developing anaemia upon exposure to certain industrial chemicals (*e.g.* aromatic amines). Decreased levels of α_1-antitrypsin predispose to developing emphysema. This protease inhibitor protects the connective tissue of the lung from destruction by proteases, and exposure to proteases is increased when noxious stimuli attract inflammatory cells to the lung. Therefore, workers with partial α_1-antitrypsin deficiency who inhale dusts and particulates may be at an increased risk, as are smokers with this deficiency, of developing lung disease. Individual differences in metabolizing chemicals can also determine relative sensitivities. About half the population acetylates aromatic amines more rapidly than the other half, rendering these amines less carcinogenic. Slow acetylators may therefore be at greater risk of developing aromatic amine-induced bladder cancer. The different acetylation rates are due to polymorphisms in the *N*-acetyltransferase gene, *NAT2*.

There is also a wide variation in the rate at which blood from different people can detoxify paraoxon (a toxic metabolite of the insecticide parathion), because of differences in paraoxonase activity coded by the *PON1, 2,* and *3* genes. Although it appears to function physiologically in lipoprotein metabolism, paraoxon influences susceptibility to organophosphates, and *PON1* polymorphism has been suggested to be linked to Gulf War Syndrome in relation to alleged chemical warfare. Cytochrome P450 isoforms are also responsible for several instances of altered susceptibilities. *CYP2D6* polymorphisms underlie differences in rates of debrisoquine, an anti-hypertensive drug, metabolism, dividing the population into poor, extensive, and ultrarapid metabolizers. *CYP2D6* is involved in metabolism of numerous drugs. The aryl hydrocarbon receptor (AHR) controls expression of several genes, including *CYP1A1, CYP1A2,* and *CYP1B1*. Higher affinity AHR may increase the risk of developing cancer, *e.g.* following exposure to polyaromatic hydrocarbons in cigarette smoke, or to dioxins, as the increased *CYP* activity may increase the activation of carcinogenic metabolites.

In none of these cases does there yet appear to be sufficient data on the levels of a particular occupational exposure that will unmask the potential increased risk. It should also be kept in mind that, although biomarkers of susceptibility are often thought of as identifying a hypersusceptible portion of the population, they should also be recognized as potentially identifying individuals with increased resistance to workplace hazards.

7.5   ETHICAL CONSIDERATIONS

Medico legal aspects of monitoring and exposure control vary among jurisdictions and only certain general ethical concerns are discussed here without specifics. Any human sampling represents an invasion of privacy and requires the consent of the worker to participate in the monitoring programme. The extent to which such consent can be made a condition of employment is a difficult ethical issue. The monitoring test should be made as non-invasive as possible. Urine collection is relatively non-invasive. At the other extreme, few people will consent to tissue biopsy without very strong indications. Minor risks and discomforts associated with procedures of intermediate invasiveness (*e.g.* blood collection, X-ray exposure) must always be considered. It is however, the employers' responsibility to ensure that any samples collected from workers are meaningful and will allow the goals of the monitoring programme to be met. As a rule, the least invasive procedure that meets the goals of the monitoring programme should be used. Interpretation must be based on sufficient knowledge, and adequate analytical facilities for obtaining meaningful results should be in place. Of course, worker privacy and anonymity must be preserved and the samples must only be used for the purpose for which consent has been given.

The decision to add biological to ambient monitoring is at times contentious. The advantages of biological monitoring have been discussed above, but the worker must never be considered a convenient sampling device. Biological monitoring must not be used as a substitute for appropriate measures of exposure control and practises of industrial hygiene.

A frequent concern of labour organizations is whether measuring absorbed doses known to be associated with health risks will result in job loss. Monitoring programmes should be proactive and preventive. The worker should not be placed in a position where fear of loss of employment interferes with prompt identification of health risks or an honest assessment of adverse effects. The decision to accept a certain level of risk for secure employment is a personal one, but should never be made under duress. The same concern arises in assessing individual susceptibilities. Screening for sensitivities that will result in denial of employment is not generally considered acceptable and, as noted above, there is not yet sufficient dose–response information to consider this approach in any case known to this author. The decision to employ women of child bearing years in jobs where there is a risk of exposure to teratogenic or genotoxic substances is particularly difficult. Finally, if biomarkers of increased resistance are identified, they should not be used to recruit a selected workforce as a substitute for a cleaner workplace.

BIBLIOGRAPHY

American Conference of Governmental Industrial Hygienists, *2004 TLVs*® *and BEIs*®, ACGIH, Cincinnati, OH, 2004.

H. Bartsch, K. Hemmink and I. O'Neill (eds), *Methods for Detecting DNA Damaging Agents in Humans: Applications in Cancer Epidemiology and Prevention*, IARC Scientific Publication No. 89, International Agency for Research on Cancer, Lyons, 1988.

Deutsche Forschungsgemeinschaft, *List of MAK and BAT Values 2003: Maximum Concentrations and Biological Tolerance Values in the Workplace*, Report 39, John Wiley/Wiley Europe, New York, 2004.

R.R. Lauwerys and P. Hoet, *Industrial Chemical Exposure. Guidelines for Biological Monitoring*, 2nd edn, Lewis, Boca Raton, FL, 1993.

M.L. Mendelsohn, L.C. Mohr and J.P. Peeters (eds), *Biomarkers: Medical and Workplace Applications*, Joseph Henry Press, Washington, DC, 1998.

S.H. Wilson and W.A. Suk (eds), *Biomarkers of Environmentally Associated Disease*, Lewis, Boca Raton, 2002.

World Health Organization, *Biological Monitoring of Chemical Exposure in the Workplace*, Vols 1 and 2, WHO, Geneva, 1996.

Chapter 8

Genetic Toxicology

DOUGLAS MCGREGOR

8.1 INTRODUCTION

Genetic toxicology – the study of toxic effects on the genetic material – originated with the experiments of Müller (1927), who observed 'artificial transmutation of the gene' by ionising radiation on the fruit fly, *Drosophila melanogaster*. The study of chemically induced mutation also has a long history, the first scientific publication by Auerbach *et al.* (1947), using Müller's fruit fly model, described mutations arising from exposure to sulfur mustards. Deep concern over mutagenesis was first expressed in the mid-1960s with the discovery of 'supermutagens', chemicals such as ICR-170, AF-2, hycanthone and β-propriolactone that induce high levels of mutation at high levels of survival. Several leading geneticists were concerned that supermutagens might be widely distributed (Crow, 1968) either because they had passed through traditional toxicity screens without showing adverse effects or because they had never been tested at all. In spite of these concerns, however, the major impetus given to mutagenesis as a branch of toxicology came from the belief that carcinogens could be predicted from assessment of their potential for interacting with DNA. Thus, Ames *et al.* (1973) pronounced 'carcinogens are mutagens'. While this may have been the spark that became a blaze of activity in developing and validating new tests for genetic toxicity, the concern that the human germ line should be well protected in its own right also benefited. The result is that tests for genetic toxicity are used to identify potential human germ-line cell mutagens as well as in assessment of carcinogenicity and research.

Mutagenicity is the process by which the genetic information for determining and maintaining the integrity, functions and relationships of living cells is permanently changed. A mutagen is an agent that damages the inherited information stored in the cell's genetic material. Mutagenic damage may be caused by the absorption of high-energy radiation, reaction with chemicals or an interaction with an invading virus. Chemical mutagens come either from the environment of the cell or are produced by the cells themselves as part of their normal metabolism.

8.2 STRUCTURE OF DNA (DEOXYRIBONUCLEIC ACID)

The principal target for mutagenic damage is DNA, a helical molecule that is primarily organised into either a single filamentous, supercoiled fibre attached at one or more points to the cell membrane, or distributed between a number of chromosomes contained by a nuclear membrane. When DNA is present in the former condition, the organisms are called prokaryotes and when DNA is found in the latter condition they are called eukaryotes. DNA is not confined to nuclear chromosomes, but also occurs outside the nucleus, in mitochondria and in the chloroplasts of plants.

In typical eukaryotic cells, each chromosome contains a single, very large linear DNA molecule, commonly of the order of 10^7–10^9 bp in length. It is complexed with special DNA-binding proteins of two types (to form chromatin): histones and non-histone chromosomal proteins. There are five types of histone, all of which are small, highly basic proteins rich in lysine and arginine. These are the structural proteins of chromatin and, on a weight basis, they approximately equal the weight of DNA. Non-histone chromosomal proteins form a bewildering array of around 10^3 different proteins, such as polymerases and other nuclear enzymes, hormone receptor proteins and many kinds of regulatory proteins.

DNA (Figure 8.1) is a twisted, ladder-like structure, where the 'poles' of the ladder are chains of deoxyribose in which the 3' position of one deoxyribose is linked to the 5' position of its neighbour on the same 'pole' by phosphates. These two deoxyribose phosphate chains run in opposite directions, so that the 3'-OH of one chain is adjacent to the 5'-OH of the other chain. The 'rungs' of the ladder consist of complementary pairs of pyrimidines [thymine (T) and cytosine (C)] and purines [adenine (A) and guanine (G)]. These bases are always paired so that A is with T and G is with C, and it is these bases that form the genetic code. Base-pairing means that the two chains or strands of the helix are complimentary and, hence, permit the same sequences of bases to recur when the strands separate during DNA replication and a new chain is synthesised using the existing chain as a template.

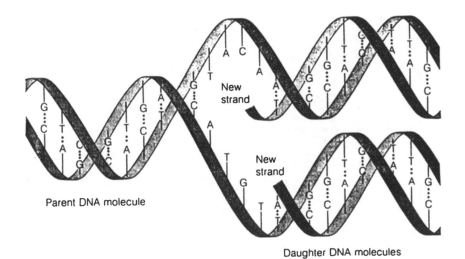

Figure 8.1 *A model of DNA replication. Each strand acts as a template for a new complementary strand. Hence, when copying is complete, there will be two double-stranded daughter DNA molecules, each identical in sequence to the parent molecule*

The fundamental unit of information in DNA is the codon or triplet of bases. Since there are four different bases, 64 combinations of a triplet code are possible. There are 20 different α-amino acids that are incorporated into proteins and hence there is sufficient information and some redundancy to code for all of them as well as start and stop signals for polypeptide synthesis (Figure 8.2).

The nucleotide sequence in a gene determines the amino acid sequence in the proteins for which it codes. In prokaryotes, there is a direct correspondence between the nucleotide sequence, the messenger RNA (mRNA) that is transcribed from it and the polypeptide chain that is translated from the mRNA. In eukaryotes, however, there exist regions of the DNA sequence that are never expressed in a polypeptide chain. These non-coding regions are called introns and they alternate with the regions that are expressed, the exons.

SECOND POSITION

<table>
<thead>
<tr><th></th><th>U(T)</th><th>C</th><th>A</th><th>G</th><th></th></tr>
</thead>
<tbody>
<tr><td rowspan="4">U(T)</td><td>Phe</td><td>Ser</td><td>Tyr</td><td>Cys</td><td>U(T)</td></tr>
<tr><td>Phe</td><td>Ser</td><td>Tyr</td><td>Cys</td><td>C</td></tr>
<tr><td>Leu</td><td>Ser</td><td>Stop</td><td>Stop</td><td>A</td></tr>
<tr><td>Leu</td><td>Ser</td><td>Stop</td><td>Trp</td><td>G</td></tr>
<tr><td rowspan="4">C</td><td>Leu</td><td>Pro</td><td>His</td><td>Arg</td><td>U(T)</td></tr>
<tr><td>Leu</td><td>Pro</td><td>His</td><td>Arg</td><td>C</td></tr>
<tr><td>Leu</td><td>Pro</td><td>Gln</td><td>Arg</td><td>A</td></tr>
<tr><td>Leu</td><td>Pro</td><td>Gln</td><td>Arg</td><td>G</td></tr>
<tr><td rowspan="4">A</td><td>Ile</td><td>Thr</td><td>Asn</td><td>Ser</td><td>U(T)</td></tr>
<tr><td>Ile</td><td>Thr</td><td>Asn</td><td>Ser</td><td>C</td></tr>
<tr><td>Ile</td><td>Thr</td><td>Lys</td><td>Arg</td><td>A</td></tr>
<tr><td>Met</td><td>Thr</td><td>Lys</td><td>Arg</td><td>G</td></tr>
<tr><td rowspan="4">G</td><td>Val</td><td>Ala</td><td>Asp</td><td>Gly</td><td>U(T)</td></tr>
<tr><td>Val</td><td>Ala</td><td>Asp</td><td>Gly</td><td>C</td></tr>
<tr><td>Val</td><td>Ala</td><td>Glu</td><td>Gly</td><td>A</td></tr>
<tr><td>Val</td><td>Ala</td><td>Glu</td><td>Gly</td><td>G</td></tr>
</tbody>
</table>

FIRST POSITION (5' end) — THIRD POSITION (3' end)

Figure 8.2 *The genetic code. The table is arranged so that it is possible to find quickly any amino acid from the three letters (written in the 5' to 3' direction) of the codon. Phe – phenylalanine, Leu – leucine, Ile – isoleucine, Met – methionine, Val – valine, Ser – serine, Pro – proline, Thr – threonine, Ala – alanine, Thyr – tyrosine, His – histidine, Glu – glutamine, Asn – asparagine, Lys – lysine, Asp – aspartic acid, Glu – glutamic acid, Cys – cysteine, Trp – tryptophan, Arg – arginine, Ser – serin and Gly – glycine*

DNA has a special need for metabolic stability because its information content must be transmitted virtually intact from one cell to another during cell replication or during the reproduction of an organism. Stability is maintained in two ways. Firstly, there are mechanisms that ensure high replication accuracy, *e.g.* 3'-exonucleolytic proof reading, which corrects errors made by DNA polymerases, and the uracil-DNA *N*-glycosylase pathway, which prevents mutations that might result from de-amination of cytosine to uracil in DNA. Secondly, there are mechanisms for repairing genetic information when DNA suffers damage. This damage may be caused by replicative errors that are not corrected or by environmental damage. The latter can result from chemical modification of nucleotides or from photochemical changes following the absorption of high-energy radiation.

8.3 TYPES OF GENETIC DAMAGE

Chromosomal damage is defined as microscopically visible modification of the number or structure of chromosomes. Variations in the number of chromosomes may involve the complete complement (polyploidy) or only some of the chromosomes (aneuploidy). The loss of a chromosome is a lethal event, but the gain of a chromosome can be viable and create significant genetic imbalances. Structural changes are mainly the result of breaks in the chromatid arms. Some of these are unstable and are not transmitted through successive cellular generations, *e.g.* achromatic gaps, breaks of one or both chromatid arms, chromatid interchanges, acentric fragments, ring and dicentric chromosomes. Stable structural modifications that are transmissible are inversions, translocations and some small deletions. These genetic factors clearly play important roles in many human diseases.

Great toxicological importance attaches to the alkylation of DNA (commonly by methylating and ethylating agents) since this can cause many base modifications. The target bases are primarily purines (although phosphate oxygen is also a target). While N^7 is frequently the quantitatively dominant alkylated product, O^6-alkylguanine is the most mutagenic because this interferes with the normal hydrogen bonding of G with C and there is a very high probability of G with T pairing when the modified strand replicates. Should this occur, there is said to have been a GC $\rightarrow$ AT *transition* mutation. Many of these alkylations (particularly methyl, but also ethyl and hydroxyethyl) can be removed by the protein O^6-methylguanine-DNA methyltransferase (MGMT), which is capable of functioning as an 'enzyme' only once, since it is inactivated when the alkyl group is transferred from the guanine to a cysteine on the protein. Haloethylation can be particularly damaging to cells unless removed by MGMT, since inter-strand crosslinks can be formed which prevent strand separation during DNA replication.

Adducts with the bases of DNA can also be formed by much larger molecules, including polycyclic aromatic hydrocarbons, aromatic amines and heterocyclic aromatic amines (which may be produced from certain amino acids during cooking). Usually, these molecules are not immediately reactive with DNA but must first be metabolised, frequently by oxidative systems, to electrophilic intermediates. This metabolism provides a large number of products, but, as in the case of benzo [*a*] pyrene, which is a carcinogenic and mutagenic constituent of any burnt carbonaceous material, there may be a high degree of specificity in the metabolite that seems to be responsible for the carcinogenic/mutagenic effects *in vivo*. Damage to DNA is also a common result of oxidative metabolism of xenobiotics during which molecular oxygen is reduced in a series of four one-electron steps through superoxide anion and hydrogen peroxide to the hydroxyl radical and hydroxyl ion. The hydroxyl radical in particular is highly reactive and this is also the most active mutagen generated by ionising radiation.

A particularly well-studied mutagen is UV radiation of about 260 nm. Such UV radiation produces a number of DNA photoproducts, prominent among which are intra-strand dimers joined by a cyclobutane structure involving carbons 5 and 6 of thymidine. These can be completely removed by a photoreactivating enzyme that, in the presence of light (especially with a wavelength of about 370 nm), binds to the cyclobutane region of DNA. However, another intra-strand dimer, the 6-4 photoproduct, is not repaired in this way and is the cause of mutations when the DNA replicates.

8.4   REPAIR OF DAMAGED DNA

Processes in addition to those already mentioned are frequently involved in the repair of DNA that has been damaged. These processes have been largely studied in prokaryotes and while excision repair is also important in mammalian cells, the status of SOS repair (see below) and error-prone repair in mammalian cells is not yet clearly defined.

8.4.1   Excision Repair

Excision repair can be initiated by DNA alkylation, arylation, the production of pyrimidine dimers by UV radiation or the production of apurinic or pyrimidinic sites. Apurinic sites can arise as a result of glycosylase activity removing abnormal bases or by non-enzymatic depurination following labilisation of the glycosyl linkage due to alkylation of N^3 or N^7 positions. The excision repair enzyme system can also repair cross-linking damage.

There are two main modes of DNA excision repair: 'short-patch' or apurinic repair, and 'long-patch' or nucleotide excision repair. Short-patch repair involves the removal and replacement of only a few (perhaps three or four) nucleotides. Long-patch repair results in the removal of the damaged DNA site and up to about 100 adjacent nucleotides. It is initiated by large distortions in the double helix, *e.g.* pyrimidine dimers or adducts with large ring systems (polycyclic aromatic amines or hydrocarbons).

Excision repair can occur throughout the cell cycle. In the bacterium *Escherichia coli*, the uvrA protein detects a distortion due to a dimer or some bulky adduct and binds to DNA distant from the damaged site. The uvrB protein binds to the DNA–uvrA protein complex and, by DNA gyrase activity, unwinds the DNA strands down to the damaged area. UvrC, with endonuclease activity, cleaves the damaged DNA strand on either side of the lesion. The damaged segment is unwound from the undamaged strand by helicase II (uvrD gene product), after which DNA polymerase 1 fills the gap and DNA ligase seals the remaining nick.

8.4.2 Post-Replication Repair

Post-replication DNA repair is an error-prone mechanism that occurs only during the DNA synthetic phase (S-phase) of the cell cycle. During DNA replication, if the polymerase enzyme encounters a large site of damage on the template strand, then that portion of DNA cannot be used as a template. The result is a newly synthesised DNA that contains gaps of up to 1000 nucleotides, as well as the possibility of shorter gaps. These gaps and nicks are eventually filled by chain elongation (*i.e.* post-replication repair) during S-phase and are ligated. Because the repair polymerase must use a damaged DNA template, it is probable that the newly synthesised DNA strand will contain errors.

If repair is not error-free and the cell survives to replicate, the genetic information may be altered. Although these changes are essential to the evolutionary process leading to the diversity of life, the great majority of them are harmful to the cell. The severity of the harm depends, to some extent, upon whether the affected cell is a unicellular organism or one of many cells in a multi-cellular organism. If the affected cell of the latter is a germ cell that is involved in the reproductive process, then the resulting offspring will carry in their genetic material the potentially harmful information (although it is not necessarily expressed). If the affected cell in a multi-cellular organism is a somatic cell (*i.e.* one which does not give rise to gametes, spermatozoa, or ova), then it may experience impaired function or impaired susceptibility to homeostatic, regulatory controls. In this case, the cell may have taken a step along the path leading to the emergence of a cancer.

8.4.3 Base Replacement

Replacement of one base by another has several possible consequences. There may be no effect at all, either because the base change is in an intron or because it results in a new codon that codes for the same amino acid, a result of the redundancy in the DNA triplet code mentioned above. Alternatively, base change may result in the coding of a different amino acid. This type of change is a miss-sense mutation. Occasionally, the codon for an amino acid residue within the original polypeptide will be changed to a stop codon. This is a nonsense mutation that results in a truncated polypeptide. A mutation in a pre-existing stop codon may code for another amino acid and the continuation of translation into a polypeptide that is elongated up to the next stop codon.

8.4.4 Deletions and Insertions

Deletions or insertions of bases in a gene may be large or small. If large, and in a codon region, the almost inevitable result is the prevention of the production of a useful polypeptide. The effects of short deletions or insertions depend upon whether or not they involve multiples of three bases. If whole codons are involved, then the consequence is the deletion or insertion of the corresponding number of amino acid residues. The deletion or insertion of any number of bases other than a multiple of three causes a shift in the reading frame during translation. Such a frameshift mutation results in a complete change in the amino acid sequence in the C-terminal direction from the point of mutation. Nonsense and frameshift mutations almost always result in the destruction of protein function. If the protein function lost is essential for cell survival, the cell will die.

In addition to the base changes that can result from the interaction of a mutagenic agent with DNA, the development of a discontinuity in the DNA imposes an extra strain on the structure of the chromosome. This can result in lesions that are microscopically visible (*e.g.* at 400 × magnification or greater) after appropriate staining of the chromosomes. Such lesions are particularly prone to develop if the DNA damage involves DNA inter-strand cross-linking or DNA–protein cross-linking agents. Numerical changes in the chromosomes can also occur if there has been interference with their movements during cell division by agents that have formed adducts with cytoskeletal proteins, particularly (but not only) tubulin. What a cytogeneticist refers to as a chromosome consists of two chromatids, one having been derived from each germ-line cell when they come together during sexual reproduction to form the zygote. Each pair of chromatids is joined by the centromere, which is also the point of attachment of the cytoskeletal proteins that push and pull the chromosomes during cell division. Each of these chromatids is what has been referred to as a chromosome elsewhere in this and other texts.

8.5   CHROMOSOMAL CHANGE

Chromosomal damage is defined as a microscopically visible modification of the number or structure of chromosomes. Variations in the number of chromosomes may involve multiples of the complete complement (polyploidy) or the elimination or multiplication of only some of the chromosomes (aneuploidy). The loss of a chromosome is a lethal event, but the gain of a chromosome can be viable and create significant genetic imbalances. Structural changes are mainly the result of breaks in chromatid arms (Figure 8.3). Some of these are unstable and are not transmitted through successive cellular generations, *e.g.* achromatic gaps, breaks of one or both chromatid arms, chromatid interchanges, acentric fragments, ring and dicentric chromosomes. Stable structural modifications that are transmissible are inversions, translocations and some small deletions.

8.6   TRANSMISSIBLE HUMAN GENETIC DAMAGE

Genetic factors clearly play important roles in many human diseases. Adverse effects upon germ-line cells are important to our descendants, but at present we cannot readily identify most adverse changes and predict what their effects will be. Even in the future, epidemiologists will have great difficulty in identifying chemically induced genetic change as the cause of any increases in sickle-cell anaemia, phenylketonuria, Down's syndrome, diabetes, atherosclerosis or any other disease with a genetic component. The fact that we cannot deduce most effects of genetic change and that these effects may not be recognised as such by our descendents is no reason for ignoring the possibilities of harm and neglecting to take preventative action.

There is little evidence that any agent has increased the frequency of any human mutation that can be transmitted from one generation to the next. That this may have

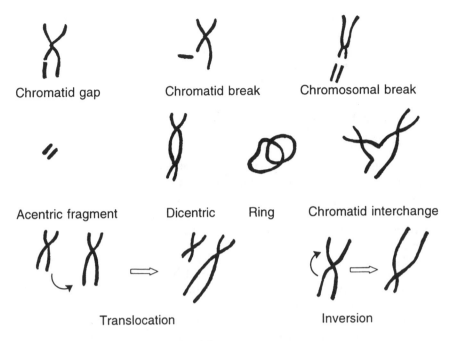

Figure 8.3 *Categories of chromosomal damage*

happened, however, is suggested by results from a study of germ-line mutation at minisatellite loci among children born in heavily polluted areas of the Mogilev district of Belarus after the Chernobyl accident (Dubrova *et al.*, 1996). Minisatellites or tandem repeat elements are particularly useful for monitoring human germ-line mutations because the very high rate of spontaneous mutation that alters allele length (repeat copy number) provides a system capable of detecting induced mutations in relatively small populations. Mutation rates in the 79 Mogilev families examined were higher in areas of high surface ^{137}Cs contamination than in areas where contamination was lower.[1] A similar minisatellite study of children from 50 families exposed to atomic bomb radiation in Hiroshima and Nagasaki (gonadal dose >0.01 Sv) and 50 control families did not show any effect (Kodaira *et al.*, 1995). The hypervariable loci examined in these two studies have spontaneous mutation rates at least 1000 times higher than in most protein-coding loci.

In most cases, those agents that induce heritable mutations in experimental animals should have been classified as hazardous chemicals because of other toxicological properties that are more easily monitored (often because of their carcinogenicity – see the next chapter). Further, it appears that there are no mutagens or clastogens (agents which break chromosomes) that are uniquely active in germ-line cells (through which their adverse effects can be passed from generation to generation). Therefore, from the view of toxicological evaluation, if carcinogens are being adequately

[1]Athough acute exposure was mainly to ^{131}I, this isotope has a half-life of only 8 days, so, when ^{131}I has decayed, more stable isotopes become the main source of radiation exposure. ^{137}Cs is the most important of these isotopes.

controlled by risk management procedures, then so are mutagens and clastogens with heritable effects. The main toxicological interest in mutagenicity therefore comes down to what mutagenic activity can tell us about carcinogenicity. Indeed, the prediction of carcinogenicity is the objective that led to the development of mutagenicity assays and their validation has always been against the yardstick of carcinogenicity.

8.7 TESTS FOR GENETIC TOXICITY

8.7.1 Test Categories

To address the need for identifying all aspects of genetic toxicity, more than 100 different test methods have been developed. Few, however, are commonly used. It is the diversity of potentially damaging events that has encouraged this development and, for this reason, it should not be expected that there will always be consistency among the results of different classes of assays. The tests can be grouped as:

(i) Genetic toxicity tests that detect types of DNA damage known to be precursors of genetic alterations (*e.g.* DNA adducts or DNA strand breakage) or cellular responses to DNA damage (*e.g.* unscheduled DNA synthesis (UDS))

(ii) Tests for resolved genetic damage (*e.g.* gene mutations, chromosomal rearrangements or deletions and loss or gain of chromosomal segments or whole chromosomes, the last also known as aneuploidy).

There are many categories of genetic damage that are open to investigation, but only the principal categories will be described briefly here.

A forward mutation is a gene mutation from the parental type to a mutant form which gives rise to an alteration or a loss of enzymic activity or other function of an encoded protein. This type of mutation is observable in prokaryotes and in mammalian cell culture assays. On the other hand, reverse mutations are the basis of the very commonly used tests in the bacteria *Salmonella typhimurium* and *E. coli* that are used to detect mutation in an amino acid requiring strain (histidine or tryptophan, respectively) to produce a strain that can grow independently of an outside supply of the essential amino acid. Such gene mutations can be induced by base pair substitution mutagens that cause a change in one or several base pairs in DNA. In a reversion test, this change may occur at the site of the original mutation, or at a second site in the bacterial genome. These mutations can also be induced by frameshift mutagens that cause the addition or deletion of one or more base pairs in DNA, thereby resulting in a shift in the three-base reading frame code.

Structural aberrations in chromosomes are changes in chromosome structure that are manifest as deletions, fragments or as intrachanges or interchanges detectable by microscopic examination of the metaphase stage of cell division. Numerical aberrations in chromosomes are also detectable by microscopic examination and are changes in the number of chromosomes (aneuploidy) from the normal number characteristic of the cells used. Special cases of such events are: polyploidy, which is a multiple of the haploid chromosome number (*n*) other than the diploid number (*i.e.* 3*n*, 4*n* and so on); and endoreduplication, a process in which cell division does not follow a phase of

DNA replication, but is followed by another phase of DNA replication. The result is chromosomes with 4, 8, 16, *etc.* chromatids, rather than the normal 2.

8.7.2 Commonly Used Tests

Genetic toxicity tests are conducted both *in vitro* and *in vivo* (see Table 8.1). These include tests for gene mutation in bacteria, *in vitro* tests with mammalian cell lines for gene mutation, chromosomal aberrations (including micronuclei) and aneuploidy, tests with primary cultures of mammalian cells (most commonly rat hepatocytes) and *in vivo* tests for chromosomal damage (including micronuclei) that normally target mammalian haematopoietic cells. Less commonly, results of tests for cell transformation may be presented that can provide useful information, although positive results obtained with them are not necessarily indicative of genetic toxicity in the form of reactivity with DNA; they may also represent a consequence of epigenetic events.[2]

No attempt will be made here to describe these tests in any detail; protocols are available in the mutagenesis literature. However, a number of them have been thoroughly discussed in an international forum and agreed general protocols have been produced. These are available for bacterial mutation tests (OECD TG471), mammalian cell mutation tests *in vitro* (OECD TG476), mammalian cell chromosomal aberration tests *in vitro* (OECD TG473), mammalian erythrocyte micronucleus tests *in vivo* (OECD TG474) and mammalian bone marrow cell chromosome aberration tests *in vivo* (OECD TG475). In addition, there is a discussion available on the performance of mammalian cell transformation tests (OECD Environmental Health and Safety Publications Series on Testing and Assessment No. 31). All of these publications, and related documents, are available from http://www.oecd. org/home/. No OECD test guideline exists for UDS assays, which are considered useful for the identification of repairable DNA damage, but a multi-national group has published recommendations for the performance of UDS tests both *in vitro* and *in vivo* (Madle *et al.*, 1994).

8.7.3 Data Assessment

With so many different types of assay, so many standard protocols and so many protocol variations possible for special studies, it is not practicable, other than in very general terms, to describe the process of data assessment. This process should, of course, include a judgment of the standards to which the experiments were conducted. A review that can be helpful in this process has been published (IARC, 1999).

Following this initial quality assessment of individual studies, the second scientific requirement is that any observation that is made should be reproducible. While this principle applies to any kind of study, the financial and time restraints that hinder replication of many mammalian toxicity and carcinogenicity experiments are less important in genetic toxicology. The OECD Test Guidelines referred to previously often state that a clearly positive result does not require repetition within a laboratory, whereas a clear-cut negative result is expected to require confirmation in a second

---

[2]An epigenetic event is any heritable influence in the progeny of cells or of individuals on chromosome or gene function that is not accompanied by a change in DNA nucleotide sequence.

Table 8.1 *Some assays for genetic toxicity*

DNA repair	Gene mutation	Chromosomal aberration
Microbial tests	*Microbial tests*	*Microbial tests*
Differential growth of repair competent and deficient cells using: *S. typhimurium* *E. coli* *Bacillus subtilis* *Saccharomyces cerevisiae*	Prophage induction Reversion to a specific nutrient independence using: *S. typhimurium* *E. coli* *Saccharomyces cerevisiae* *Aspergillus nidulans* *Neurospora crassa*	Gene conversion and mitotic recombination using: *Saccharomyces cerevisiae* *Schizosaccharomyces pombe*
Mammalian tests		Chromosome loss/aneuploidy using: *Saccharomyces cerevisiae* *Aspergillus nidulans* *Neurospora crassa*
UDS using: Primary cultures (often hepatocytes) *In vivo* exposure of many tissues	Forward mutation to resistance to a specific antimetabolite using: *S. typhimurium* *E. coli* *Saccharomyces cerevisiae* *Aspergillus nidulans* *Neurospora crassa*	*Insect tests* Somatic cell chromosomal aberrations and aneuploidy and using: *D. melangaster*
DNA strand breakage alkali labile sites monitored by sucrose gradient or filter elution using: Cell cultures *In vivo* exposure of many tissues	*Insect tests* Somatic cell mutation and recombination test (SMART) using: *D. melanogaster*	Germ-line cell dominat lethals and heritable translocation using: *D. melanogaster*
DNA strand breakage and alkali labile sites monitored by single-cell gel electrophoresis (COMET assay) using: Cell cultures *In vivo* exposure of many tissues	Sex-linked recessive lethal test using: *D. melanogaster* *Mammalian tests* *In vitro* assays	*Mammalian tests* *In vitro* assays Sister-chromatid exchange (SCE) Chromosomal aberrations, micronuclei and aneuploidy using: CHO and V79 cell lines and human lymphocytes
	Hypoxanthine-guanine phosphoribosyl transferase (*hprt*) gene using cell lines such as: Chinese hamster ovary (CHO) Chinese hamster lung (V79) Human lymphoblastoid	*In vivo* assays Somatic cell assays SCE Chromosomal aberrations, micronuclei and aneuploidy using: Bone-marrow cells (rodent), lymphocytes (rodent and human)
	Thymidine kinase (*tk*) gene using cell lines such as: Mouse lymphoma L5178Y Mouse lymphoma P388F Human lymphoblastoid	Germ-line cell assays Chromosomal aberrations, heritable translocations and dominant lethals using mice and rats

Table 8.1 *Some assays for genetic toxicity (continued)*

DNA repair	Gene mutation	Chromosomal aberration
	In vivo assays	
	Somatic cell assays: Mouse coat 'Spot' test *LacZ* (Muta™ Mouse) or *lacI* (BigBlue™ mouse)	
	Germ-line cell assays: Specific locus tests in mice	

(not normally identical) experiment. Nevertheless, the strength of evidence is increased if a particular finding can be demonstrated in a number of laboratories. Indeed, where an observation is made in a single laboratory – even if made on a number of occasions – it is generally viewed with suspicion if other laboratories fail to achieve the same result. This suspicion may not be justified in some cases, but it is nevertheless an understandable view in data evaluation.

The third step is to look for a plausible hierarchy in the pattern of results. Such hierarchical patterns can only be used as general guides, because there can always be exceptions. It is expected that a substance that is clastogenic *in vivo* will also be clastogenic *in vitro*; and that, *in vivo*, a germ-line cell clastogen will also be clastogenic to somatic cells. Deviations from this pattern may occasionally occur, but these should be scrutinised with special care. The basis for suggesting this procedure is as described below for cytogenetic assays. Unfortunately (apart from a short comment, below), a similar basis cannot be presented for gene mutation induction, because of the current lack of *in vivo* test data. This situation may be expected to change as more data accumulate from the increasing use of the transgenic mouse models. These have been used on a small number of well-established germ-line cell mutagens (EMS, MMS, iPMS, MNU) (Douglas *et al.*, 1997; van Delft *et al.*, 1997). Such a meagre database is not, however, a firm basis for using transgenic mouse models in place of the more arduous, traditional specific locus germ-cell tests.

8.7.3.1 Cytogenetic Assays In Vivo and In Vitro. Thompson (1986) reviewed the literature to find 216 chemicals that had been tested both *in vitro* and in rodent bone marrow tests for clastogenicity. Definitive results were obtained with 181 of them among which there was an agreement for 126 chemicals. Of the 55 for which the results did not agree, 53 were positive *in vitro* and negative *in vivo*. Only D-ascorbic acid and ethynyloestradiol were negative *in vitro* while inducing significant results in bone marrow. This leads to the conclusion that a chemical that fails to induce a significant response in an *in vitro* clastogenicity assay is unlikely to be clastogenic *in vivo*, in bone marrow assays.

8.7.3.2 Germ-Line and Somatic Cell In Vivo Cytogenetic Assays. Holden (1982) reviewed the available literature and found 76 compounds that had been tested for chromosomal effects *in vivo* in both somatic and germ-line cells. Concordant results were obtained with 58 of these chemicals. The remaining 18 chemicals for which

Table 8.2   *Mouse-specific locus test-positive chemicals (Russell et al., 1981; Shelby et al., 1993)*

Acrylamide	Methyl methane sulfonate (MMS)
Chlorambucil	Methyl nitrosourea (MNU)
Coal liquid A	Mitomycin C
Cyclophosphamide	Procarbazine
Ethyl methane sulfonate (EMS)	*iso*Propyl methane sulfonate (iPMS)
Ethyl nitrosourea (ENU)	Triethylene melamine (TEM)
Ethylene oxide	
Melphalan	

there were discordant results were all positive (*i.e.* induced damage) in somatic cells only. At that time, therefore, the available evidence suggested that a negative somatic cell effect is highly predictive of a negative germ-line cell response. Subsequently, it was suggested by the U.S. EPA GeneTox Workshops that six chemicals could be uniquely germ-line cell mutagens (Auletta and Ashby, 1988), but a re-evaluation of the GeneTox Program literature on these compounds indicates that they were misclassified (Adler and Ashby, 1989). There was, therefore, as of 1989, no reason to change the presumption that all germ-line cell clastogens are also somatic cell clastogens. This kind of thinking led the European Chemical Industry Ecology and Toxicology Centre (ECETOC) to propose a testing strategy in which agents without somatic cell genotoxicity *in vivo* could be assumed to have no potential for germ-line cell genotoxicity (Arni *et al.*, 1988).

8.7.3.3   Germ-Line and Somatic Cell In Vivo Mutation Assays.   Data reviewed by Russell *et al.* (1981) and Shelby *et al.* (1993) showed for the mouse-specific locus test only 13 chemicals as clearly positive. In addition, after review of data initially considered negative by formal statistical methods, Russell *et al.* (1981) considered that one chemical mixture was positive. The chemicals studied (Table 8.2) are all highly reactive and induce a wide variety of toxic effects in man and other animals; they are also clastogenic in germ-line cells.

8.8   CONCLUSION

The concept of genetic toxicology in safety evaluation has been accepted in academic, industrial and regulatory circles. While it is debatable whether specific concern for the induction of germ-line cell genetic damage is needed, it is clear that genetic toxicology does have a role to play in hazard identification and risk assessment. The tests that are used – at the very least – provide evidence that the chemical under study can react with biologically important molecules, either directly or after metabolism. Validation studies, using rodent carcinogenicity data as the yardstick, have shown that any prediction of carcinogenic hazard will be imperfect. Similarly, evidence of a particular mode of carcinogenic action that might be derived from positive results of genetic toxicity tests will always have an element of uncertainty about them. As long as these weaknesses are not forgotten and the strengths are not over-emphasised, however, the results can provide useful guidance in chemical risk assessment.

BIBLIOGRAPHY

I.-D. Adler and J. Ashby, The present lack of evidence for unique rodent germ-cell mutagens, *Mutat. Res.*, 1989, **212**, 55–66.

B.N. Ames, W.E. Durston, E. Yamasaki and F.D. Lee, Carcinogens are mutagens: a simple test system combining liver homogenates for activation and bacteria for detection, *Proc. Nat. Acad. Sci. USA*, 1973, **70**, 2281–2285.

P. Arni, J. Ashby, S. Castellino, G. Engelhardt, B.A. Herbold, R.A.J. Priston and W.J. Bontinck. Assessment of the potential germ cell mutagenicity of industrial and plant protection chemicals as part of an integrated study of genotoxicity *in vitro* and *in vivo*, *Mutat. Res.*, 1988, **203**, 177–184.

C. Auerbach, J.M. Robson and J.G. Carr, The chemical production of mutations, *Science*, 1947, **105**, 243–247.

A. Auletta and J. Ashby, Workshop on the relationship between short-term information and carcinogenicity; Williamsburg, Virginia, January 20–23, 1987. *Environ. Mol. Mutagen.*, 1988, **11**, 135–145.

J.F. Crow, Rate of genetic change under selection, *Proc. Natl. Acad. Sci. USA*, 1968, **59**, 655–668.

J.H.M. van Delft, A. Bergmans and R.A. Baan, Germ-cell mutagenesis in λ *lacZ* transgenic mice treated with ethylating and methylating agents: comparison with specific-locus test, *Mutat. Res.*, 1997, **388**, 165–173.

G.R. Douglas, J.D. Gingerich, L.M. Soper and J. Jiao, Toward an understanding of the use of transgenic mice for the detection of gene mutations in germ cells, *Mutat. Res.*, 1997, **388**, 197–212.

Y.E. Dubrova, V.N. Nesterov, N.G. Krouchinsky, V.A. Ostapenko, R. Neumann, D. Neil and A.J. Jeffreys, Human minisatellite mutation rate after the Chernobyl accident, *Nature*, 1996, **380**, 683–686.

H.E. Holden, Comparison of somatic and germ cell models for cytogenetic screening, *J. Appl. Toxicol.*, 1982, **2**, 196–200.

IARC, in *The Use of Short- and Medium-term Tests for Carcinogens and Data on Genetic Effects in Carcinogenic Hazard Evaluation*, D.B. McGregor, J.M. Rice and S. Venitt (eds), IARC Scientific Publications No. 146, Lyon, 1999, 536.

M. Kodaira, C. Satoh, K. Hiyama and K. Toyama, Lack of effects of atomic bomb radiation on genetic instability of tandem-repetitive elements in human germ cells, *Am. J. Hum. Genet.* 1995, **57**, 1275–1283.

S. Madle, S.W. Dean, U. Andrae, G. Brambilla, B. Burlinson, D.J. Doolittle, C. Furihata, T. Hertner, C.A. McQueen and H. Mori, Recommendations for the performance of UDS tests *in vitro* and *in vivo*, *Mutat. Res.*, 1994, **312**, 263–285.

H.J. Müller, Physiological effects on "spontaneous" mutation rates in *Drosophila*, *Genetics*, 1926, **31**, 225.

OECD/OCDE, http://www.oecd.org/home/.

L.B. Russell, P.B. Selby, E. von Halle, W. Sheridan and L. Valcovic, The mouse specific-locus test with agents other then radiation, *Mutat. Res.* 1981, **86**, 329–354.

M.D. Shelby, J.B. Bishop, J.M. Mason and K.R. Tindall, Fertility, reproduction, and genetic disease: studies on the mutagenic effects of environmental agents on mammalian germ cells, *Environ. Health Perspect.* 1993, **100**, 283–291.

E.D. Thompson, Comparison of *in vivo* and *in vitro* cytogenetic assay results, *Environ. Mutagen.* 1986, **8**, 753–768.

Chapter 9

Carcinogenicity

DOUGLAS MCGREGOR

9.1 INTRODUCTION

Tumours are, literally, swellings. They have, however, come to mean a special kind of swelling that, popularly, is called a 'growth' or, if it is a particularly dangerous growth, it is called a 'cancer'. These tentative definitions are made with deference to a highly respected German pathologist, Rudolf Virchow (1821–1902), who once remarked that no man, even under torture, could say exactly what a tumour is. Nevertheless, a workable definition is that a tumour is a tissue mass formed as a result of abnormal, excessive and inappropriate cell proliferation, the growth of which continues indefinitely and regardless of the mechanisms that control normal cellular proliferation.

Nineteenth century pathologists divided tumours into benign, simple or innocent tumours and malignant tumours. Benign tumours (which can still be dangerous in some situations) remain localised, forming a single mass that frequently produces no symptoms and can usually be excised completely. If symptoms are experienced, these are usually due to pressure on adjacent tissues or to excessive hormone production. Malignant tumours invade adjacent tissues and their cells may dissociate and spread through blood and lymphatic vessels to other parts of the body, where they may lodge, divide and give rise to secondary tumours or metastases. These are difficult to control, and account for most of the approximately 25% of the deaths due to cancer in most developed countries.

Cancer as a cause of death is not negligible at any age, but it is primarily a terminal illness in aged populations. Carcinogenesis is the process involved in the development of malignant tumours. A carcinogen, therefore, is a risk factor causally related to an increase in cancer incidence or prevalence. Cancer epidemiologists study all the factors by which people develop or die of cancer. To an experimental oncologist (someone who performs experiments in his or her studies of any kind of tumour), however, a carcinogen is not necessarily an agent causing a pathological phenomenon resulting in death. A carcinogen in this context is an agent that increases the incidence of any neoplasm (a new growth), a term applied to all types of tumour, irrespective of whether it is lethal, potentially lethal or benign. In extreme circumstances, a carcinogen may be defined also as an agent that increases the incidence of certain types of pre-neoplastic change.

Carcinogens may be physical, chemical or biological agents. Important physical phenomena that are carcinogenic are ultraviolet (UV) and ionising radiation. Circumstances involving biological agents that are judged to be carcinogenic are chronic infection with the hepatitis viruses B and C (hepatocellular carcinoma), human papilloma virus types 16 and 18 (cervical cancers), human immuno-deficiency virus-I and, possibly, -II (Kaposi's sarcoma, non-Hodgkin's lymphoma), human T-lymphotropic virus-I (adult T-cell leukaemia/lymphoma in Japan, Caribbean and West Africa), Epstein-Barr virus (non-Hodgkin's lymphoma) and infestation with the blood parasite, *Schistosoma haematobium* and the liver fluke, *Opisthorchis viverrini*. However, there are many more chemicals that have been recognised as human carcinogens and some circumstances that entail exposures to carcinogens (without, necessarily, identifying the actual carcinogens involved). Those identified by the International Agency for Research on Cancer may be found on their website, http://www.iarc.fr/.

9.2 MECHANISMS OF CARCINOGENICITY

9.2.1 Genes

Currently, it is thought that genes of three main types may be involved in carcino-genesis: oncogenes, tumour suppressor genes and DNA repair enzyme genes. The concept that there are genes capable of causing cancer (oncogenes) is based largely on studies with transplantable tumours in chickens, mice and rats. In 1911, Rous found that a transmissible sarcoma in chickens could be induced by a cell-free filtrate of the tumour. Similar results were found for some rodent sarcomas and the responsible agent was later demonstrated to be an RNA virus. There is, however, little evidence that any human tumour is induced by RNA viruses (except in the case of hepatitis C virus, chronic infection with which does increase the risks of liver cancer). In 1965, however, it was proposed that the cells of many, if not all, verte-brates possess the information in their DNA to produce RNA viruses. This informa-tion (the virogene) could code for a virus, but a portion of it (the oncogene) was responsible for malignant transformation. Normally, this DNA is not expressed, but it could be activated as a result of exposure to chemical carcinogens or radiation. During the following years, it was found that while some transforming RNA viruses carry the information to induce tumours quickly and directly (oncogenes), others induced transformation slowly, apparently by insertion of reverse-transcribed DNA adjacent to host cellular genes and in some way altering the transcription of them.

9.2.2 DNA Analysis

Analysis of DNA from many different organisms, including yeast, insects, birds and mammals (including man), revealed that it contains base sequences with a high degree of homology to the oncogene region of the viruses. These normal cellular DNA sequences are called proto-oncogenes, or c-onc sequences. It is highly probable that the viral oncogenes (or v-onc sequences) are derived from the proto-oncogenes. Over 20 oncogenes have been identified and, in normal cells, their expression is well controlled. Their functions have been determined in some cases and they appear to play a role in growth and development. The known proto-oncogene products include growth factors (*e.g.*, sis), protein kinases (*e.g.*, src) which phosphorylate serine, threonine or tyrosine, including membrane receptors with protein kinase activity, GTP-binding proteins (*e.g.*, ras) and nuclear proteins (*e.g.*, c-myc), which seem to be involved in the control of cellular proliferation and differentiation. For some proto-oncogenes, there is a close association between their expression and cell proliferation.

Stimulation of a non-malignant cell into division often depends upon a signal initiated by an extracellular chemical. This chemical interacts with a receptor on the cell membrane, which transfers the signal through the membrane and into the cytoplasm and, ultimately, to the nucleus where DNA synthesis is initiated. Proto-oncogene products have been found to function at each step of this pathway. Alteration of the proto-oncogenes may result in a development along a pathway leading to neo-plasia. Inappropriate activation of a signal transduction pathway (*e.g.*, by sustained exposure to certain hormones or exogenous chemicals such 3,4,7,8-tetrachlorodibenzo-*p*-dioxin) may have a similar effect. Activation may occur by mutation (*e.g.*, *RAS* family of oncogenes), by gene amplification (*e.g.*, *MYC, ERBB2*) or by rearrange-ments such as chromosomal translocation. Activated proto-oncogenes (oncogenes) have been detected in malignant cells of human origin by their ability to transform normal cells after transfection of malignant cell DNA and by using labelled DNA probes that are complementary to known viral oncogenes. The most commonly activated oncogenes in human cancers are *ERBB2* (in breast and ovarian cancers), members of the *RAS* family (in particular *KRAS* in lung, colorectal and pancreatic cancers) and *MYC* (in a large variety of tumours, such as cancers of the breast, oesophagus and some forms of acute and chronic leukaemia).

Tumour-suppressor genes code for products that control aspects of normal cell turnover within a tissue. Loss of the effective gene product by mutation of the gene therefore results in reduced control, *e.g.*, of cellular proliferation, differentiated state or programmed cellular death (apoptosis). However, only one active gene is necessary for suppression. Since there are normally two genes (except on the sex-chromosomes), both copies (alleles) need to be lost before tumour suppression is lost. The concept of tumour suppressor-genes arose from observations on two human, childhood tumours, retinoblastoma, which affects the eye, and nephroblastoma or Wilms' tumour, which affects the kidney.

9.2.3 Hereditary and Nonhereditary Forms of Cancer

Retinoblastoma occurs in both hereditary (40%) and nonhereditary forms. The nonhereditary form is always unilateral, whereas about 80% of the hereditary form cases have bilateral disease. In most families carrying the germ-line mutation that predisposes to retinoblastoma, the penetrance of the mutation is 90–95% and affects both sexes (*i.e.*, it is an autosomal dominant trait). The germ-line mutation has been traced to the long arm (*i.e.*, q) of chromosome 13 (*i.e.*, 13q) and localised to a particular staining band of the chromosome that is identified as 13q 14.11, reading from the centromere (which separates the short, p, arm from the long, q, arm). This locus has been named RB1 (although it seems to be the only locus at which muta-tion can produce retinoblastoma). The normal cells that are able to transform into retinoblastoma are probably embryonic cells that differentiate into retinoreceptors early in life. Once differentiated, they are no longer at risk of transformation by mutation (because the mature cells no longer divide). For nonhereditary retinoblastoma, both mutations at the RB1 locus are somatic and therefore must occur in the same cell. The probability of two independent mutations occurring in this way is very low indeed. However, when the first mutation is in the germ line, then every potential target cell in the retina has undergone the first mutation and the probability of the required second mutation occurring in just one cell of the total target population is very much higher and so bilateral disease is much more likely.

Wilms' tumour has many of the characteristics of retinoblastoma, in this case the tumour suppressor function being located on the short arm of chromosome 11 (at 11p13.05–06). A number of other tumour-specific suppressor genes have been identified, but much attention has been drawn to p53, which is located on chromosome 17q12–13.3, because it seems to be associated with several different tumours, including astrocytomas, osteocytomas and carcinomas of the breast, colon, liver, oesophagus and lung. Its behaviour is, however, complex and much remains to be learned about its true significance.

The importance of deficiencies in DNA repair in carcinogenesis is suggested by the observation of highly elevated cancer incidences in people with well-characterised autosomal recessive diseases that produce high sensitivity to DNA-damaging agents. These diseases include xeroderma pigmentosum, Fanconi's anaemia, ataxia telangiectasia and Bloom's syndrome. Xeroderma pigmentosum carries with it a deficiency in the excision repair of DNA lesions induced by UV-irradiation. A very high proportion of such people develop benign and/or malignant skin tumours, particularly if they are exposed to the sun or other sources of UV-irradiation. Fanconi's anaemia patients repair enzyme-induced incision of DNA at cross-links inefficiently. They have an increased risk of developing liver tumours, skin carcinomas and leukaemias. Ataxia telangiectasia patients' DNA is highly susceptible to damage by ionising radiation and chromosome-breaking chemicals. About 10% of them develop malignancies, mainly of the lymphoid system, but another factor contributing to their increased cancer risk is the commonly associated immunodeficiency. Bloom's syndrome patients also show an increased frequency of chromosomal damage and it appears that the disease may result from a mutation in the gene coding for DNA ligase I, which is an important enzyme during DNA replication. Approximately 25% of these patients develop a leukaemia, lymphoma or carcinoma.

The research areas described above influence our understanding of carcinogenicity as a disease process and may suggest new approaches to therapy. Further, clarification of the role of heredity is important in the way in which changes recorded in experimental and epidemiological studies are interpreted and evaluated. Such evaluation helps to establish the appropriate level of concern when an excess of tumours of a particular type is observed.

9.2.4 Genotoxic and Non-genotoxic Mechanisms

In evaluation of chemicals for carcinogenic hazard and risk, there is a dogma that cancer is a genetic disease. This dogma has led to the default assumption that if a substance is mutagenic or clastogenic, it must be carcinogenic. The origins of the dogma are simple to understand. All neoplasms studied contain mutations. There is a single copy of DNA in every cell and therefore, it is reasoned, there can be no threshold below which DNA damage (mutation) has no consequence. Hence, there can be no safe exposure level to a carcinogen that is genotoxic. Currently, there are growing doubts about the default assumption, but it remains useful to consider mechanisms of carcinogenesis as being either genotoxic or non-genotoxic (Bolt *et al.*, 2004).

Cell proliferation is a common requirement for carcinogenesis. This is true whether the basic mechanism leading to the appearance of a tumour is genotoxic or not. Another general feature is that there is genetic damage in the emergent tumour. This raises the question as to how these changes can occur if the inducing agent has no genotoxic activity. Another question that may be asked is whether a carcinogen that is genotoxic necessarily acts by a genotoxic mode of action, as asserted by the default assumption stated above. If not, then what modes of action might account for neoplasia and how do the genetic changes arise?

9.2.4.1   Carcinogens with a Genotoxic Mode of Action.   Cancer is a collection of multi-step and factorial diseases (Figure 9.1). An important factor in our under-standing of carcinogenicity was the realisation of the significance of metabolism as a process that can convert a large number of apparently very different chemical structures into a common form of chemical reactivity. While the metabolism of a chemical may be complex, a frequent step (called an activating reaction) is the gen-eration of an electrophile that can readily react with nucleophilic centres in complex biological molecules, such as nucleic acids and proteins. An example of such an activating metabolism is presented for benzo[a]pyrene (Figure 9.2). Although metabolism can occur by a variety of routes, the activation pathway *in vivo* proceeds as a consequence of oxidation by the cytochrome P450 family of enzymes via the 7,8-oxide and 7,8-dihyrodiol to the 7,8-diol 9,10-oxides, the so-called bay-region diol-epoxides, which are generally regarded as the ultimate, carcinogenic and DNA-reactive species of benzo[a]pyrene. Both the 7,8-oxide and the 7,8-dihydrodiol exist as pairs of enantiomers, while the 7,8-diol 9,10-oxides are formed as two diasteriomers, each comprising a pair of enantiomers (Figure 9.3). Of these four diol-epoxides, the (+)anti-7,8-diol 9,10-oxide is the most biologically active and adducts derived from the covalent binding of anti-7,8-diol 9,10-oxides to DNA have been identified as the major DNA adducts following the exposure of cultured cells or tissues to benzo[a]pyrene. It is tumourigenic to mouse skin and to the lung of newborn mice, while the (−)anti-7,8-diol 9,10-oxide is inactive in these systems. There are several potential conse-quences of adduct formation, but the important one as far as carcinogenesis is concerned is that the cells containing adducts survive and there is some mis-repair of the molecular lesion resulting in a change in the proliferation and differentiation control of the cells, so that a subpopulation develops with an increased susceptibility to neoplasia. The cells are said to have been initiated.

Cell division is necessary at all stages of neoplasia. In the early stages, this process allows the formation of a clone of initiated cells, within which further genetic changes may occur, leading to a yet more altered cell population. Out of this promotion phase, histologically recognisable pre-neoplastic lesions emerge. Most of these develop no further, but a small number (perhaps only one) may experience more genetic changes, as a result of which a cell population can arise that are no longer

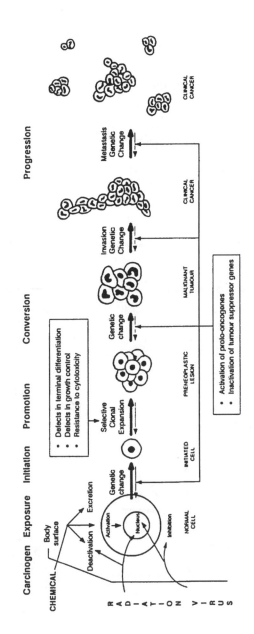

Figure 9.1 *Carcinogenesis shown as a multistage process involving multiple genetic and epigenetic events in proto-oncogenes, tumour suppressor genes, and antimetastasis genes (Adapted with kind permission from C. Harris, 'Tumour suppressor genes', in Mechanisms of Carcinoenesis in Risk Identification, H. Vainio, P. Magee, D. Mcgregor and A.J. McMichael (eds), IARC Scientific Publications, Lyon, 1992.)*

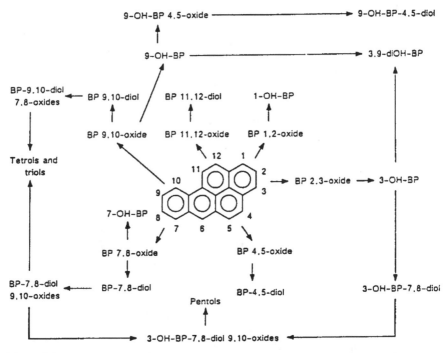

Figure 9.2 *Pathways involved in the metabolism of benzo[a]pyrene (BP) (Adapted with kind permission from D.H. Phillips and P.L.Grover, Polycyclic hydrocarbon activation: bay regions and beyond, Drug Metab. Rev., 1994, 26, 443–467)*

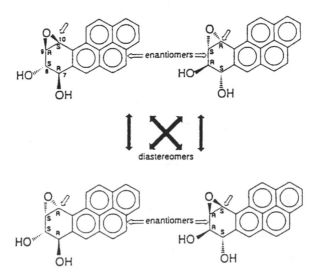

Figure 9.3 *The configurational, isomers of the bay region 7,8-dihydrodiol 9,10-oxides of benxo[a]pyrene (Adapted with kind permission from A. Dipple, 'Reactions of polycyclic hydrocarbons with DNA', in DNA Adducts: Identification and Biological Significance, K. Hemminki, A. Dipple, D.E.G. Shuker, F.F. Kadlubar, D. Segerback and H. Bartsch (eds), IARC Scientific Publications, Lyon, 1994)*

susceptible to the usual cell population size controls. This conversion to autonomous growth marks the emergence of a tumour, but yet more changes are necessary if the tumour is going to progress to a spreading type of neoplasm (metastasis).

9.2.4.2 Carcinogens with Nongenotoxic Modes of Action. The agents that induce genetic changes and produce selective environments favouring the clonal expansion of cell populations are not necessarily the same throughout the neoplastic process and may not act directly on DNA. It is notable that 3,4,7,8-tetrachlorodibenzo-*p*-dioxin, one of the most potent rodent carcinogens known, shows no sign of being able to damage DNA in a wide range of assays. This and some other chemical carcinogens (including certain hormones and hormone-like substances) appear to act primarily through receptor mechanisms. Many mechanistic schemes based on observations on rodents and man involve disturbance of hormonal regulating networks. Although this may be important, it is clear that a property common to all proposed mechanisms of hormone-like action is the persistent stimulation of cell populations to divide, either as a consequence of a toxic effect or of induced mitogenesis (often associated with apoptosis).

While some rodent carcinogens are hormones or hormone-like substances, other non-genotoxic mechanisms are possible. Examples of compounds with such mechanisms are:

(a) blood lipid-lowering compounds that induce a specific proliferation of cytoplasmic organelles, called peroxisomes, in the rat liver;
(b) compounds that cause an accumulation of a specific protein, α_{2u}-globulin, in the kidney of male rats;
(c) compounds that interfere with the feedback control of thyroid, pituitary and ovarian hormones;
(d) dietary components that enhance calcium absorption, which leads to hyperplasia of the adrenal medulla; and
(e) cytotoxic substances that cause damage at the entry portal (*e.g.*, forestomach or nose), resulting in reparative hyperplasia (increased cell proliferation).

The importance of these observations in the mechanism of human carcinogenicity is not yet clear and may range from negligible to substantial. However, irrespective of whether a carcinogen has or has not any direct effect upon the cell's DNA, changes in gene expression always occur during carcinogenesis.

The normally assumed mechanism has been that a shortened cell-cycle time reduces the time for repair of 'background' damage to DNA and increases the probability of mutation. While that may be part of the process, another possibility is that interactions of carcinogens with proteins in chromatin have interfered with normal gene regulatory function, causing inappropriate phenotypic changes, which could include the silencing of repair genes or of tumour-suppressor genes.

9.2.4.3 Chromatin Modification. At the molecular level, mechanisms of epigenetic carcinogenesis are beginning to emerge, particularly from studies that have focussed on gene regulation by chromatin modification. Two important types of chromatin modification are recognised – DNA methylation and histone modification (including methylation, acetylation, phosphorylation and ubiquitylation).

These processes are important for changes in genomic expression during early embryogenesis and differentiation and for tissue-specific gene expression and global gene silencing. Aberrant promoter hypermethylation is associated with inappropriate gene silencing and this affects almost every step in tumour development. In general, modification of histones can modify gene activity with currently unpredictable consequences.

9.2.4.4 DNA Methylation. In the mammalian genome, enzyme-mediated methylation takes place only at cytosine bases that are located 5′ to a guanosine in a CpG dinucleotide. These dinucleotides cluster in the so-called CpG islands. Most CpG islands are found in the proximal promoter regions of almost all of the genes in mammalian genomes and are, generally, unmethylated in normal cells at all stages of development and in all tissue types (Antequera and Bird, 1993). A small but significant proportion of all CpG islands becomes methylated during development and, when this happens, the associated promoter is stably silent, as in genomic imprinting and X-chromosome inactivation.

De novo methylation can also occur in adult somatic cells. A significant fraction of all human CpG islands are prone to progressive methylation. This may occur in certain tissues during aging (reviewed by Issa, 2002) or in abnormal cells, such as cancers (reviewed by Baylin and Herman, 2000), as well as in permanently cultured cell lines. It appears, however, that methylation does not usually silence promoters of genes that are active, but may permanently silence currently silent genes.

In cancer, the hypermethylation of promoter regions is now the most well-recognised epigenetic change to occur in tumours. It is found in, virtually, every type of human neoplasm and is associated with the inappropriate transcriptional silencing of genes. Indeed, in human cancer, promoter hypermethylation is at least as common as the disruption of tumour-suppressor genes by mutation. Aberrant promoter methylation is associated with a loss of gene function that can provide a selective advantage to neoplastic cells, as do mutations. For example, the von Hippel–Lindau syndrome (*VHL*), breast cancer 1, early onset (*BRCA1*) and serine/threonine kinase 11 (*STK11*) genes – germ-line mutations of which cause familial forms of renal, breast and colon cancer, respectively – are often epigenetically silenced in the sporadic forms of these tumour types.

A very important consequence of methylation is when certain DNA repair enzymes are silenced, particularly *MLH1*, which results in increased microsatellite instability, and *MGMT*, which results in mutations. Thus, interference with DNA repair enzymes clouds the distinction of genetic and epigenetic mechanisms.

Although gene silencing is an important function of DNA methylation, it is also important to recognise that cytosine methylation can influence tumourigenicity by other mechanisms (Figure 9.4). These effects occur because 5-methylcytosine is itself mutagenic; it can undergo spontaneous hydrolytic de-amination to cause C → T transitions. This enhanced mutagenesis is seen in the germ line of all organisms that methylate their DNA.

The presence of the methyl group in the CpG dinucleotides, in the coding region of the *TP53* gene, strongly increases the rate at which mutations are induced by UV radiation during the development of skin cancers. This occurs because the methyl group shifts the UV absorption spectrum for cytosine to a region that is prevalent in sunlight.

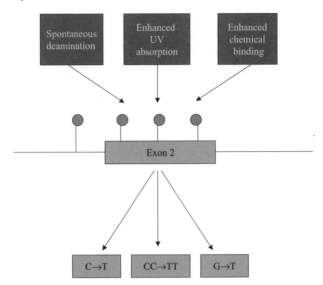

Figure 9.4 *Cytosine methylation and mutation (Adapted from Jones and Baylin (2002))* Reproduced with permission from *Nature Review Genetics*, Vol. 3, 418–428, Copyright 2002, Macmillan Magazines Ltd.

Furthermore, methylated CpG dinucleotides are the preferred targets of G $\rightarrow$ T transversions that are induced in mammalian cells by benzo[a]pyrene diol epoxide. Thus, the methylation that occurs in the transcribed region of *TP53* increases its susceptibility to spontaneous de-amination, UV-and hydrocarbon-induced mutation.

9.2.4.5 Histone Modification. It is quite clear that DNA methylation alone is insufficient to explain the suppression or activation of gene transcription. One function of methylated DNA is to recruit methyl-binding proteins, which in turn recruit a chromatin-remodelling complex that includes histone deacetylases. These enzymes, by deacetylating histones, silence DNA transcription.

Histone modification seems to be a universal regulatory mechanism among eukaryotes from yeast to man. These are small, highly conserved basic proteins, found in the chromatin of all eukaryotic cells, which associate with DNA to form a nucleosome. Each nucleosome contains two copies of each of the core histones H2A, H2B, H3 and H4 wrapped by 146 bp DNA. Modification of core histones at the lysine, arginine and serine residues that lie in their amino-terminal tails is far more complex than DNA methylation. Histone acetylation and deacetylation have been shown to determine the transcriptional activity of chromatin. Precisely how these processes might be involved in carcinogenesis remains, for the most part, unclear. It can be at least speculated that if such changes in protein structure induced by a xenobiotic lead to changes in gene expression, structural or conformational changes brought about by this may lead to inappropriate gene expression.

The study of gene regulation is now firmly in the area of protein chemistry. This study has led to the proposal that there is a 'histone code' that is generated by modification of the amino-terminal tails of the histones protruding from the nucleosomes, a code that is read by other proteins. This hypothesis suggests that it is these interactions that determine whether genes are transcribed or not.

In conclusion, the common property of mitogenesis or hyperplasia is but one step in the epigenetic processes leading to neoplasia. Modification of protein structure, and therefore function, by persistent exposure to certain endogenous or exogenous chemicals may be at least as important as mutation in the neoplastic process. The neoplastic process may commence with an increase in a cell population, but sooner or later must involve a transformation step, which could be an inappropriate re-programming of the genome, with mutational events occurring secondarily to increased genomic instability and the silencing of key repair and tumour-suppressor genes.

9.3 TESTS FOR CARCINOGENS

9.3.1 The Standard Test

The currently accepted design for carcinogenicity testing is to expose rats and mice to the agent for about 2 years. Rarely is any other species used. For each species, 50 male and 50 female animals per group are dosed with a vehicle or the test agent in that vehicle. Daily observations are made and if any animals become moribund during the experiment, they are killed so that tissues are not lost due to autolysis. All animals, including those that survive to the scheduled end of the experiment, are subjected to autopsy and almost 40 different tissues are taken from each animal for histological examination. All observations are recorded, summarised and analysed. While this is the currently accepted basic design for regulatory purposes, it has not always been so and special experimental designs may be used in particular circumstances. If properly justified, these also may be acceptable to regulatory authorities.

The objective is to expose a statistically acceptable number of animals to the highest dose that they will tolerate without reducing their life span for reasons other than tumour development. A series of preliminary toxicity tests is conducted, the purpose of which is to identify a minimally toxic dose, as manifested by reduced body weight gain, reduced food intake, altered appearance or altered behaviour patterns and non-neoplastic histology observed at the end of the experiment. This is the so-called maximum tolerated dose (MTD) determination. There are many ways in which the MTD can and has been defined, but the most commonly used definition states: the MTD is the highest dose that can be predicted not to alter the animals' normal longevity from effects other than carcinogenicity. In practical terms, the MTD is the dose that, in a 3-month preliminary study, causes no more than a 10% weight decrement as compared to the appropriate control group and does not produce mortality, clinical signs of toxicity or pathological lesions (other than those that may be related to a neoplastic response) that would be predicted to shorten an animal's natural life span. Ideally, this is what should happen in the carcinogenicity experiment itself, but in reality the prediction may be incorrect for various reasons. In the carcinogenicity test, animals may die too early, thereby invalidating the study, or the predicted toxicity may not be realised, which invites the criticism that higher dose levels could have been used.

Other considerations in establishing the MTD value are metabolism and pharmacokinetics. Disproportionate changes in these parameters with increasing dose may signal saturation of metabolic pathways that are dominant at lower and therefore more probable human dose levels.

The MTD level is clearly needed for the experimental demonstration of a carcinogenic response in many instances. The basic problem is that the rodent carcinogenicity bioassay is insensitive. In comparison with human populations, the numbers of individuals that are exposed are low. Hence, it is readily conceded that some means must be used to compensate for the low statistical power inherent in the assay and that the use of high dose levels may be one way of compensating for this weakness.

For those substances that are known to be human carcinogens, the rodent carcinogenicity tests may be able to demonstrate their potential at doses lower than the MTD. For a sample of 13 human carcinogens for which there were also experimental carcinogenicity data, the rodent carcinogenicity results indicated that the lowest observed effective doses ranged from 0.005 × MTD to 0.5 × MTD. It would be wrong to conclude, however, that a substance showing a carcinogenic effect only at the MTD in the rodent bioassay is not a human carcinogen. Such a substance may not be a recognisable human carcinogen only for reasons of limited statistical power in the epidemiological studies, or because there is no significant human exposure under normal circumstances. At the same time, however, the use of very high dose levels in carcinogenicity tests raises legitimate concerns about the mechanisms and relevance of any tumour induction observed in these assays.

9.3.2 Alternative Tests

9.3.2.1 Transgenic Animals in Carcinogenicity Testing. Advances in molecular biology in recent years have opened up the possibility of approaches to carcinogenicity testing differing from the more traditional methods. It is now possible to transfer new or altered genes to the germ-line of mammals. These transgenic animals can be used in short-term *in vivo* tests for carcinogenicity and may be useful for research into the characterisation of critical, genotoxic events in carcinogenesis.

In looking for alternatives to life-span studies that are so expensive and often provide results that are contentious, much consideration has been given to models involving proto-oncogenes and tumour-suppressor genes. These classes of genes have been highly conserved in evolution and are therefore likely to be similar in man and inbred strains of rats and mice. Also, they are known to have key roles in the early and late stages of carcinogenesis in both man, where the aetiology is usually unknown, and in rats and mice, where there is a high probability that the causative agent is the test substance administered.

Consequently, transgenic and knockout models have been constructed for the purpose of carcinogen identification. They include the c-Ha-*ras* proto-oncogene, activation of which alters signal transduction and growth control, and the *p53* tumour suppressor gene, inactivation of which seriously disturbs control of the cell cycle and DNA repair.

The Tg.AC mouse is derived from the FVB/N strain and carries multiple copies of the v-Ha-*ras* transgene, which already contains point mutations, and the site of integration, fused to the promoter of the mouse zeta-globin gene, confers on these mice the characteristics of genetically initiated skin as a target for tumourigenesis. The untreated skin appears normal in comparison with the skin of the wild-type FVB/N parent strain. Thus, it can be used to demonstrate promotion, or late stage, activity, but it cannot demonstrate a difference between carcinogens that are genotoxic or are not genotoxic.

Skin papilloma development is the most commonly reported response, but target sites other than skin also can be affected. Specific activation of the expression of the transgene is the critical event underlying responses in the model. Since induction of transgene expression can be achieved either by treatment with known carcinogens or tumour promoters or by full-thickness wounding of the skin, "false-positive" responses might be induced by a chemical that induces some epidermal toxicity; such responses have not been observed, but the possibility should be recognised.

CB6F1-Tg.H*ras*2 mice carry five or six copies per cell of the human proto-oncogene c-Ha-*ras*. The transgene is constitutively expressed in normal tissues. Mutations of the gene have, so far, been found only in tumour cells. Pre-neoplastic lesions or "spontaneous" tumours do not appear before 6 months of age.

Mice that are heterozygous for the *p53* tumour suppressor gene are viable and have a low incidence of tumours during the first year. Mice that are homozygous for the null allele generally have a high incidence of spontaneous tumours within the first 3–6 months of life. Thus, induced deletion of the active allele in the heterozygote will lead to an increase in tumour incidence. This model responds to carcinogens that are genotoxic. Indeed, it could be considered as an *in vivo* test for mutagens.

9.3.2.2 Initiation and Promotion Models. Based on an assumed two-stage process, in which there is initiation followed by promotion, models designed to detect neoplastic end points in single organs have been in use for many years. These include the induction of lung adenomas in strain A mice, the induction of skin papillomas in mice, the induction of mammary tumours in female rats and initiation and promotion models in several organs in mice and rats. Similar, but less-extensive, data are available for other systems, *e.g.* rat stomach, kidney, urinary bladder and colon. Such studies can provide results within 1 year and, frequently, a much shorter period. These systems could be considered appropriate for identifying carcinogens in rodents. However, the possibility of false-positive results has yet to be tested rigorously.

9.3.2.3 Assays with Pre-Neoplasia as the End Point. Focal pre-neoplastic lesions appear prior to the occurrence of neoplasms and generally there are more of these lesions than there are subsequently developing neoplasms. Three widely studied models are aberrant crypt foci in the colon, pepsinogen-altered pyloric glands in the rat glandular stomach and foci of hepatocellular alteration in the liver. While pre-neoplastic lesions can often be identified in conventionally stained sections, lesions that are otherwise difficult to detect can be readily identified in several tissues by special staining techniques, such as staining for placental-type glutathione *S*-transferase (GST-P) activity in liver foci. Such lesions have been extensively used as surrogates for neoplasms to assess the carcinogenic activity of chemicals. The use of histologically defined pre-neoplastic lesions for this purpose has the advantage that the end point occurs within 20–40 weeks, which is before the development of other age-related pathological changes, including spontaneous pre-neoplastic lesions.

9.4 EPIDEMIOLOGY

Epidemiology has many complexities, but it offers, as its major strength, immediate relevance to human health as it uses human data in an attempt to relate human disease to its causes, thereby avoiding the greatest weakness of the experimental approach, its necessary concentration on data from experimental animals. The two main techniques used are cohort and case-control studies. Although most epidemiology concerns causality, it should be clearly understood that epidemiological evidence by itself is often insufficient to establish causality and that causality is never established on the basis of a single study.

Some differences between the epidemiological and experimental approaches to carcinogen identification may not be immediately obvious. They arise because of differences in the pathology studies and data collection. As mentioned above, all experimental animals are subject to extensive histopathology as well as to autopsy. Such detailed study is almost never undertaken on people, although there are national and regional policy differences on this matter. While experimental histopathology is restricted to small samples of each of the (up to) 40 tissues taken, those samples are very much more representative of a mouse or rat than they would be of the much larger human body. Thus, there is a higher probability of discovering small, benign neoplasms in 2-year old rodents, and these are all taken into account in the subsequent analysis. They are not looked for in any but a few human autopsies, and epidemiological studies commonly take into account only the neoplasm that was either clinically diagnosed (with or without histological verification), or was the cause of death or contributed to it, *i.e.*, they are dependent upon what is clinically important or information in death certificates or cancer registries. Data used for the analysis of experimental studies are collected directly from the experiment and are not filtered according to their clinical significance. Common incidental human tumours (which do not find a mention on death certificates) frequently include tumours of the prostate, thyroid, adrenal and kidney.

9.5 CONCLUSION

This brief excursion into carcinogenicity can be no more than an introduction (especially the epidemiology section). The mechanisms by which cancers arise are diverse, complex and only partially understood. Procedures have been adopted for the detection of carcinogens and while they clearly have the ability of demonstrating carcinogenicity in rodents, it is frequently unclear how relevant these findings are for the prediction of human cancers. Statistical limitations have led to the development of exaggerated testing methods that may often detract from simple applicability of the results. As a consequence, data evaluation for carcinogenic risk assessment has become a complex art in which disagreements abound. There are, however, certain undeniable facts about human carcinogens. The most important single human carcinogen, in terms of its effects upon human mortality, is tobacco smoke. In comparison, all other fatal carcinogenic risks are small. Unburnt tobacco also is carcinogenic, but its burning creates yet more carcinogens, *e.g.*, benzo[a]pyrene. Sunlight exposure certainly produces more tumours, but only a small proportion of these is lethal. However, sunlight is more difficult to avoid than tobacco smoke should be, and low exposures to sunlight are beneficial. Occupational carcinogens have a much smaller community impact, although they can be significant in specific populations. Exposure controls have improved significantly in the developed countries during the latter half of this century, but the conditions in some developing countries are a matter of growing concern.

BIBLIOGRAPHY

F. Antequera and A. Bird, Number of CpG islands and genes in human and mouse, *Proc. Natl. Acad. Sci. USA*, 1993, **90**, 11995–11999.

A. Balmain and K. Brown, Oncogene activation in chemical carcinogenesis, *Adv. Cancer Res.*, 1988, **51**, 147–182.

S.B. Baylin and J.G. Herman, DNA hypermethylation in tumorigenesis: epigenetics joins genetics, *Trends Genet.*, 2000, **16**, 168–174.

H.M. Bolt, H. Foth, J.G. Hengstler and G.H. Degen. Carcinogenicity categorization of chemicals – new aspects to be considered in a European perspective, 2004, *Toxicol. Lett.*, **151**, 29–41.

H.C. Grice and J.L. Ciminera, (eds), *Carcinogenicity: The Design, Analysis and Interpretation of Long-term Animal Studies*, Springer, New York, 1988.

International Agency for Research on Cancer web site: http://www.iarc.fr/.

J.P. Issa, Epigenetic variation and human disease, *J. Nutr.*, 2002, **132**, 2388S–2392S.

P.A. Jones and S.B. Baylin, The fundamental role of epigenetic events in cancer, *Nature Rev. Gen.*, 2002, **3**, 415–428.

A.J. Levine, The tumor suppressor genes, *Ann. Rev. Biochem.*, 1993, **62**, 623–651.

D.E. Lilienfeld and P.D. Stolley, Foundations of Epidemiology, 3rd edn, Oxford University Press, New York, Oxford, 1994.

H. Vainio, P.N. Magee, D.B. McGregor and A.J. McMichael (eds), *Mechanisms of Carcinogenesis in Risk Identification*, IARC Scientific Publication No. 116, IARC Scientific Publications, Lyon, 1992.

B. Vogelstein, E.R. Fearon, S.E. Hamilton, S.E. Kern, A.C. Preisinger, M. Leppert, Y. Nakamura, R. White, A.M Smits and J.L. Bos, Genetic alterations during colorectal-tumor development, *N. Engl. J. Med.*, 1988, **319**, 525–532.

Chapter 10

Introduction to Toxicogenomics

DARRELL BOVERHOF, JEREMY BURT AND TIMOTHY ZACHAREWSKI

10.1 INTRODUCTION

In order to assess fully the potential adverse health effects of chronic and subchronic exposure to synthetic and natural chemicals and their complex mixtures, a more comprehensive understanding of the molecular, cellular, and tissue-level effects of these compounds is required within the context of the whole organism. Toxicogenomics is an emerging interdisciplinary field that incorporates '-Omic' technologies into toxicology with the goal of furthering the elucidation of mechanisms of toxicity. The omic technologies have evolved into three disciplines: genomics is the study of genes and their function, proteomics is the study of the levels and states of proteins found in a cell, and metabonomics is the study of metabolite levels in biological fluids as they change in the cell, tissue, or organism. Traditional toxicological studies have generally focused on selected aspects of toxicity and response (*i.e.* expression-level change of one gene). By contrast, toxicogenomics attempts systematic integration and comprehensive assessment of data across all levels of biological organization with the aim of producing a complete picture of the molecular, cellular, and physiological basis of toxicity and associated disease.

10.2 MICROARRAY TECHNOLOGY

Microarray technology was the first technology to emerge in the field of toxicogenomics. It permitted profiling of gene expression levels following exposure to a toxicant. Gene expression profiling may be better referred to as 'transcriptomics', since it is simply the study of the mRNA transcript levels within a cell or tissue.

The use of microarrays is based on the principle of hybridization between complementary strands of nucleic acids, one of which is immobilized on the slide surface and the other is the labeled sample of interest. Nucleic acid hybridization is both sensitive and specific and has been used for many years in the field of molecular biology to monitor mRNA levels of a single gene. When combined with fluorescent labeling and sensitive detection instrumentation, microarrays allow for rapid and accurate characterization of the expression of hundreds to tens of thousands of genes simultaneously.

Microarray technology is well suited to studying molecular effects of toxicants both *in vivo* and *in vitro*, since changes in gene expression often precede a toxic response. Microarrays have been used to compare gene expression profiles between control and toxicant-treated animals in order to identify specific gene expression changes. The gene expression profiles identified may permit elucidation of the mechanisms of action that underlie toxicity and, as a result, improve risk assessment. In the context of drug development, this can lead to the production of molecules with fewer side effects. The following sections provide an introduction to the basic concepts of microarray technology and the subsequent analysis and interpretation of the data (Figure 10.1).

10.2.1 Array Platforms

DNA microarrays are available in two forms, namely spotted and synthesized arrays. With spotted arrays, the nucleic acid elements are presynthesized and stored in 96- or 384-well plates and are robotically spotted onto the slide surface. The spotted DNA products are typically generated by PCR (polymeric chain reaction) amplification of cloned cDNAs (100–3000 nt (nucleotides) in length). The cloned cDNAs are subsequently purified and transferred into the plates for storage and spotting. Alternatively, oligonucleotides for spotting (50–120 nt) can be synthesized directly in 96- or 384-well formats, an approach that is becoming increasingly popular owing to the expanding knowledge of genomic sequences of various organisms, the reduced cost and increased yields of synthesis, and the ability to optimize the spotted products for specificity, concentration, and purity when compared to PCR amplified cDNAs. The products of these reactions (PCR or oligonucleotide) are robotically spotted at a high density onto a substrate that allows for chemical attachment of the sequence to the slide surface. Typically, this is a glass slide coated with substrates, such as reactive aldehyde, amine, or epoxide groups, to which the DNA products are electrostatically or covalently attached.

Synthesized arrays involve the *in situ* synthesis of short oligonucleotides (25–60 nt) directly onto the slide surface using photolithography or ink-jet technologies (www.affymetrix.com, www.chem.agilent.com). This allows oligonucleotides to be specifically designed for optimal detection without the need to maintain extensive collections of cDNA clones or oligonucleotides. However, short oligonucleotide arrays suffer from reduced specificity in detecting a single gene product owing to the shorter sequences for binding. To compensate for this limitation, a tiling approach is used in which multiple short oligonucleotides that span across the gene of interest are represented in order to determine more accurately its true expression level. Owing to the technological implications of *in situ* synthesis, these arrays are provided mainly by commercial vendors, and as a result, the cost associated with these arrays can be significant, often limiting their use in academic laboratories. Consequently, spotted microarrays, which are cheaper to produce but more labor intensive to maintain, are often utilized in academic laboratories that have centralized facilities providing amplification, robotic, and printing services for array production. Spotted arrays also offer greater flexibility in the choice of represented genes, and allow for the preparation of customized arrays for answering specific research questions.

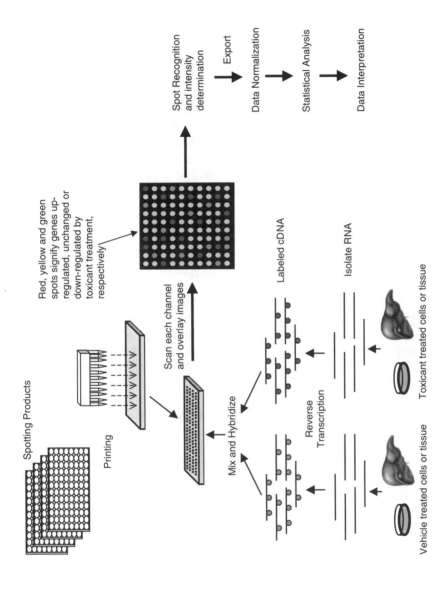

Figure 10.1 *General overview of the microarray process using spotted arrays with a two-color labeling approach*

10.2.2 Sample Labeling and Hybridization

Expression profiling involves the labeling of cDNA derived from mRNA and subsequent hybridization to the microarray, followed by detection and quantification to determine the relative expression levels of the various transcripts in the sample. The starting material in this process is RNA, which can be in the form of highly purified poly(A) mRNA or total RNA from cells or tissues. It is imperative that the RNA be pure and of high quality in order to allow for the efficient production of labeled target. The target sequence used in hybridization is prepared by reverse transcription of RNA in the presence of fluorescently labeled deoxynucleotides producing fluorescently labeled cDNA, which will competitively hybridize to cDNAs or oligonucleotides on the array. This approach to target-production is often referred to as direct incorporation because the fluorescence is incorporated directly during the reverse transcription reaction. However, owing to the steric hindrance of 'bulky' dyes during the enzymatic reaction, direct incorporation generally reduces the efficiency of the reverse transcription reaction. Direct incorporation can also introduce fluor-dependent labeling bias (referred to as dye bias) owing to differences in the incorporation rate of fluor-labeled nucleotides.

Indirect incorporation methods are also used. In these methods, nucleotides or primers with less 'bulky' modified chemical groups are utilized in the reverse transcription reaction, and fluorescent dyes are conjugated to these groups in a subsequent reaction. Owing to its increased efficiency, indirect incorporation allows for the use of less starting material and, therefore, may be appropriate in situations when RNA is limiting. Regardless of the approach, the fluorescently labeled samples are washed, purified, and dissolved in buffers that allow for efficient hybridization between labeled samples and the immobilized sequences on the surface of the microarray. The samples are then applied to the surface of the array and incubated (16–24 h) at a temperature that facilitates the discriminate hybridization of the labeled probes to the immobilized sequences on the array (42–55 °C).

Following hybridization, the slides are washed and the images for each fluorescent label (and therefore each sample RNA) are obtained by laser-induced fluorescence imaging using either a confocal detector or a charged coupled device (CCD) camera. The output of the detector is a tagged-image file format (TIFF), which is a two-dimensional (2D), 16-bit numerical representation of the microarray surface with intensity values ranging from 0 to 65,536 (2^{16}). Although fluorescent labeling methods provide a simple and sensitive approach for detection, fluorescent dyes are light sensitive and prone to quenching during the scanning process. Furthermore, some fluorescent dyes are extremely sensitive to oxidation and, as a result, weather conditions such as high ozone levels have been reported to have a detrimental effect on fluorescent signals from the array. Therefore, the acquisition of high-quality images is not only dependent on the quality of the RNA-starting material, but also on the timely and appropriate scanning of the slides.

Labeling and hybridization can be done in one- or two-color formats. In the one-color approach, an expression profile for each sample is generated on separate arrays using a single fluorescent label (typical for commercial slides such as Affymetrix GeneChips). The differential expression of transcripts from two different samples is then assessed by comparing the intensity levels obtained from the individual array images.

Alternatively, the two-color format involves the independent labeling of two different RNA samples with different fluorescent dyes (*e.g.* cyanine 3 and cyanine 5), which fluoresce at different wavelengths. These samples are then mixed in the hybridization buffer and applied to a single array. This approach has been referred to as competitive hybridization and typically involves the co-hybridization of a sample of interest (*e.g.* the toxicant-treated sample) to a reference sample (untreated or vehicle control). After hybridization and washing, the slide is scanned using two different wavelengths, corresponding to the optimal excitation wavelength of each fluorescent dye. The resultant images are then overlaid and the signal intensities are used to calculate ratios to determine the relative expression levels of the transcripts in each sample.

10.2.3 Image Analysis

After image acquisition, microarrays are generally imported into software programs for spot recognition, a process referred to as gridding. In this process, the image files (TIFFs) are imported into an image analysis software application (*e.g.* Genepix (www.axon.com), Spotfire (www.spotfire.com), or Molecularware (www.molecularware.com)), which determines the foreground and background intensities for each spot on the array. The output files generated by these programs are then subjected to a normalization strategy. Normalization refers to the adjustment of the microarray data, both within and across the arrays in an experiment, for effects that could arise from variation in the technology rather than from real differences between the biological samples of interest. There are numerous reasons why the data must be normalized including differences in the labeling or detection efficiencies of the different fluorescent dyes, quality differences between scanned arrays, and differences in sample quantity and/or integrity. Numerous normalization strategies have been proposed including log centering, rank invariant methods, and locally weighted linear regression (LOWESS) analyses. Regardless of the approach utilized, normalization of the data is essential to facilitate comparison of arrays and to select genes that are differentially expressed between samples in order to ensure the generation of meaningful results in downstream data analysis and interpretation.

10.2.4 Data Analysis and Interpretation

After the data have been fully processed, the researcher is left with the daunting task of interpreting the vast amounts of information that have been generated. This process is aided by first utilizing statistical approaches to identify those genes that are active or that differ significantly from the reference samples. Various statistical approaches have been proposed in the literature from those as simple as the *t*-test to the more complicated ANOVA and Bayesian analysis approaches. Importantly, as with any experiment, these statistical analyses require biological replication within the experimental design, an important aspect that is often overlooked owing to the costs and resources associated with microarray technology. However, replication in microarray experiments has been shown to reduce markedly the number of potential false positives and, when combined with a robust statistical analysis, experimental results can be interpreted with high confidence.

For purposes of data interpretation, the final gene list is typically converted to a form that is organized and readily viewable. This often involves a graphical representation of the data using various microarray analysis software packages (*e.g.* GeneSpring (www.silicongenetics.com), Bioconductor (www.bioconductor.org)). Various forms of data representation are available including scatter plots, hierarchical clustering, K-means clustering, principle components analysis, self organizing maps, neural networks, and other complex algorithms. Researchers are also interested in assigning functional information to the data. This may be done using tools, such as GoMiner (http://discover.nci.nih.gov/gominer/) or OntoExpress (http://vortex.cs. wayne.edu/ research.htm), which utilize Gene Ontology annotation (http://www. geneontology.org/) of the genes to help classify the data and to offer insight into the particular biochemical process, cellular component, or molecular function that may be altered by the treatment. Various software programs also exist for compiling complex biological signaling pathways through the reverse-engineering of the data with the aim of identifying and developing global gene regulatory networks.

10.2.5 Microarray Standards

Microarrays have become a very powerful and widely used tool as evidenced by the exponential increase in the number of publications using this technology since 1997. With the rapid growth and increased availability of microarray technology, many researchers have realized the need to implement standards for conducting and reporting data from these experiments. Without such standards, the ability to interpret accurately a microarray experiment or judge the validity of the reported results is significantly compromised. As a result, the microarray gene expression data (MGED) Society (www.mged.org) has proposed the minimum information about a microarray experiment (MIAME) standards, which define the minimum information required to ensure that the microarray data can be unambiguously interpreted and potentially repeated by an independent investigator. Many prominent scientific journals, including Nature, Cell, Lancet, and the New England Journal of Medicine, now require that microarray data be MIAME compliant prior to publication. In addition, a toxicology version of these standards, referred to as MIAME-Tox, is being developed and outlines the minimum information required to interpret fully and, if required, repeat a toxicogenomics study. These guidelines adopt the MIAME standards and extend them to include documentation specifically relevant to toxicological experimental designs and end points (*e.g.* dose–response experiments, clinical observations, and histopathological end points). To aid in these efforts and assure open access to published data, various groups have developed public repositories for storing microarray data in MIAME compliant formats (*e.g.* Chemical Effects in Biological Systems (CEBS) – http:// cebs.niehs.nih.gov/microarray/index.jsp, Gene Expression Omnibus (GEO) – http:// www.ncbi.nlm.nih.gov/geo).

10.2.6 Microarrays in Toxicology

Microarrays provide toxicologists with a valuable tool to define precisely the molecular responses that precede and accompany the effect on a biological system to a chemical or drug insult. It is hoped that these gene expression responses will further elucidate mechanisms of toxicity and also lead to the development of biomarkers for environmental health research and high-throughput chemical screening, which may be particularly valuable in drug development. Profiling gene expression responses between *in vitro* and *in vivo* models of toxicity and between human and rodent models will also help to define the limits of animal-to-human extrapolation. This will improve risk assessment for environmental chemicals and for preclinical safety assessment of new drugs.

In order to exploit fully the power of microarray technology, it is important to determine the relationships between the gene expression data and traditional toxicological end points. Understanding these relationships will allow the vast quantity of information contained within the gene expression data to be harnessed fully, allowing for a greater understanding of the mechanisms of action of chemicals in biological systems.

10.3 PROTEOMICS

Microarray technology provides gene expression profiling at the mRNA level. While this approach provides researchers with a powerful tool for elucidating gene function, it suffers from the fact that gene expression may not correlate directly with protein expression and does not consider the multiple forms a protein may adopt from a single transcript. To examine these issues, proteomics, the large-scale study of the proteins in a cell, has emerged as an important complementary field. As with microarray experiments, much of the pioneering proteomic research has focused on examining altered protein expression profiles found in cells following exposure to various toxicants.

Characterizing protein expression levels is only one part of proteomics. Proteomics includes the study of the multiple forms of proteins caused by alternative splicing of transcripts, post-translational modifications, gene mutations, protein–protein interactions, and protein–small molecule interactions. Proteomics also includes the study of protein localization. The ultimate goal of proteomics in toxicology is the integration of all these separate components in order to further our understanding of the various pathways that contribute to the manifestation of complex adaptive and toxic responses observed in cells, tissues, and whole organisms following toxicant exposure. This will lead to a better understanding of the onset of various toxic disease states. In addition, it is expected that proteomics will lead to the identification of novel protein biomarkers that are indicative of toxicant exposure.

Examining and analyzing the whole proteome, all the proteins contained in a cell, has become a viable discipline owing to the deciphering of the sequence of entire genomes and the development of bioinformatics. Having the genetic blueprint for an organism has enabled the creation of putative gene and protein sequence databases (such as Swiss-Prot (http://us.expasy.org/sprot/)), which has facilitated the digital prediction of many components of the proteome, including protein molecular weights and p*I* values, post-translational modifications and alternative splicing forms, and potential protein structures and functions. This information has been integral to the design of proteomic studies and the analysis of the vast amounts of data generated. Most of the current proteomic research is directed toward those organisms with highly annotated genomes and proteomes, such as yeast, while the research on other toxicologically relevant species, such as the rat, lags behind.

The methods used in proteomic studies tend to be a hybrid between traditional biochemical and molecular biology approaches and the new large-scale, high-throughput methods. While techniques such as yeast two-hybrid systems have been used successfully to answer questions about the proteome, what follows in this section is a review of some of the technologies that currently define proteomics.

10.3.1 Mass Spectrometry-Based Proteomics

Mass spectrometry (MS)-based technologies are the techniques most commonly used to examine the expression of proteins as a function of cell or tissue state. In these techniques, a biological sample is analyzed by separating and identifying as many proteins as possible, usually with an emphasis on those proteins with altered abundances relative to a control sample (Figure 10.2). There are two methods commonly used to identify proteins, namely 2D gel electrophoresis followed by matrix-assisted laser desorption ionization (MALDI)–MS and liquid chromatography (LC) followed by electrospray ionization (ESI)–MS. However, in the rapidly growing field of proteomics, many more methods are being developed and refined as the field matures and the equipment becomes more sensitive.

2D gel electrophoresis followed by MALDI–MS has been the method of choice for analyzing protein mixtures. 2D gels first separate proteins on the basis of their charge by isoelectric focusing prior to further separation by molecular weight using sodium dodecyl sulfate polyacrylamide gel electrophoresis. The proteins in the gel are visualized using a protein dye, such as coomassie blue or silver stain, and individual protein bands are excised and digested using trypsin to yield peptides that are then subjected to MS analysis. The proteins to be examined are usually chosen by comparing the protein patterns of two separate proteomes (such as toxicant treated *vs*. reference or control samples) and examining the differentially expressed proteins. MALDI uses an ultraviolet laser to ionize peptides when they are in a solid or crystalline state. The ionized peptides enter a time-of-flight (TOF) mass spectrometer, which measures the mass of individual peptides and gives a 'peptide mass fingerprint' (PMF) for the protein of interest. The experimental PMFs can be used to identify the original protein by searching online databases of hypothetical PMFs using search engines such as PeptIdent (http://au.expasy.org/tools/peptident.html) and PeptideSearch (http://www.mann.embl-heidelberg.de/GroupPages/PageLink/ peptidesearchpage.html).

An alternate MS-based approach for proteomic analysis uses LC coupled with ESI–MS. The columns used in LC are capable of separating proteins based upon size, charge, or affinity depending on the matrix. ESI ionizes the proteins while they are in solution and thus is well suited to high-throughput approaches as ESI–MS is able to analyze the proteins immediately they are eluted from the columns. With highly complex mixtures of proteins, multiple tandem chromatography columns are used to achieve the necessary separation for subsequent analysis. As with other high-throughput technologies, a challenge for this technology is the analysis of the large number of peptide spectra that are generated. Peptide assignment using database search engines has an inherent rate of false identification; thus, filtering and statistical methods must be utilized to identify the proteins accurately. The ultimate goal of this technology is to evolve a form of shotgun proteomics in which it is possible to analyze rapidly very complex mixtures of proteins and determine accurately their identity.

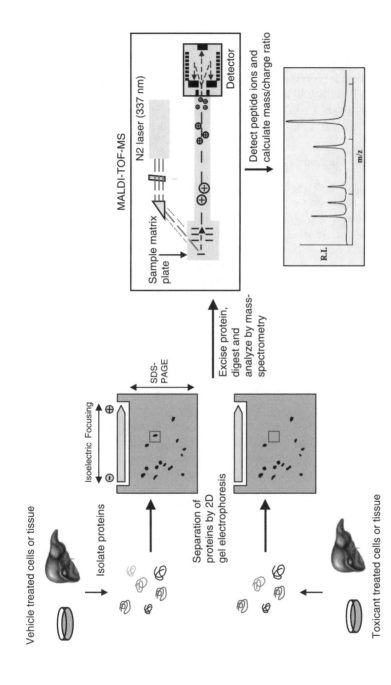

Figure 10.2 *MS-based proteomics; identifying differentially expressed proteins using 2D gel electrophoresis and MALDI-TOF–MS*

While these methods have proved to be very effective in identifying a subset of the proteins in a sample, they are not so effective in coping with the complex nature of the proteome because of the large dynamic range of protein expression and the diverse physical properties of the proteins themselves (*e.g.* size and/or charge). Proteins found in very low abundance are difficult to detect using MS-based methods, although with a large amount of starting material and considerable effort it is possible to detect them with certain LC–ESI–MS methods. Another problem with these technologies is their inability to quantitate accurately the proteins found in a sample. Unless other methods are utilized (stable-isotope dilution for one), it is impossible to quantitate the amount of protein present, only to detect it.

10.3.2 Protein Microarrays

While MS-based methods are currently the method of choice, microarray technology is also being considered for proteomics. Protein arrays can be categorized into two types: protein-detecting arrays and protein-function arrays. In protein-detecting arrays, protein-specific affinity reagents (usually in the form of antibodies) are conjugated to a solid support (usually a glass slide) and allowed to capture proteins out of a mixture (usually a cell lysate). The bound proteins are then visualized by being bound to a second antibody that is fluorescently labeled, or by fluorescently labeling the protein extract. One drawback of such an experiment is that it requires prior knowledge of what is to be studied and highly specific antibodies with comparable affinities for the antigens. Although this allows the researcher to direct the experiment, it limits true 'discovery'-based experiments. This method also suffers from the difficulty in detecting proteins of low abundance and is compromised by the cross-reactivity of many antibodies.

Protein function arrays attempt to use microarray technology to assign an activity to a protein or group of proteins. This technology is currently evolving and is being used to investigate protein–protein interactions, protein–small molecule interactions, protein modifications, and protein activity. With these arrays, the proteins themselves are immobilized on the glass slide and used to assay for function, by adding other proteins to look for interactions or substrates to test for enzyme activity. These arrays are useful as they allow the researcher to control the conditions of the reaction/interaction, but require purified proteins to spot on the arrays. This is possible for smaller, precisely directed experiments, but is a limitation that must be overcome for larger global studies that require large numbers of pure proteins.

10.3.3 Proteomics Standards

The ultimate goal of proteomics is to elucidate the protein networks involved in a toxicant response. However, with the data currently generated being reported in different formats using different databases, it is difficult to compare and integrate data from a variety of independent sources. Following the lead of the MGED Society and their MIAME standards, several groups are currently working on developing a similar set of standards that will allow for such functionality. The SASHIMI project (http://sashimi.sourceforge.net), the PEDRo model, and the proteomics standards initiative (PSI) group, each have the goal of establishing a set of required parameters necessary for the reporting of data and deposition in public databases that will facilitate the integration of data from different sources.

10.4 METABONOMICS OVERVIEW

The latest emerging 'omic' technology involves the profiling of drug and toxicant-induced changes in the metabolite composition of biofluids such as blood or urine. This approach, referred to as metabonomics, has been defined as 'the quantitative measurement of the dynamic multiparametric metabolic response of living systems to pathophysiological stimuli or genetic modification'. Metabonomics takes advantage of the fact that drug and toxicant-induced alterations in cell and tissue function result in disruption of the dynamic equilibrium (steady state) of cellular metabolites resulting in changes in the composition of the biofluids within an organism. Thus, analysis of metabolite levels within biofluids has the potential to assess and predict the severity of target-organ toxicity. Biofluids available for metabonomic analysis are diverse and include plasma, whole blood, milk, saliva, and seminal fluid. However, urine and blood are often the fluids of choice owing to the ease of sampling and the ability to obtain sufficient quantities for analysis.

Although both genomics and proteomics technologies capture the global responses of cells or tissues in response to a toxicant, they do not assess the integrated toxic response of the organism as a whole. By contrast, metabonomics monitors the *in vivo* multiorgan response and the functional integrity of the biological system in real time. Metabonomics also has the advantage of being noninvasive. A single animal can be profiled for a toxic response for the duration of the study, thereby removing uncertainties introduced by inter-animal variability and allowing for characterization of both the onset of toxicity and the adaptation of the organism. This is a distinct advantage over *in vivo* genomics and proteomics, which require tissue removal and the sacrifice of an animal for each designated time point.

The use of nuclear magnetic resonance (NMR) technology is currently the most common approach to metabonomic analysis of altered metabolite profiles. This technology can quantitate a wide range of metabolites at one time and thus is amenable to high-throughput analysis. Although more sensitive detection strategies, such as high performance liquid chromatography (HPLC) or LC coupled with MS, can be used, NMR analysis requires very little sample preparation and is non-destructive and the sample can be recovered for additional analyses. Published reports have indicated that NMR-based metabonomics is robust, with same-sample analysis yielding comparable results in different laboratories. With recent technological developments, NMR can now be utilized for the analysis of whole tissue samples using high-resolution magic angle spinning (MAS). This approach allows for the metabolic analysis of specific cellular or tissue environments and will provide invaluable data in the assessment of toxicants and their mechanisms of action in target tissues.

10.4.1 Applications and Approaches

Metabonomics has the potential to be applied to a wide range of experimental studies including the phenotyping of laboratory animals, the diagnosis of disease states, the rapid assessment of toxicity, and the monitoring of toxic effects in people at risk.

Like genomics and proteomics, metabonomics holds tremendous promise, and studies have already demonstrated the ability of NMR metabolite patterns to identify accurately both the location and severity of toxic lesions. In addition, various metabolite biomarkers of toxicity have already been identified through metabonomic analysis of urine, *e.g.* the metabolites taurine and creatine have been associated with acute liver and testicular toxicity, respectively.

The general approach to metabonomics in toxicology requires the collection of the biofluid of interest from control and drug- or toxicant-treated animals, the preparation of the sample and subsequent NMR analysis (Figure 10.3). The resulting spectrum is used as a 'fingerprint' of the metabolic state of the organism. Subsequent comparison of this 'fingerprint' to that of control animals and to a database of other compounds with known toxicity allows for the classification of the test compound of interest for its toxic potential.

For metabonomics to be most useful, an extensive reference database of toxicants with known target site toxicities and mechanisms of action will need to be developed. This database will require establishment of strong associations between metabolite profiles of known toxicants and traditional toxicological parameters such as histopathological changes and clinical chemistry biomarkers in order to achieve general acceptance.

In an effort to evaluate comprehensively the applicability of metabonomics for the toxicological assessment of drug candidates, the Consortium for Metabonomic Technology (COMET) has been assembled. The consortium consists of six pharmaceutical companies and the Imperial College of Science, Technology and Medicine in London. Their goal is to develop an effective approach for metabonomic analysis,

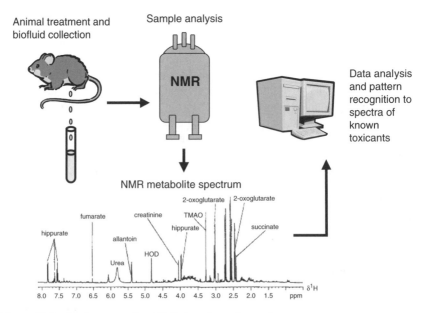

Figure 10.3 *Basic approach to NMR-based metabonomics*

which includes establishing baseline data on the normal physiological spectrum of metabolites, the characterization of known toxicants, the identification of biomarkers of toxicity, and the development of analytical approaches for pattern recognition of metabolite profiles of drug candidates. Reports of round robin studies of toxicant-treated rodent serum and urine samples indicate remarkable consistency between laboratories. Future reports will provide a detailed evaluation of prototype-pattern recognition systems that promise to identify samples from animals experiencing liver or kidney toxicity. Although still in its infancy, metabonomics will eventually complement both transcriptomic and proteomic technologies in providing a better understanding of toxic effects in living organisms.

10.5 CONCLUSIONS

Toxicogenomics has resulted in the accumulation of vast quantities of data dealing with the profiles of transcripts, proteins, and metabolites found in cells, tissues, and organisms in response to natural and synthetic chemicals as well as their mixtures. Ideally, results from omic technologies will be integrated with other disparate data (*e.g.* histopathology, clinical chemistry) to develop computational models that comprehensively describe the molecular and cellular responses contributing to the etiology of observed physiological or pathological effects. This strategy, termed systems biology or systems toxicology, is an iterative process that elucidates the underlying network structures of toxicity in order to develop predictive computational models of response for living organisms. The advent of high-throughput technologies and the continuing development of associated computational resources necessary for data storage, analysis, and modeling are consistent with current computational toxicology efforts (http://www.epa.gov/comptox/), which propose the development of predictive models for use in human health risk and preclinical drug development safety assessments. Thus, omic technologies not only further our understanding of the molecular basis of toxicity, but also contribute to preclinical drug development safety assessments, science-based human health risk assessments, and the potential identification of biomarkers of exposure for environmental health and clinical drug research.

BIBLIOGRAPHY

R. Aebersold and M. Mann, Mass spectrometry-based proteomics, *Nature*, 2003, **422**, 198–207.

P. Brazhnik, A. de la Fuente and P. Mendes, Gene networks: how to put the function in genomics, *Trends Biotechnol.*, 2002, **20**, 467–472.

A. Brazma, P. Hingamp, J. Quackenbush, G. Sherlock, P. Spellman, C. Stoeckert, J. Aach, W. Ansorge, C.A. Ball, H.C. Causton, T. Gaasterland, P. Glenisson, F.C. Holstege, I.F. Kim, V. Markowitz, J.C. Matese, H. Parkinson, A. Robinson, U. Sarkans, S. Schulze-Kremer, J. Stewart, R. Taylor, J. Vilo and M. Vingron, Minimum information about a microarray experiment (MIAME)-toward standards for microarray data, *Nat. Genet.*, 2001, **29**, 365–371.

S.P. Gygi, Y. Rochon, B.R. Franza and R. Aebersold, Correlation between protein and mRNA abundance in yeast, *Mol. Cell Biol.*, 1999, **19**, 1720–1730.

H. Hermjakob, L. Montecchi-Palazzi, G. Bader, J. Wojcik, L. Salwinski, A. Ceol, S. Moore, S. Orchard, U. Sarkans, C. von Mering, B. Roechert, S. Poux, E. Jung, H. Mersch, P. Kersey, M. Lappe, Y. Li, R. Zeng, D. Rana, M. Nikolski, H. Husi, C. Brun, K. Shanker, S.G. Grant, C. Sander, P. Bork, W. Zhu, A. Pandey, A. Brazma, B. Jacq, M. Vidal, D. Sherman, P. Legrain, G. Cesareni, I. Xenarios, D. Eisenberg, B. Steipe, C. Hogue and R. Apweiler, The HUPO PSI's molecular interaction format – a community standard for the representation of protein interaction data, *Nat. Biotechnol.*, 2004, **22**, 177–183.

T. Ideker, T. Galitski and L. Hood, Building with a scaffold: emerging strategies for high- to low-level cellular modeling, *Annu. Rev. Genomics Hum. Genet.*, 2001, **2**, 343–372.

T. Ideker and D. Lauffenburger, A new approach to decoding life: systems biology, *Trends Biotechnol.*, 2003, **21**, 255–262.

T. Ito, T. Chiba and M. Yoshida, Exploring the protein interactome using comprehensive two-hybrid projects, *Trends Biotechnol.*, 2001, **19**, S23–S27.

M. Kaern, W.J. Blake and J.J. Collins, The engineering of gene regulatory networks, *Annu. Rev. Biomed. Eng.*, 2003, **5**, 179–206.

A. Keller, A.I. Nesvizhskii, E. Kolker and R. Aebersold, Empirical statistical model to estimate the accuracy of peptide identifications made by MS/MS and database search, *Anal. Chem.*, 2002, **74**, 5383–5392.

J.C. Lindon, J.K. Nicholson, E. Holmes, H. Antti, M.E. Bollard, H. Keun, O. Beckonert, T.M. Ebbels, M.D. Reily, D. Robertson, G.J. Stevens, P. Luke, A.P. Breau, G.H. Cantor, R.H. Bible, U. Niederhauser, H. Senn, G. Schlotterbeck, U.G. Sidelmann, S.M. Laursen, A. Tymiak, B.D. Car, L. Lehman-McKeeman, J.M. Colet, A. Loukaci and C. Thomas, Contemporary issues in toxicology the role of metabonomics in toxicology and its evaluation by the COMET project, *Toxicol. Appl. Pharmacol.*, 2003, **187**, 137–146.

Y.F. Leung and D. Cavalieri, Fundamentals of cDNA microarray data analysis, *Trends Genet.*, 2003, **19**, 649–659.

G. MacBeath, Protein microarrays and proteomics, *Nat. Genet.*, 2002, **32**(Suppl), 526–532.

J.K. Nicholson, J.C. Lindon and E. Holmes, 'Metabonomics': understanding the metabolic responses of living systems to pathophysiological stimuli via multivariate statistical analysis of biological NMR spectroscopic data, *Xenobiotica*, 1999, **29**, 1181–1189.

P.H. O'Farrell, High resolution two-dimensional electrophoresis of proteins, *J. Biol. Chem.*, 1975, **250**, 4007–4021.

J. Peng, J.E. Elias, C.C. Thoreen, L.J. Licklider and S.P. Gygi, Evaluation of multidimensional chromatography coupled with tandem mass spectrometry (LC/LC-MS/MS) for large-scale protein analysis: the yeast proteome, *J. Proteome Res.*, 2003, **2**, 43–50.

J. Quackenbush, Microarray data normalization and transformation, *Nat. Genet.*, 2002, **32**(Suppl), 496–501.

L.C. Robosky, D.G. Robertson, J.D. Baker, S. Rane and M.D. Reily, *In vivo* toxicity screening programs using metabonomics, *Comb. Chem. High Throughput Screen*, 2002, **5**, 651–662.

A. Schulze and J. Downward, Navigating gene expression using microarrays–a technology review, *Nat. Cell Biol.*, 2001, **3**, E190–E195.

D.W. Selinger, M.A. Wright and G.M. Church, On the complete determination of biological systems, *Trends Biotechnol.*, 2003, **21**, 251–254.

R.L. Stears, T. Martinsky and M. Schena, Trends in microarray analysis, *Nat. Med.*, 2003, **9**, 140–145.

C.F. Taylor, N.W. Paton, K.L. Garwood, P.D. Kirby, D.A. Stead, Z. Yin, E.W. Deutsch, L. Selway, J. Walker, I. Riba-Garcia, S. Mohammed, M.J. Deery, J.A. Howard, T. Dunkley, R. Aebersold, D.B. Kell, K.S. Lilley, P. Roepstorff, J.R. Yates III, A. Brass, A.J. Brown, P. Cash, S.J. Gaskell, S.J. Hubbard and S.G. Oliver, A systematic approach to modeling, capturing, and disseminating proteomics experimental data, *Nat. Biotechnol.*, 2003, **21**, 247–254.

M. Waters, G. Boorman, P. Bushel, M. Cunningham, R. Irwin, A. Merrick, K. Olden, R. Paules, J. Selkirk, S. Stasiewicz, B. Weis, B.V. Houten, N. Walker and R. Tennant, Systems toxicology and the chemical effects in biological systems (CEBS) knowledge base, *Environ. Health Perspect.*, 2003, **111**, 811–824.

D.A. Wolters, M.P. Washburn and J.R. Yates III, An automated multidimensional protein identification technology for shotgun proteomics, *Anal. Chem.*, 2001, **73**, 5683–5690.

Chapter 11

Reproductive Toxicity

FRANK M. SULLIVAN

11.1 INTRODUCTION

The term "reproductive toxicity" is defined as any adverse effect on any aspect of male or female sexual function or fertility, or on the developing embryo or foetus, or postnatally, which would interfere with the production or development of a normal offspring which can be reared to sexual maturity, capable in turn of reproducing the species. This is a very wide definition and includes various types of toxicity that are often considered separately. The two main subdivisions of reproductive toxicity are (i) sexual function and fertility in males and females and (ii) developmental toxicity to the embryo and foetus. "Sexual function and fertility" refers to effects on the male and female sexual behaviour and gonads. This includes any effects on spermatogenesis or oogenesis through puberty to conception and on development of the fertilised ovum up to the stage of implantation in the uterine wall. "Developmental toxicity" includes adverse effects on embryofoetal development from the stage of implantation through parturition and postnatal development up to the stage of puberty. Examples include reduced intrauterine embryofoetal growth and developmental retardation, organ toxicity, death, abortion, structural (teratogenic) defects resulting in congenital malformations, and functional defects such as impaired postnatal mental or physical development. It may also include adverse effects on lactation that could interfere with normal postnatal development, either by altering the quality or quantity of milk produced, or by passage of chemicals into the milk to affect neonatal development.

Various other terms are used which have specific meanings. For example, "teratogenicity" is the ability to cause gross structural (anatomical) malformations in the developing embryo/foetus; "behavioural teratogenicity" is a term which has been used to describe the ability to affect the developing embryo/foetus in such a way as to result in abnormal nervous system development either to impair the neurological or intellectual development or the behaviour of the offspring after birth. Since a good deal of central nervous system functional and biochemical development occurs not only *in utero* but also in the early postnatal period up to about 2 years of age in humans, it is possible that chemicals which affect lactation or which are transferred to the infant via the milk may also affect the normal nervous system development of the offspring in such a way as to produce permanent effects. The term "embryofoetal toxicity" is commonly used to describe all of the different types of toxicity that may affect the conceptus without defining whether these are induced in the embryonic period or in the foetal period. These include death, which is followed in rodents by resorption of the foetal remains, and in primates and humans by abortion. Also included are intrauterine growth retardation with decreased weight and retardation of ossification of the foetuses, as well as an increase in the minor variations that commonly occur in foetuses such as changes in the proportion of foetuses with 12, 13 or 14 pairs of ribs. The term "developmental toxicity" is used now to include all of these different aspects of reproductive toxicity that can result in abnormal structural or functional development of the offspring following exposure of pregnant or lactating females.

11.2   RISK ASSESSMENT FOR REPRODUCTIVE TOXICITY

"Risk Assessment" is a process that involves three phases: (i) hazard identification, (ii) dose–response extrapolation and (iii) exposure assessment. This then allows an estimate to be made of the risk from a defined exposure to humans. Each one of the phases in the assessment is complex and open to many variations and highlights problems that have to be overcome. Hazard identification usually involves the detection of reproductive toxic effects in animals using screening tests that are described below. The precise characterisation of these effects is determined by further, more detailed, animal experiments and assessment in humans. Dose–response analysis attempts to define the range of doses which cause, or which do not cause adverse effects. It also involves extrapolation within and between species and the use of modifying factors. Exposure assessment involves examination of the actual levels of a chemical to which humans are exposed and may need to take into account special features of relevance such as route of exposure.

11.3   THRESHOLDS IN REPRODUCTIVE TOXICITY

It is generally accepted in reproductive toxicity that a threshold will exist for each and every adverse effect of chemicals observed, and that exposure to chemicals at levels below the thresholds will not produce any adverse effect on reproduction. In this respect, reproductive toxicity differs fundamentally from carcinogenicity where no thresholds for genotoxic carcinogens are usually accepted. The important consequence of this is to understand that a chemical that may produce gross structural malformations, *i.e.* teratogenic, at high doses in animals, or which may affect fertility by producing testicular atrophy at high doses in animals, may be completely safe in humans exposed to low doses in the workplace or in the environment. This explains why the observation of adverse effects in animals, *i.e.* hazard identification, has to be coupled to dose–response analysis, in the process of identifying whether any risk will apply to humans exposed in defined circumstances, *i.e.* risk assessment.

11.4   SCREENING TESTS IN ANIMALS FOR REPRODUCTIVE TOXICITY

Hazard identification is normally done by means of animal experiments. Many different designs of study have been used to assess adverse effects of chemicals on fertility and development and these will be discussed in four sections dealing with (i) drugs, (ii) pesticides, (iii) food additives and (iv) industrial chemicals.

11.4.1 Drug Testing

Thalidomide is a hypnotic drug which, if used by pregnant women in the first few weeks of pregnancy, was found to produce severe congenital malformations, usually short or missing limbs as well as other less obvious defects in the baby. One important consequence of its disastrous use in 1957–1960 was that virtually every country in the world introduced legal requirements for the safety testing of potential new drugs before they could be released for use by the general public. A major aim was to prevent a repetition of the teratogenic effects of thalidomide so not unnaturally, detailed reproductive tests were included as part of the testing battery. In 1966 under the direction of Dr Lehmann, the United States Food and Drug Administration (FDA) published guidelines for a three-segment study for drug testing for adverse effects on fertility and pregnancy. This proved to be a classic design, and forms the basis of most of the more recent designs introduced worldwide for drugs and other types of chemicals (Sullivan, 1988). The Segment 1 of the test was the fertility and general reproductive performance segment in which male and female rats were treated to cover the whole of spermatogenesis in males and several oestrous cycles in females. The animals were then mated and treatment continued in females throughout the whole of pregnancy and lactation up to weaning of the offspring. If an adverse effect is observed on fertility or pregnancy, then separate mating of treated males with untreated females and vice versa may be necessary to demonstrate if the effect is on one sex only, and other studies may be necessary to investigate the exact nature of the effects observed. Segment 2 of the study was the "teratogenicity" or embryotoxicity segment and was normally carried out in two species, usually rats and rabbits. Mated animals were treated during the period of organogenesis and the foetuses delivered by caesarean section for examination of the foetuses for malformations. Segment 3 of the study was the "peripostnatal" segment, carried out on pregnant rats that were treated during the last part of gestation, not covered by treatment in the Segment 2 study, through parturition until weaning. This was to examine whether the drug has any adverse effects on parturition or lactation that had not been detected in the Segment 1 study.

11.4.2 International Harmonisation of Drug-Testing Guidelines

This basic study design was used worldwide for nearly 30 years, but there has been a gradual change in recent years from studies designed mainly to detect teratogenic effects to studies with a much wider range of end points. In particular, there has been increased interest in adverse effects on postnatal development following exposure during pregnancy. This necessitated modification of the older study designs and, since most drugs are tested for worldwide markets, resulted in a huge increase in the amount of work and numbers of animals required to satisfy the requirements of the different regulatory authorities. This led to a major revision in the drug-testing guidelines by the International Conference on Harmonisation, ICH (1993) and the new guidelines have been readily accepted worldwide, leading to a reduction in duplication of studies, with reduced costs in terms of both finance and in the numbers of animals used. The new guidelines are essentially based on the three-segment design, still testing for effects on fertility, embryofoetal development and postnatal development, but allow variations in the timing of dosing and various combinations of segments to be used, giving more scope for the toxicologist to produce studies tailored to specific chemicals. Increasing the flexibility of guidelines is regarded by most toxicologists as very desirable, but it does place an extra responsibility on the toxicologist to justify his choice of any particular study design that will depend on the anticipated uses of the new drug. All stages of the reproductive cycle must be studied from conception in one generation through to conception in the following generation. The study may be divided into sections so long as no gaps are left untested, suffering should be avoided and the minimum number of animals used. In general, rats are used for the majority of the investigations, but a second species, usually the rabbit, is required in addition when testing for teratogenicity. Normally, three dose levels of the drug are administered by the same route as will be used clinically, in groups of 20–25 males and females. The relevant guideline for the conduct of the studies ICH (1993) contains an extensive discussion of the many factors to be given consideration in designing and conducting the studies.

Rats have high sperm counts compared with humans, and drugs can dramatically reduce the count in rats without any adverse effect being detected in fertility studies. This is in contrast to humans, in which relatively small reductions in sperm count can result in infertility. It has been shown that histopathology of the testis is the most sensitive test for detecting the adverse effects of chemicals on spermatogenesis in rats, and this has led to an addendum to the 1993 guidelines giving revised suggestions for testing drugs for toxicity to male fertility (ICH, 2000). This puts emphasis on the use of the repeat dose toxicity studies to detect adverse effects on the testis (or ovaries) by careful histopathological examination. If no effects are observed, then male animals may be treated for only 2 weeks prior to mating (instead of 10 weeks prior) to detect any effect not studied by histopathology, *e.g.* libido, mating behaviour or functional effects on sperm. In addition, sperm analysis for count, morphology and motility may be performed.

When positive results are observed in any of the segments in the reproduction studies, these will normally lead to further studies being carried out to investigate the mechanism of action. The effects observed may be extensions of the pharmacological actions of the drug, which might be expected at the high doses used in the toxicity tests, but would not be expected at the lower doses used clinically. Sometimes the actions may be exerted on rodent-specific aspects of reproductive physiology and would not be a problem in humans. Differences in results between rats and rabbits may be due to metabolites, which are produced in one species but not in another. It is then important to know which metabolites are produced in humans and to be sure that these are adequately tested in the animal studies. It may be necessary sometimes to synthesise adequate quantities of human-specific metabolites and for animal tests to be repeated using these.

11.4.3 Safety Testing of Other Chemicals, Namely Pesticides, Food Additives, Industrial Chemicals

In the past, regulatory requirements for most other types of chemicals, food additives, pesticides and industrial chemicals have been based on the drug-testing guidelines. However, the work of the ICH has been specifically directed towards the testing of drugs. The other regulatory bodies have therefore tended to follow the guidelines for toxicity studies published by the Organisation for Economic Co-operation and Development (OECD) and these have been increasingly adopted during recent years.

11.4.4 Pesticide Testing

The safety evaluation of pesticides for reproductive toxicity has to take into consideration several factors that are not relevant for drug testing. By their nature, pesticides are toxic compounds since their use is to kill target species, which may be plants, insects, fungi, *etc.* though they may have a very low toxicity for mammals. Consideration has to be given to the safety of workers manufacturing and handling these chemicals. Exposure of operators using pesticides in the field may be considerable and this also has to be taken into account in the risk assessment of safety in use. In addition, for chemicals used on food crops, residue data have to be obtained so that exposure of the general public consuming the food and via drinking water can be calculated. For chemicals used on amenity lands such as parks, gardens and playing fields, consideration has to be given to exposure of chemicals to the public using these facilities. Exposure of workers either in the manufacture of the pesticides or as field operators tends to be for relatively short periods, but with possibly high exposures. On the other hand, the general public exposures tend to be low and potentially for more prolonged periods. These two different scenarios have to be taken into account in the safety testing of pesticides. The short high exposures may be covered by tests for fertility and developmental toxicity along similar lines to those used for drug testing, whereas the long-term exposures are studied using a multigeneration test, which involves treatment of parental animals prior to mating and through gestation, followed by continuous treatment throughout lactation and rearing of the offspring until sexual maturity and production of a subsequent generation. The environmental ecological impact must also be examined so that studies may have to be carried out to look for effects on beneficial species such as earthworms and bees, as well as on fishes, birds, predators and so on. Reproduction tests on these latter species are outside the scope of this chapter but are of great environmental interest.

There is a good international agreement now on the design of studies required for most pesticides. The two main study designs accepted are a test for developmental toxicity known as the OECD 414 guideline, and a two-generation multigeneration study known as the OECD 416 guideline. These guidelines were extensively revised in 2001 (OECD, 2001a,b) respectively, and cover possible effects on developmental toxicity as well as male and female fertility. The developmental toxicity study is normally performed in both rats and rabbits and differs from the older versions of the tests in that the animals are now required to be exposed to the test chemical throughout gestation from implantation until just before delivery at term, instead of just during embryogenesis. The idea is to cover not just the period when gross malformations may be induced, but in addition later periods of fetogenesis when sexual organ and central nervous system development is occurring, as well as organ functional development. For effects on fertility, postnatal development and maturation, the two-generation study OECD 416 is used. This is also useful for pesticides leaving significant residues on food. In such tests, male and female animals (usually rats) are treated continuously with the chemical administered in the diet, through two generations. This involves treating young male and female rats with three dose levels of the test substance plus a control group and allowing them to produce a litter. Some of the pups of the litters are killed for examination at weaning at 3 weeks of age, and selected pups are reared to maturity, still under treatment, and in turn are allowed to produce a litter. Again, selected pups from the second generation litters are reared to maturity, still under treatment, and are then subjected to full pathological and histopathological examination for any effects of long-term exposure. In addition to the fertility tests by mating, the parental animals of the first two generations have detailed histological examination of the testis and epididymis, and examination of homogenisation-resistant spermatids and epididymal sperm reserves, as well as sperm evaluation for motility and morphology. In the females, oestrous cycle length and normality are assessed by vaginal smears. If any adverse effects are observed in the first generation, then additional tests may have to be performed in the second generation. Similarly, if effects are observed in the first litter, a second litter may be required in each generation. An immense amount of information is obtained in such studies, not only on the different aspects of fertility and pregnancy, but also on possible effects on growth and sexual and neurological development of the young animals. The tests also indicate when endocrine disruptor activity may be present. A modified test design may also be used where an additional litter is produced in one or two of the generations with the dams killed just before delivery for examination of the foetuses, so that the test may include a teratology element as well.

11.4.5 Food Additive Testing

The reproductive toxicity test requirements for the testing of food additives were defined in Europe by the Scientific Committee for Food (SCF) 2001, and these are still used although this Committee has now been replaced by the European Food Safety Authority (EFSA). In the USA, the FDA has published an extensive account of the types of test that may be required for food additives depending on the use, amount, frequency and duration of exposure expected (US FDA, 1993, 1999a,b). In general in most countries of the world, toxicity to reproduction and fertility are now tested using the OECD 416 guideline discussed above and developmental toxicity is tested using the OECD 414 guideline in two species, though postnatal developmental tests may be added to the OECD 414 test by the use of additional animals allowed to deliver and rear their offspring. Consideration has to be given also to whether the metabolism of the chemicals may alter on chronic administration, so that more than one dosing regimen may have to be used.

11.4.6 Industrial Chemicals Testing

In most industrialised countries there are now testing requirements for industrial chemicals when these are produced in significant amounts. In Europe, for example, under the requirements of the laws generally known as the Dangerous Substances Directive, there is a stepwise system that permits regulatory authorities to require sequential testing for fertility, teratogenicity and multigeneration effects as the tonnage of chemical produced per year, or *in toto*, reaches certain critical values. This is described in detail below. The methods used worldwide are essentially the OECD 414 and 416 developmental and reproductive toxicity testing guidelines for chemicals, as well as for pesticides and food additives, which have been described above.

11.5   EXTRAPOLATION OF RESULTS OF ANIMAL STUDIES
TO HUMANS

When adverse effects are detected in any of the above screening tests, then extrapolation to humans involves analysis of the mechanisms and sites of action of the chemicals causing the effects. This requires a good knowledge of the comparative physiology of reproduction in the different animal species.

The most satisfactory situation is when one can define the precise mode of action of the reproductive toxicant in the animal model and then state whether such a mechanism would operate in humans at the exposure levels concerned. The differences in reproductive physiology between species are sufficiently large for there to be cases where a chemical may affect fertility or development in rodents or rabbits by acting on hormones or systems that do not operate in humans. In such a situation, the animal studies can be discounted, though the possibility of a different effect in humans is still possible. If similar mechanisms to those responsible for the adverse effects in animals do exist in humans, then some confidence can be achieved by determining the no effect levels for the underlying changes in human studies where this is possible.

The more usual situation, however, is that the underlying mechanisms of toxicity in animals are not known, and then one has to resort to the use of safety factors in arriving at the acceptable levels of exposure for humans. This is the least scientific part of risk assessment, and is regarded by some as a "numbers game" and by others as a scientific judgement. In Europe, a safety factor of 100 is usually applied to the lowest "no adverse effect level" (NOAEL) determined in the animal studies to define the exposure level tolerable to humans. In the USA, an extra factor of tenfold, giving an overall safety factor of 1000 may be applied when a reproductive effect is observed with a chemical that may contaminate food.

11.6   THE EUROPEAN COMMUNITY CLASSIFICATION OF
CHEMICALS FOR REPRODUCTIVE TOXICITY

The classification and labelling of dangerous substances was first introduced in 1967 in the European Community with Council Directive 67/548/EEC known as the Dangerous Substances Directive. The "sixth amendment" to this directive in 1979 introduced a notification procedure and a requirement for labelling chemicals for toxicity. Three special categories for labelling were for "carcinogenicity, mutagenicity and teratogenicity". The teratogenicity classification was restricted to chemicals inducing "teratogenic" effects in the classical sense of the word, *i.e.* producing only gross structural malformations. Discussions by expert advisors to the European Commission over several years lead to a widening of concern in this area of toxicology, and under the current "seventh amendment" in 1992 (European Commission, 1992) the classification of teratogenicity has been changed and expanded to "toxic to reproduction". This includes adverse effects on fertility, pre- and postnatal development and lactation, and encompasses not only structural malformations but also functional deficits. This has resulted in a major change in the testing requirements to allow adequate classification of chemicals for these other aspects of reproductive toxicity.

11.7 THE SEVENTH AMENDMENT TO EC DIRECTIVE 67/548/EEC 1992

This directive relates to the classification, packaging and labelling of dangerous substances so as to reduce barriers to trade within the common market and to approximate the laws of the member states to permit this. An important function of the directive is to protect man and the environment, and concerns itself with the health and safety of the population and especially with the safety of workers exposed to potentially dangerous chemicals. The directive requires the notification of new chemicals to appropriate authorities for assessment with a view to classifying and labelling them with appropriate health and safety warnings. The definition of "dangerous", which is used includes the following properties: explosive, oxidising, flammable, corrosive; toxic including death, acute or chronic damage, irritant or sensitising, carcinogenic, mutagenic and toxic for reproduction. The toxicity categories refer to exposure by the inhalation, oral and dermal routes.

There is also a requirement for production of safety data sheets for chemicals when sold relating to health and safety at work, and manufacturers have to review their own data in order to provide adequate safety warnings and to obtain toxicity data on their chemicals in cases where such data do not exist at present. As there is also legislation within the community and in the USA concerning the health and safety of pregnant women at work, there will be increasing pressure to obtain data on the reproductive hazards of chemicals. Many large companies currently have policies relating to the safety of women workers during pregnancy, but the majority of workplaces are still not prepared to deal with this type of situation. Most countries have legislation relating to prevention of unfair sexual discrimination in the workplace and many countries have legislation to protect the rights of the pregnant worker. The converse side of this, however, is the increased risk of litigation in the event that the developing foetus has a malformation that may be claimed to have arisen from exposure of the mother to dangerous chemicals at work.

An example of the effect of changing the classification from "teratogenicity" to "toxic to reproduction" is given by consideration of methyl mercury. Exposure of humans to this chemical may result in severe adverse effects on central nervous system development, causing blindness, deafness and mental retardation in the absence of obvious structural defects. This chemical would not have been eligible for classification under the previous legislation as a teratogen, but would now be classified as "toxic to reproduction". In addition, the potential of chemicals to induce adverse effects at any stage of the reproductive cycle, in males as well as females now results in classification. Substances classified for carcinogenic, mutagenic or reproductive toxicity are now commonly referred to as CMR substances.

11.8 CLASSIFICATION OF CHEMICALS AS TOXIC FOR REPRODUCTION

11.8.1 Effects on Male or Female Fertility

Effects on male or female fertility include adverse effects on libido, sexual behaviour, any aspect of spermatogenesis or oogenesis, hormonal activity or physiological response that would interfere with the capacity to fertilise, fertilisation itself or the development of the fertilised ovum up to and including implantation.

11.8.2 Developmental Toxicity

Developmental toxicity is taken in its widest sense to include any effect interfering with normal development, which is induced prenatally and may be manifest either pre- or postnatally. This includes embryofoetal toxicity, death, abortion, retarded development, structural (teratogenic) effects, functional defects, impaired postnatal mental and physical development up to and including normal pubertal development.

The basis of the classification system is "hazard" and not "risk". Chemicals are only classified as toxic to reproduction when they have specific intrinsic toxic potential to affect reproduction adversely. That is, the effects should not merely be secondary to other toxic effects. For example, it is well known that any treatment with high toxic doses of chemicals that cause severe inanition will reduce fertility or produce adverse effects on the developing foetus. Effects of this type would not lead to classification. This does not exclude substances that affect reproduction at around the same dose levels that cause other signs of toxicity, if it is thought that the effects on reproduction are not secondary to the other toxicity. Because of the above restrictions, information obtained from *in vitro* tests are only regarded as supportive data and would not normally be a basis for classification.

11.9 CATEGORISATION

Chemicals are categorised into three categories. Allocation to Category 1 is made on the basis of human data, with evidence of reproductive toxicity to humans. Allocation to Category 2 is usually on the basis of animal studies, with results suggesting that adverse effects would be likely with human exposure. For Category 2 that relates to fertility, results in one species plus evidence on mechanism of action, site of action or structure–activity relationships to known active compounds, and evidence that the results are likely to be relevant to humans would be expected. With less data, Category 3 or no classification would be appropriate. For Category 2 that relates to developmental toxicity, good animal data in one or more species, in the absence of marked maternal toxicity, and administered by a relevant route would be expected. If the data are less convincing, if there is the possibility of non-specific effects, if only small changes in common variants are reported, or if there are only small changes in postnatal developmental tests, then it may be that Category 3 or no classification would be appropriate. Chemicals that are classified have to carry certain risk phrases on the labels such as "May impair fertility" or "May cause harm to the unborn child".

11.10 LACTATION

When pregnant women are exposed to chemicals at work, there is concern not only with respect to embryo–foetal development, but also for the baby during the lactation period. With improved creche facilities being provided by some employers, there is the possibility of women returning to work after the birth of a baby, but while the child is still being breastfed. In general, chemicals that have high fat solubility, and especially chemicals that are also poorly metabolised, may be expected to accumulate in breast milk and pass to the baby in potentially toxic amounts. There is also concern for the general public about contamination of breast milk from environmental pollution. If such chemicals have toxic potential for the baby, they are labelled with a risk phrase "May cause harm to breastfed babies". Chemicals are not labelled in this way simply because they are present in milk. There must be harmful potential as well.

11.11 TESTING REQUIREMENTS UNDER THE DANGEROUS SUBSTANCES DIRECTIVE

When new substances, or mixtures, are notified under this directive, the amount of testing required to be submitted to the appropriate national authorities depends on the amount of chemical being sold. Amounts less than 10 kg per year used for research and similar purposes may be exempt. With less than 1 ton per year, a simple "base set" comprising chemistry, production, uses, precautions; acute, skin and eye toxicity; mutagenicity *in vitro* and simple ecotoxicity (degradation), is required.

At 1 ton per year (or 5 ton total) the "base set" plus a repeat dose 28-day toxicity test, second mutagenicity test, screen for toxicity to reproduction, kinetics and ecotoxicity are required. When the amount sold reaches 10 ton per year (50 ton total) the authority *may* require "Level 1" tests, and when the amount reaches 100 ton per year (500 ton total) the authority *shall* require "Level 1" tests. When the amount sold reaches 1000 ton per year (5000 ton total) the authority *shall* require "Level 2" tests. Depending on the results obtained at all the different stages, the possibility exists for the authority to require tests to be carried out at different times from the above basic plan.

The "Level 1" tests include a fertility study (single generation), a teratogenicity study (*e.g.* OECD 414), sub/chronic toxicity, as well as further mutagenicity, kinetic and ecotoxicity tests. The "Level 2" tests include a further fertility (*e.g.* OECD 416) study if appropriate from the earlier tests, a teratology test in a second species, a developmental toxicity test (on peri-postnatal effects), as well as chronic toxicity, carcinogenicity, kinetics and ecotoxicity tests.

11.12 DOWNSTREAM CONSEQUENCES RELATING TO THE CLASSIFICATION OF CHEMICALS (CMR SUBSTANCES)

Originally, it was hoped that classification of chemicals for toxicity to reproduction and consequent labelling of the chemicals with warnings would substantially reduce hazards for men and women exposed to these chemicals at work and in the environment. The basis of the classification, as described above, is 'hazard' and not 'risk' so that if a chemical is classified as toxic to reproduction, this merely implies that at some dose level and route of administration, the chemical is capable of producing an adverse effect. The intention was that people exposed to such chemicals, usually in the workplace, should have a risk assessment carried out to ensure that a toxic level of exposure does not occur. However, in the years following the introduction of the classification system, about 70 new directives have been published by the EU, which place restrictions on the sale, use and handling of CMR substances without any risk assessment being carried out. For example, it is forbidden to supply or sell CMR Category 1or 2 substances to the general public. CMR Category 1 or 2 pesticides may not be sold for home use, and in many EU countries may not be sold for non-professional use. Recent changes to the cosmetics laws (European Commission, 2003) have banned the use of CMR substances in cosmetics. Such restriction may be valid for genotoxic carcinogens for which no threshold effect is generally assumed to exist. However, for substances toxic to reproduction, it is widely accepted that thresholds do exist. The majority of chemicals, if administered to animals in large amounts will cause some developmental toxicity. At lower exposure levels, however, no adverse effects may be seen, and the chemical may be perfectly safe for use under specified conditions. Because the classification for reproductive toxicity, in the way the classification is currently performed, is based on hazard and does not take account of actual exposure levels, this has led to many chemicals being classified and so restricted in use, in a way that was never intended in the original classification system. Currently there are initiatives to change the whole chemicals safety system within the EU and this may result in changes to the current classification system.

BIBLIOGRAPHY

S.M. Barlow and F.M. Sullivan, *Reproductive Hazards of Industrial Chemicals*, Academic Press, London, 1982.

European Commission, Council Directive of 5th June 1992 amending for the 7th time Directive 67/548/EEC. O.J. of the European Communities No. L. 154/1, 1992.

European Commission (2003): Directive 2003/15/EC of the European Parliament and of the Council of 27th February 2003. O.J. of the European Union, L 66/26, 11.3.2003.

ICH, Harmonised Tripartite Guideline: Detection of Toxicity to Reproduction for Medicinal Products S5A, *International Conference on Harmonization*, 1993. May be downloaded from www.ich.org. Also published in USA FDA, Federal Register Vol. 59, Food and Drug Administration, 1994.

ICH (2000) Maintenance of the ICH Guideline on Toxicity to Male Fertility S5B(M), *International Conference on Harmonization*, 2000, www.ich.org.

J.C. Lamb and P.M.D. Foster (eds), *Physiology and Toxicology of Male Reproduction*, Academic Press, San Diego, 1988.

D.R. Mattison (ed), *Reproductive Toxicology*, Alan R. Liss Inc, New York, 1983.

H. Nau and W.J. Scott (eds), *Pharmacokinetics in Teratogenesis*, CRC Press Inc, Boca Raton, FL, 1987.

OECD, 'Guideline for the testing of chemicals', *414 Prenatal Developmental Toxicity Study*, Organization for Economic Cooperation and Development, 2001a, www.oecd.org.

OECD, 'Guideline for the testing of chemicals', *416 Two-Generation Reproduction Toxicity Study*, Organization for Economic Cooperation and Development, 2001b, www.oecd.org.

SCF, *Guidance on submissions for Food Additive Evaluations by the Scientific Committee on Food*, European Community Scientific Committee on Food, 2001, http://europa.eu.int/comm/food/fs/sc/scf/out98_en.pdf.

Ch. Schaefer (ed), *Drugs during Pregnancy and Lactation*, Elsevier, Amsterdam, 2001.

J.L. Schardein, *Chemically Induced Birth Defects*, 3rd edn, Dekker, New York, 2000.

A.R. Scialli, *A Clinical Guide to Reproductive and Developmental Toxicology*, CRC Press, Boca Raton, FL, 1992.

F.M. Sullivan, Reproductive toxicity tests: Retrospect and prospect, *Hum. Toxicol.*, 1988, **7**, 423–427.

J.G. Wilson and F.F. Clarke (eds), *Handbook of Teratology*, Plenum Press, New York, 1977.

US FDA, Toxicological principles for the safety assessment of direct food additives and color additives used in food, *Redbook Two, Draft*, Centre for Food Safety and Applied Nutrition, Food and Drug Administration, Washington, 1993.

US FDA, Food and Drug Administration proposed testing guidelines for reproduction studies, *Regul. Toxicol. Pharmacol.*, 1999a, **30**, 29–38.

US FDA, Food and Drug Administration proposed testing guidelines for developmental toxicity studies, *Regul. Toxicol. Pharmacol.*, 1999b, **30**, 39–44.

Chapter 12

Immunology and Immunotoxicology

GRAEME WILD

12.1 INTRODUCTION

Immunity describes the body defence mechanisms against harmful substances. These include bacteria, viruses, parasites and malignant cells. The word immunity is derived from the Latin 'immunitas', which means free from disease. The immune response can be divided into three parts; recognition, activation and removal. These functions are achieved by cellular, humoral and innate immune systems.

12.2 INNATE IMMUNITY

Innate immunity is the part of the immune system that is present in an individual from birth; it is also referred to as non-specific immunity. Innate immunity does not require previous exposure to an antigen to be effective and does not result in specificity. The innate immune system is the first line of defence against microorganisms and a major component is the physical barrier provided by intact skin or mucous membranes (see Section 13.2). Sebum is a skin secretion that has a low pH (3–5) and inhibits microorganism growth, whereas mucous on membranes traps pathogens. Tears, mucous secretions and saliva all act to wash away organisms or have bactericidal properties *e.g.* lysozyme in tears. Soluble factors such as interferon, complement and acute phase proteins are also important in the innate immune system. Interferon is released by virus-infected cells and helps provide protection against the virus to surrounding cells. Various microbial proteins can activate the alternative complement pathway resulting in cell lysis, recruitment of phagocytic cells and opsonisation. Acute phase proteins include C-reactive protein (CRP) and antitrypsin, and serve to enhance the immune response (CRP) or to limit its site of activation (antitrypsin).

Removal of the pathogen or cell debris triggering the innate system is achieved by phagocytic cells, macrophages and neutrophils. Organisms may also be ingested and killed by phagocytes either by an oxygen-dependent pathway involving the production of superoxides, chlorates(I) and nitrogen(IV) oxide or an oxygen-independent pathway using defensins, lysozyme and tumour necrosis factor. Defensins are antimicrobial peptides acting against gram-negative and -positive bacteria, fungi and viruses. In addition, polymorphs have cytoplasmic granules containing lytic enzymes such as peroxidase, elastase and collagenase.

Natural killer (NK) cells also form part of the innate system and kill virally infected or tumour cells by releasing perforins, which perforate cell walls, and induce apoptosis of the target cell.

Innate immunity is an effective system but can be breached by various organisms. Some pathogens do not activate complement whereas others are resistant to phagocytosis or phagocytic killing. Other organisms can gain access to the inside of cells and so evade innate immune system removal.

12.3 ADAPTIVE IMMUNITY

Adaptive immunity results in an acquired or specific response to an antigen and can be divided into humoral or cellular responses. The adaptive response confers specificity against individual antigens, generates diversity in that an unlimited number of antigens can be reacted against, results in memory where the system responds faster on subsequent exposure and enables self to be differentiated from non-self.

An antigen is a protein or polysaccharide that can induce a specific antibody or cellular response. An antigenic determinant is a specific epitope on an antigen, which reacts with a lymphocyte receptor or antibody. Multiple epitopes on an antigen result in a polyclonal response, a heterogenous population of antibodies or lymphocytes is produced. Haptens are molecules that are too small to generate an immune response, but when bound to a protein molecule can become antigenic.

Antibodies or immunoglobulins are Y-shaped molecules consisting of two heavy (long) chains and two light (short) chains. The heavy chains are divided into five types, which determine the class of the antibody molecule, IgG, IgA, IgM, IgD and IgE. There are two types of light chain, κ or λ. An individual immunoglobulin molecule has two identical heavy chains and two identical light chains, *e.g.* two G heavy chains and two λ light chains. The arms of the Y-shaped molecule are termed as the Fab region or antigen binding and the rest of the molecule forms the Fc region and has various receptors, *e.g.* for complement. IgG, IgD and IgE antibodies occur as single molecules whereas IgA is a dimer and IgM is pentamer of the basic Y shape. Heavy and light chains are composed of constant regions, where the amino acid sequence is similar for each of the individual types, and variable regions, where the sequence differs considerably even in the same antibody class. These variable regions allow the adaptive immune system to exhibit diversity and specificity in its response to pathogens.

Antibodies are themselves capable of functioning as antigens with epitopes that can be classed as isotypic and idiotypic. Isotype relates to the heavy chain class and the idiotype relates to the heavy and light chain variable regions. There are a number of antigenic determinants within the variable region and these are termed idiotopes and collectively form the idiotype. An idiotype can be the antigen-binding site of the variable region and in this case is referred to as the paratope.

Antibodies help eliminate antigens or pathogens in a variety of ways including complement activation, opsonisation and toxin neutralisation.

12.4 HUMORAL IMMUNITY

Humoral immunity is a response against extracellular or soluble antigens and is initiated by antigen-presenting cells (APC). These cells can be either macrophages, dendritic cells or B lymphocytes that ingest antigen and then express antigen peptides bound to major histocompatibility complex (MHC) class 2 molecules on the cell surface. Cells known as T helper lymphocytes, CD4 cells, interact with the MHC 2–antigen complex (MHC 2 class restricted) via the T-cell receptor and after co-stimulation from the APC the T helper cells secrete cytokines that activate B cells and results in antibody production. Lymphocytes are formed in the bone marrow, T cells mature in the thymus, whereas B cells mature within the bone marrow. During thymus development, T cells start to express the T-cell receptor and then show specificity and memory. T cells are only activated by APC bearing the same MHC molecules. When an immature B cell is activated and contacts antigen, it differentiates into memory cells and plasma cells that secrete antibody. Plasma cells do not have membrane-bound antibody whereas memory cells express antibody on the membrane which has the same specificity as that of the original B cell from which the clone was formed. Mature B cells encounter antigen, which cross-links the surface immunoglobulin, and is then internalised by receptor-mediated endocytosis. Antigen peptides in association with MHC class 2 molecules are expressed on the cell surface. This MHC 2–peptide complex binds to T helper cells having specificity for that peptide and induces cytokine production that causes the B cell to proliferate. B cells can also secrete antibody without the help of T cells, *i.e.* are T-cell independent, but most B cell responses are T-cell dependent. The first antibody produced by a B cell is IgM, later class switching can occur following which IgG, IgA or IgE classes may be expressed or secreted. Mature T cells circulate round the blood and lymphatic system. Some of these lymphocytes become seeded into peripheral lymphoid tissue that is adjacent to external surfaces, *e.g.* gut-associated lymphoid tissue (GALT) and mucosal-associated lymphoid tissue (MALT). First contact with antigen results in a primary immune response, subsequent contact with that antigen will give a secondary response and results in much faster antibody production than the primary response. The cells have acquired memory and are ready primed for specific antibody production in the secondary response.

12.5   CELL-MEDIATED IMMUNITY

Cell-mediated immunity describes the immune response generated by intracellular antigens. Infected cells contain proteins arising from the organism causing the infection and cancerous cells contain proteins due to malignant growth. These proteins are degraded to peptides that are then complexed with MHC class 1 molecules and expressed on the cell surface. MHC 1 molecules are contained within all nucleated cells. Cytotoxic T lymphocytes (CTL), CD8 lymphocytes, recognise the MHC 1 peptide complex (MHC class 1 restricted) via the T-cell receptor and undergo proliferation and differentiation into CTL. These CTL release lytic enzymes and induce apoptosis in the target cell.

12.6   DELAYED TYPE HYPERSENSITIVITY

Delayed type hypersensitivity is mediated by a population of T lymphocytes called T delayed type hypersensitivity cells (Tdth cells). The production of these cells follows initial contact with MHC 2–antigen complexes resulting in sensitisation. The APC associated with this type of response are macrophages, dendritic cells and epidermal Langerhans cells in particular. Subsequent contact with the antigen causes CTL activation in which various cytokines are released. These cytokines cause monocytes to accumulate in the surroundings and mature into activated macrophages. These macrophages destroy surrounding infected cells by releasing lytic enzymes. In situations where the antigen persists, a granuloma can develop. This is an accumulation of activated macrophages that fuse to form multinucleated giant cells surrounded by Tdth lymphocytes. Granulomas are under continual antigenic stimulation and release lytic enzymes that can destroy surrounding healthy tissue.

Delayed type hypersensitivity responses can also occur against small molecules, such as nickel, insect venoms and anilines, that can penetrate the skin and bind to various proteins, *e.g.* albumin as haptens. These substances are then ingested by APC and produce a delayed type hypersensitivity response. The route of antigen entry for sensitisation includes the skin, gut and respiratory tract.

12.7 COMPLEMENT

The complement system comprises a number of proteins acting in a cascade amplification type manner, *i.e.* for an individual molecule generated at the beginning of the pathway many molecules are generated at the end. Complement can be activated by three different mechanisms; the classical pathway, alternative pathway and mannan binding lectin pathway. The classical pathway is activated by immunoglobulin bound in immune complexes and consists of components C1, C4, C2 and C3 in sequence. The alternative pathway is activated by bacterial cell wall proteins and starts at C3. The mannan binding lectin pathway activates C1 but does not require antibody interaction. All three pathways converge at C5 and form the membrane attack complex (MAC) with C6, C7, C8 and C9 binding in sequence to form a C56789 complex. The MAC forms holes through cell membranes and causes lysis. The complement system is kept in tight control by a number of inhibitory molecules to prevent host-cell damage. Activated complement components also play a role in the inflammatory response, opsonisation and clearance of immune complexes.

12.8 HYPERSENSITIVITY

An immune response that results in excessive tissue damage or is inappropriate in response to an antigen is called hypersensitivity. Prior exposure to the antigen or sensitisation is required. Gell and Coombes classified hypersensitivity reactions according to their pathogenic mechanism.

12.8.1 Type I: IgE-Mediated (Immediate) Hypersensitivity

IgE bound to mast cells is cross-linked by antigen (allergen) causing degranulation and release of mediators such as histamine and tryptase. This reaction gives the typical allergic response seen in hay fever and eczema and can also result in anaphylaxis. Type I reactions can be detected by skin-prick test or *in vitro* allergy test.

12.8.2   Type II: Antibody-Mediated Hypersensitivity

Antibody directed against cell surface antigens causes destruction of that cell by complement activation or antibody-dependent cellular toxicity via NK cell activation. This type of response occurs in blood transfusion reactions and haemolytic disease of the newborn.

12.8.3   Type III: Immune Complex-Mediated Hypersensitivity

Preformed immune complexes are deposited in tissues causing complement activation and localised inflammation with resulting tissue damage. Examples are glomerulonephritis, serum sickness and rheumatoid arthritis.

12.8.4   Type IV: Delayed Type Hypersensitivity

Sensitised Tdth lymphocytes on second contact with an antigen cause cytokine release, which activates macrophages and cytotoxic T cells and results in localised inflammation and tissue damage. Examples of this type of response include contact allergy and transplant rejection. The antigen responsible for contact allergy can be determined by patch testing.

12.9   IMMUNODEFICIENCY

Immunodeficiency diseases may be primary in origin or acquired from drug treatment or malnutrition, both types result in a lack of response of part of the immune system. These diseases can be classed as antibody, cellular or combined defects and manifest as recurrent infections especially from low-grade pathogens. Antibody deficiencies can result in a total lack of immunoglobulin as in X-linked agammaglobulinaemia. An absence of individual classes of antibody can also occur. For example, in hyper IgM syndrome, class switching cannot occur, thus no IgG, IgA or IgE is produced. There is a selective IgA deficiency disease, which is self-explanatory. Antibody deficiency syndromes are susceptible to infection from pyogenic-encapsulated bacteria because opsonisation or complement activation by antibody does not occur.

Cellular immunodeficiencies can be associated with recurrent infections by intracellular pathogens and may be due to defects in lymphocytes and neutrophils. Leukocyte adhesion deficiency occurs when neutrophils or macrophages are unable to adhere to blood vessel walls for adiapedesis to occur. Defects in neutrophil-killing ability can lead to chronic granuloma disease. A congenital immune deficiency known as complete Di George syndrome is associated with thymic aplasia and low T-cell numbers, giving rise to multiple infections and hypocalcemia sometimes leading to tetany. Combined immunodeficiences have defects in both antibody and cellular immune systems. Defects in T-cell development can inhibit both cellular and antibody immunity resulting in low T and B lymphocyte numbers and low immunoglobulin levels leading to the condition of severe combined immunodeficiency (SCID). This disorder manifests soon after birth and can have a number of causes ranging from X-linked SCID to enzyme defects such as adenosine deaminase deficiency.

Immune competence can be assessed by the numbers of lymphocytes and subsets, immunoglobulin quantitation and cellular function. Neutrophil activity can be measured by the presence or absence of an oxidative burst when stimulated. Lymphocyte function can be determined by measuring cell proliferation after stimulation of these cells with various mitogens such as phytohaemagglutinin, cononcanavalin A or anti CD3 antiserum for T lymphocytes and pokeweed mitogen for B cells. Specific antigens such as purified protein derivative may also be used.

12.10 AUTOIMMUNITY

Autoimmunity occurs when the immune system reacts against host antigens and causes chronic inflammation and tissue damage. Autoimmune disease can be systemic, *e.g.* systemic lupus erythematosus (SLE) and rheumatoid arthritis, or organ specific, *e.g.* Goodpastures syndrome and Hashimotos thyroiditis. Self-reacting lymphocytes are normally removed during the development phase, but any breakdown in this process leads to autoimmunity. The manifestation of this disorder may be due to both environmental and genetic factors causing a loss of tolerance to self antigens. Autoimmunity may arise from infection through a process of molecular mimicry where some microbial epitopes are similar to self epitopes, and any antibodies produced due to the microbial infection can cross-react with self tissue. Some areas of the body are not seen by the immune system and thus are not subject to tolerance. If lymphocytes gain access to these antigens then an immune response can occur.

12.11 TRANSPLANTATION

An organ transplanted from one individual to another will stimulate the recipient's immune system and lead to graft rejection. Rejection occurs by an adaptive response and can be antibody or cellular mediated. Graft survival is much greater if recipient and donor are MHC compatible. Rejection is due to T cells being activated by alloantigens present on cells in the transplanted organ, *i.e.* the graft is recognised as non-self. Hyperacute rejection occurs within minutes of a transplant, acute rejection 7–10 days post-transplant and chronic rejection several months after transplant.

Graft *vs.* host disease occurs where a bone marrow transplant reacts against and rejects the recipient's tissues. Graft survival is enhanced by using immunosuppressant drugs to control the host immune response, although this treatment places the transplant patient at a greater risk of potential infection. Immunosuppression can be used in treating autoimmune disorders as well as preventing transplant rejection and is achieved using drugs such as corticosteroids, cytotoxic agents, cyclosporins and various monoclonal antibodies. Some diseases can also result in immunosuppression either by depressing the immune response or causing loss of antibody through enteric or renal damage. Conversely, immunoenhancement causes an increase in immune system activity and may be a result of the inflammatory response or artificially induced by treating with levamisole or thymopentin.

12.12   VACCINATION

Antibodies are formed after a process of immunisation, which can be an active or passive process. Passive immunisation occurs when antibodies formed in a donor are transferred to a recipient. The passage of maternal antibodies to a foetus or injection of preformed antibody into an individual are examples of passive immunisation. Active immunisation is achieved by infection with a pathogen and can occur by natural exposure or introduction of the pathogen through vaccination. The organism used for vaccination may be a live attenuated source or a killed source. Purified antigen and synthetic microbial proteins may also be used. Attenuated organisms retain antigenicity but are avirulent; a problem that can be encountered with this type of vaccine is a reversion where virulence is recovered. The immune response to a vaccine can be enhanced by the addition of an adjuvant. This is a substance that, when mixed with an antigen, enhances antigenicity and gives a superior immune response. Different adjuvants have different effects on the immune system and can produce different antibody classes directed against an antigen or cause sensitisation of different cell types. Many vaccines contain aluminium hydroxide or phosphate as adjuvant. Other vaccines include bacterial toxins, *e.g.* pertussis toxin, or microbial cell wall constituents. Freund's complete adjuvant, a well-known adjuvant, consists of a mineral oil emulsion containing killed *Mycobacterium tuberculosis*.

12.13   IMMUNOTOXICOLOGY

Immunotoxicity refers to the adverse effects various substances can exert on the immune system. A variety of agents exhibit immunotoxic properties including drugs, metals, proteins and numerous organic and inorganic chemicals.

The main symptoms caused by immunotoxic agents are hypersensitivity, autoimmunity, immune enhancement and immune suppression. Immune enhancement causes over stimulation of the immune response and can lead to gross inflammation and tissue damage. Immune suppression lowers the activity of the immune system including surveillance, *e.g.* the elimination of tumour cells. This type of response can give recurrent infections with increased severity and neoplastic cell growth. Autoimmunity and hypersensitivity responses can also be a feature of immunotoxicity and occur as a result of tolerance loss or induction of sensitivity. Drug hypersensitivity is a cause of morbidity and to a lesser extent mortality. The potential for a drug to induce an allergic response is thoroughly investigated before marketing. All new drugs are tested for their potential immunotoxicity on rodents but interpretation of the results may be difficult. The trigger for immunotoxicity to occur is variable and can be due to structure/activity type relationships, length of exposure or the amount of drug encountered. However, there are some responses where the concentration or length of exposure may not always be applicable as the effects seen may not be directly due to the immunotoxic agent. Metals such as gold or mercury can induce immune-complex formation, which can be deposited in organs such as the kidney or skin. Drugs containing gold have been used to treat autoimmune disorders and thus have the potential to enhance the formation of immune complexes in susceptible individuals. Many drugs can induce the production of autoantibodies, particularly antinuclear antibodies and cause conditions such as drug-induced SLE.

Cessation of treatment with the particular drug removes the stimulus and both antibody level and clinical signs subsequently disappear. The symptoms associated with immunotoxicity are varied and are dependent on the part of the immune system that is activated or suppressed. Animal models may not necessarily indicate that a substance is immunotoxic to humans.

The immune system serves as a powerful resource in preventing infection and tumour cell development. Chaotic activation or loss of control in the system can have catastrophic consequences. Many substances are able to modulate the immune response to a certain degree, but some individuals are capable of reacting more severely than others and show clinical signs of immunotoxicity. This response may be dependent on environmental or genetic factors acting together with the immunotoxin. Immunotoxicology is not only concerned with therapeutic agents but also toxic substances encountered in the general environment and from prosthetic devices.

BIBLIOGRAPHY

D.R. Germolec, Sensitivity and predictivity in immunotoxicity testing: immune endpoints and disease resistance, *Toxicol. Lett.*, 2004, **149**, 109–114.

C.A. Janeway, T. Travers, M. Walport and M. Shlomchik, *Immunobiology*, 6th edn, Churchill Livingstone, London, 2004.

J. Kuby, R.A. Goldsby and B.A. Osbourne, *Immunology*, 5th edn, W.H. Freeman & Co., New York, 2002.

K. Miller, J.L. Turk and S. Nicklin, *Principles and Practice of Immunotoxicology*, Blackwell, Oxford, 1992.

D.S. Newcombe, N.R. Rose and J.C. Bloom (eds), *Clinical Immunotoxicity*, Raven Press, New York, 1992.

I. Roitt and P.J. Delves, *Roitt's Essential Immunology*, 10th edn, Blackwell Scientific, London, 2001.

D.J. Snodin, Regulatory immunotoxicology: does the published evidence support mandatory nonclinical immune function screening in drug development?, *Regul. Toxicol. Pharmacol.*, 2004, **40**, 336–355.

Chapter 13

Skin Toxicology

DAVID A. MCKAY AND ROGER D. ALDRIDGE

13.1 INTRODUCTION

The skin is the largest organ of the human body and interfaces with the environment, maintaining the equilibrium of the internal milieu and preventing the ingress of environmental toxins. Toxicity results when physical or chemical agents breach this normally effective barrier. Advances in molecular biology have both increased our understanding of the physical and biological elements that contribute to this skin barrier function and helped in the development of tools for more reliable dermatotoxicological risk assessment.

13.2 SKIN ANATOMY

The skin consists of an inner layer called the 'dermis' that sits astride the subcutaneous fat and adjoins the overlying epidermis. The cellular content of the dermis comprises predominantly fibroblasts and mast cells. The fibroblasts produce the collagen and elastin fibres that give the skin its tensile strength. These fibres are embedded in a ground substance of sugar-containing molecules known as glycosaminoglycans, which are also produced by the fibroblasts. This extracellular matrix situated at the dermo-epidermal junction is now known to have a significant effect on growth of cells of the epidermis. The dermis also contains cells of the monocyte/macrophage system, which together with the dendritic, antigen- presenting cells of the epidermis are part of the body's cellular immune defence mechanism, and are involved with both allergic and irritant dermatitis responses.

The epidermis covers the dermis and contains a basal layer of replicating cells. Above the basal layer, several layers of terminally differentiating cells evolve into the stratum corneum. The latter is a water-impenetrable, anucleated cell layer that interfaces with the environment. This extraordinary barrier is highly resistant to mechanical and chemical assault and an understanding of its structure and development is essential to an understanding of the toxicology of the skin (Figures 13.1 and 13.2).

The stratum corneum can be thought of as a two-compartment system consisting of lipid-depleted protein-rich corneocytes bound with complex bilayer lipids. The formation of this protective structure occurs predominantly in the stratum granulosum, which is situated just under the stratum corneum. Within the stratum

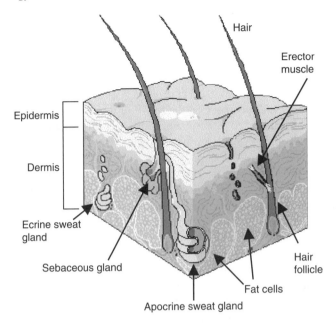

Figure 13.1 *Diagrammatic cross-section of the skin*

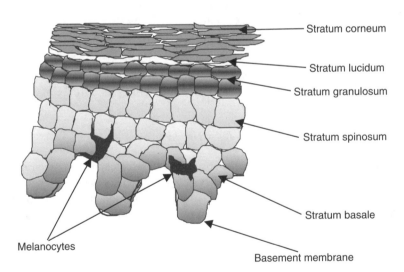

Figure 13.2 *Diagrammatic cross-section of the epidermis*

granulosum enzymes transform the lipid content of the cell into the complex bilayers that are then excreted, the remaining protein-rich cell being termed a corneocyte. The control of stratum corneum production is intimately associated with perturbation of its barrier function. Thus, damage to the stratum corneum will generate both enzyme activity within the stratum granulosum and increased replication of basal keratinocytes. Keratinocytes themselves express transport-associated and metabolising enzymes that affect xenobiotic uptake and biotransformation. However, the initiating event in skin toxicity is permeation of the stratum corneum.

13.3 PERMEATION

Penetration of the stratum corneum is the rate-limiting step in percutaneous absorption. If the barrier is damaged by vesicants such as acids or alkali, absorption will be enhanced. Permeation of toxins through the intact stratum corneum is dependent upon the lipid solubility of the toxin, the concentration of the toxin at the skin surface and the nature of the carrier vehicle within which the toxin exists. Penetration will be influenced by the thickness of the stratum corneum and the presence of skin appendages such as hair follicles and sweat glands. The study of regional variations in absorption has been limited but it is recognised that considerable differences exist, with permeation in the peri-orbital and genital skin being rapid, and in the hyperkeratotic skin of the palms and soles extremely slow. In a similar fashion, the effects of ageing and disease on factors such as stratum corneum thickness and lipid content have an influence on skin permeation.

The cutaneous blood flow primarily serves the purpose of thermoregulation and vastly exceeds the metabolic requirements of the skin. It is generally accepted that this excess flow results in rapid chemical clearance from the subcutis, preventing any build-up or reservoir effect that would inhibit penetration. Under these circumstances Fick's first law of diffusion applies, such that the flux (J) is proportional to the concentration of the penetrant (C_v) thus:

$$J = K_p C_v$$

where K_p is the permeability constant. K_p itself is defined as a function of the diffusivity (D), the partition coefficient (K_m), which describes the affinity of a toxin for the skin, and the skin thickness (d):

$$K_p = DK_m/d$$

The permeability constant enables the absorption to be calculated from various concentrations of any given penetrant. A variety of *in vitro* techniques have been developed for the calculation of absorption. Such pharmacokinetic modelling can be of help in risk assessment, but care must be taken in extrapolating the results of such *in vitro* studies to the dynamic *in vivo* situation. For example, studies with single toxins are unlikely to reflect the true environmental and occupational exposure to combinations of chemicals that may interact, affecting their absorption and ultimately their toxicity. Such factors may explain the known unpredictability of metal absorption, which is not dose related. In France, the use of the topical antiseptic hexachlorophene in neonates in combination with talcum powder was responsible for 40 deaths as a consequence of increased systemic bioavailablility. However, several *in vitro* systems do produce sufficiently consistent results, which in conjunction with such *in vivo* studies as are practicable, enable useful predictions to be made on dermal absorption of certain toxins.

13.4 POTENCY

The ability of a toxin to elicit a cutaneous or systemic response depends not only on its absorption but the relative potency of that toxin. Using these parameters to calculate a threshold below which toxicity would not be expected is an attractive concept and indeed this threshold effect has been extensively studied in the induction and elicitation of the contact hypersensitivity response to cutaneous allergens. Unfortunately, the calculation of a toxin threshold not only requires consideration of the individual's susceptibility but also the wide spectrum of potentially toxic effects, and for those substances with more than one biological consequence the threshold may be different for each outcome. As an example, the widely used insect repellent diethyltoluamide (DEET) may induce a spectrum of responses ranging from irritant dermatitis often at therapeutic concentrations to a corrosive effect manifested by blistering, ulceration and scarring at high concentration. However, the rare phenomena of DEET-associated neurotoxicity and encephalopathy can occur over a wide spectrum of concentrations. It is therefore necessary when considering the threshold for dermatological toxins to appreciate these thresholds will be effect specific in which case the lowest threshold for any effect is likely to become the determining threshold for that product.

13.5 DERMAL TOXICOLOGY

Dermal toxicology must therefore consider both the direct effect of environmental toxins on the skin and the systemic effects of cutaneous absorption. Substances that are absorbed through other routes may have a particular affinity for the skin, for example the antifungal agent itraconazole that preferentially accumulates in keratin. In these circumstances, the skin may be the primary target of damage from toxins that are not absorbed cutaneously. Conversely, many dermally absorbed toxins may have little direct effect on cutaneous function, but possess profound systemic effects. Thus the plasticiser triorthocresyl phosphate is a neurotoxin with little adverse cutaneous effect. For other toxins, the cutaneous and systemic effects differ. Polyvinyl chloride is a narcotic in high dosage and can cause the rare hepatic haemangiocarcinoma; in the skin, digital scleroderma may develop. In humans, dioxin is remarkably non-toxic in comparison with many other species but it can cause deranged hepatic function when absorbed in large quantities. On the skin, it can produce a characteristic abnormality known as chloracne. Toxins freely passing through the dermal barrier are generally lipophillic, a characteristic that favours neurotoxicity. Irrespective of any systemic effect, the commonest cutaneous response elicited by environmental toxins is dermatitis.

13.6 DERMATITIS

Dermatitis is an inflammatory condition that can be caused by irritants or allergens. In the latter, damage is mediated through antigen-presenting cells and sensitised lymphocytes following recognition of antigen contact with the skin. In irritant-contact dermatitis, the irritant itself inflicts damage that is compounded by the body's defence mechanism. Both these disorders give rise to a systemic response which may have effects elsewhere. Thus, in the case of allergic contact dermatitis, secondary eczematisation is often seen and in irritation the phenomenon of conditioned hyperirritability can affect responses to irritation outside the primary site of contact. Consideration of dermatitis is outside the scope of this review but many toxins can have irritant or allergic potential and, if dermatitis arises, increased absorption will occur through the inflamed skin.

13.7   TOXIN ACCUMULATION, METABOLISM AND TRANSPORT

Toxins may accumulate in the skin as a result of absorption or by dermal localisation after ingestion, inhalation or injection. Toxin accumulation is affected by the duration and frequency of exposures, which in the occupational setting may typically be multiple and short lived. Where exposure cannot be prevented by engineering solutions, personal protective equipment (PPE) may be employed. Skin occlusion by such PPE itself has effects on susceptibility to skin toxicity. Occlusion increases percutaneous absorption of mainly lipid-soluble molecules (in itself a useful therapeutic modality in the topical treatment of some cutaneous diseases) and there is increased hydration of the stratum corneum, which in turn reduces skin barrier function. Some chemicals are able to penetrate or permeate through protective clothing, and trapping of the toxin under occlusion increases the potential for absorption. Care must be taken to ensure PPE is renewed before penetration or permeation occurs.

While the stratum corneum is the critical first-line physical barrier against toxins, it is evident a second 'biochemical barrier' exists in the form of xenobiotic metabolising enzymes and transport proteins expressed by keratinocytes. The liver metabolises ingested toxins resulting in less harmful compounds. Similar detoxifying enzymes exist within the skin, which is estimated to have between 2% and 6% of the detoxifying potential of the liver. While enzyme activity has been identified in epidermis, epidermal appendages and dermis, the major site of detoxification is the epidermis.

Phase 1 metabolism (of largely lipophilic xenobiotics which penetrate the stratum corneum) involves the introduction of a polar group by hydroxylation of a carbon, nitrogen or sulfur atom using molecular oxygen. These oxidative reactions may activate or inactivate the xenobiotic chemically. As in the liver, the most important group of enzymes involved in the oxidation process are cytochrome P450 enzymes, using molecular oxygen with the formation of water as a by-product. These enzymes are the terminal component of an electron transport system that is found in the microsome by which NADPH transfers high-potential electrons to a flavoprotein, which then conveys them to adrenotoxin, a non-haem iron protein. Adrenotoxin then transfers an electron to the oxidised form of cytochrome P450. The reduced form of P450 then activates O_2.

It is clear that in the liver P450 is not a single entity and a variety of individual forms exist that exhibit differing substrate specificity. Similarly, in the skin various isoenzymes of P450 are constitutively expressed by human keratinocytes and fibroblasts. Some isoenzymes are inducible, as demonstrated *in vitro* following incubation with agents such as dexamethasone, and in animal work by the application of topical toxins. Thus, the toxin itself may modulate the detoxifying properties of cutaneously expressed P450.

Phase 2 metabolism involves a conjugation of the Phase 1 reactant to form water-soluble products that are more easily excreted. The skin possesses Phase 2 enzymes such as phenol uridine 5'-diphosphate (UDP)-glucuronosyltransferase and glutathione *S*-transferase, and thus the complete metabolism of a toxin to a water-soluble excretable form can occur within the epidermis. This process is assisted by the expression of specific transporter proteins that affect influx and efflux into and out of cells. Multi-drug resistance (MDR) transporter proteins such as MDR1 P-glycoprotein and multi-drug resistance-associated proteins (MRP1) are expressed in the membrane of human keratinocytes, with related moieties on dermal fibroblasts. Members of the organic anion transporting polypeptide (OATP) family have diverse roles in human physiology transporting large organic ions such as bile acids, toxins and drugs. Expression of isoforms of OATP has been demonstrated in human skin, suggesting a role in toxin/metabolite transport. Inevitably, there will be interaction and synergism between the processes of xenobiotic metabolism and transportation.

In the process of metabolising xenobiotics, highly reactive intermediate products can be formed from substances that are non-toxic or even inert. The effect of these toxic metabolites varies. Thus, the commonly used topical anti-acne preparation, benzoyl peroxide, does not become effective until it is metabolised. In addition to its irritant and proliferative effects, the metabolised benzoyl peroxide can act as an allergen. Most common contact allergens are easily absorbed, low molecular mass substances that become allergic by haptenisation with skin constituents into a form capable of being presented to T lymphocytes, a process influenced at least in part by isoenzymes of the P450 system. Thus inter-individual variation in expression of P450 isoenzymes may affect susceptibility to potential contact allergens. Potential chemocarcinogens may undergo, or be a product of similar biotransformational events.

13.8 CHEMICAL CARCINOGENESIS

Carcinogenesis is a multistage process involving initiation, promotion and progression. Environmental toxins and other damaging insults are being increasingly incriminated in the process of carcinogenesis. Indeed, it is difficult not to accept an association between life-long exposure of the skin to environmental assault from sun and chemical carcinogens and it being the commonest site of human malignancy.

Given that the molecular events that generate genetic changes differ at each stage, it seems unlikely that any single environmental agent is responsible for all stages of tumour development. Initiation is considered to be the first, irreversible genetic event that involves damage to DNA by covalent binding with the toxic molecule. The suggestion by Potts in 1775 that scrotal cancer in sweeps originated from their occupational exposure to soot initiated a series of investigations that culminated in the identification of the polycyclic aromatic hydrocarbons (PAHs) as the factors responsible for the carcinogenic potential in soot, oil and coal tars. These PAHs are metabolised by constitutively expressed P450 isoenzymes such as aryl hydrocarbon hydroxylase. PAH metabolites probably bind to the aryl hydrocarbon receptor (AhR) present on cells of the epidermis and dermis. Translocation of the AhR receptor–ligand complex to the nucleus results in alteration of DNA transcription, with up and downregulation of genes including those coding the isoenzymes of the P450 system. Activation of the *ras* oncogene, at least in animal work, is associated with tumour initiation.

Subsequent tumour development is dependent upon epidermal hyperplasia, and many irritants and mitogens capable of inducing epidermal hyperplasia are recognised tumour promoters, which can selectively encourage the growth of initiated cells. Variation in expression of skin lipogenases and cyclooxygenases may contribute to such tumour promotion.

Tumour progression is likely to be a consequence of further chemically induced mutagenic events. Such events may activate oncogenes or inactivate tumour suppressor genes such as p53. Mutations of p53 are detectable in human skin cancers, particularly those associated with exposure to ultraviolet radiation.

13.9 ULTRAVIOLET RADIATION

Ultraviolet radiation (UVR), and to some extent near infrared radiation, are capable of inducing skin damage. Photoageing, photocarcinogenesis, phototoxic and photoallergic reactions, and cutaneous disease such as lupus erythematosus and solar urticaria can all occur as a consequence of UVR exposure.

The most clinically obvious consequence of acute UV-B (280–315 nm) radiation exposure is the development of increased blood flow and redness (erythema) at the site of exposure, a process that can be shown to be nitric oxide dependent. Free radicals are also generated by the action of UV-B light on the skin. Indeed, it is likely that many of the adverse effects of light on the skin are mediated by free radical production. UV-B generates cyclobutane pyrimidine dimers within cellular DNA, an important initiating event in photocarcinogenesis, and also affects on the keratinocyte membrane. UV-A (315–400 nm) radiation membrane effects include hydrolysis of sphingomyelin and the generation of the second-messenger ceramide, which in turn initiates events that alter the binding of transcription factors to DNA. This dysregulation of gene transcription is crucial to the photocarcinogenicity of UVR. The effect of light in the UV-A spectrum can be enhanced by a number of plant derivatives, such as fluorocoumarines, of which psoralens are the most well characterised. These are used, both topically and systemically, and are referred to as PUVA therapy.

The excitation of a ground state molecule by energy transfer from another species is termed 'sensitisation' and the deactivation of an excited species 'quenching'. Both these phenomena are of central importance in organic photochemistry. Many chemicals are capable of producing dermal phototoxicity and the potential effects of UV light on absorbed xenobiotics are immense, and as yet, poorly understood.

13.10 ENZYMES

A number of enzymes of metabolic importance are known to exist within the epidermis. The best characterised are those responsible for dermal steroid metabolism. The activity of these enzymes differs in body sites and between persons. Thus 5a reductase, the enzyme responsible for mediating the reduction of testosterone to the more active dihydrotestosterone is believed to vary in concentration at various body sites, and this variation in distribution may be of importance in the development of secondary sexual characteristics. In a similar manner, aromase enzymes, which have been found to be present at the outer root sheath of the hair follicle, are present at five times greater concentration in women than in men. Since aromatase mediates the conversion of testosterone and androstenedione to oestradiol and oestrone, respectively, this enzyme may well be responsible for the differing pattern of hair loss in men and women. It is as yet unknown whether these enzymes play any part in the metabolism of exogenous steroids or other xenobiotics.

Many other enzyme systems exist in the skin, which are involved in the metabolism of endogenous substrates. In some cases, the absence of these may give rise to skin abnormalities, for example deficiency of cholesterol sulfate in the epidermis of those with x-linked ichthyosis. Such deficiencies may, by the alteration of the cutaneous barrier, have an effect on the fate of applied toxins, but direct participation in metabolism has not been demonstrated.

13.11 PEROXISOMES

Peroxisomes (microbodies) are, like mitochondria, a major site of oxygen utilisation. It is believed that they are present in all cells, and peroxisomes have been demonstrated in keratinocytes and fibroblasts. Catalase comprises some 40% of the total protein content of peroxisomes and it seems likely that these organelles are the source of the catalase activity demonstrated in human skin.

Peroxisomes are diverse organelles and in addition to catalase contain differing sets of enzymes that can adapt to changing conditions. Peroxisomes have broad effects on the metabolism of lipids, hormones and xenobiotics, and use molecular oxygen to remove hydrogen ions from organic substrates. This oxidation process produces hydrogen peroxide (H_2O_2) and catalase utilises the hydrogen peroxide to oxidise such substances as phenols and alcohol with the production of H_2O. It seems likely that this system may participate in the metabolism of phenol and other toxins applied to the skin. Such metabolism is likely to generate free radicals, the importance of which has already been discussed with regard to UV light. The discovery of peroxisomal proliferator activator receptors (PPARs), which when bound to their ligand act as nucler transcription factors, has led to the development of drugs which alter peroxisomal function.

Reactive oxidants can therefore be generated in the skin not only from normal endogenous processes but also from exogenous sources by redox cycling or electromagnetic radiation. Toxicity of reactive oxidants is felt to be due to non-specific attack upon structural cell constituents. These effects may vary from destruction to physiological disturbance of growth including apoptosis (programmed cell death, a p53-dependent event) and cell signalling. Oxidative stress is mitigated by scavenging systems such as catalase and superoxide dismutase. If these protective measures fail and damage occurs, repair is effected by a mixture of enzyme and non-enzyme processes. It is outside the scope of this review to discuss the complex antioxidant system in depth, but our knowledge of the system is incomplete, as is our knowledge of the susceptibility of the components of the system to damage from cutaneously applied toxins.

13.12 CONCLUSION

The factors that determine cutaneous susceptibility to toxins are complex. Physical and biological interplay, including the metabolism and transport of xenobiotics, combine to modulate the threat of any potential toxin. Undoubtedly, our understanding of these processes has increased significantly over the last decade, largely as a result of improved techniques in molecular biology. However, much remains to be understood, and caution must be used when interpreting the results of our present methods of assessing dermatotoxicological risk. Illuminating the predictable in an unpredictable world is no easy task, and as such the reader would be well served by re-reading Chapter 25 of this volume.

BIBLIOGRAPHY

M. Aumailley and B. Gayraud, Structure and biological activity of the extracellular matrix, *J. Mol. Med.*, 1998, **76**, 253–265.

O. Baadsgaard and T. Wang, Immune regulation in allergic and irritant skin reactions, *Int. J. Dermatol.*, 1991, **30**, 161–172.

P. Brookes and P.D. Lawley, Evidence for the binding of polynuclear aromatic hydrocarbons to the nucleic acids of mouse skin: relation between carcinogenic power of hydrocarbons and their binding to deoxyribonucleic acid, *Nature*, 1964, **202**, 781–784.

J.W. Cook , I. Hieger, E.L. Kennaway and W.V. Mayneord, The production of cancer by pure hydrocarbon. Part 1, *Proc. R. Soc. Lond.*, 1932, **111**, 455–484.

J.M. Coxon and B. Halton, *Organic Photochemistry*, Cambridge University Press, Cambridge, 1987.

P.M. Elias and G.K. Menon, Structural and lipid biochemical correlates of the epidermal permeability barrier, *Adv. Lipid Res.*, 1991, **24**, 1–26.

European Centre for Ecotoxicology of Chemicals, Monograph 20, Percutaneous Absorption, European Centre for Ecotoxicity and Toxicology of Chemicals, Brussels, 1995.

S.M. Fischer, Is cyclooxygenase-2 important in skin carcinogenesis?, *J. Environ. Pathol. Toxicol. Oncol.*, 2002, **21**, 183–191.

F.K. Jugert, R. Agarwal, A. Kuhn, D.R. Bickers, H.F. Merk and H. Mukhtar, Multiple cytochrome P450 isozymes in murine skin: induction of P450 1A, 2B, 2E, and 3A by dexamethasone, *J. Invest. Dermatol.*, 1994, **102**, 970–975.

B.N. Kemppainen and W.G. Reifenrath (eds), *Methods for Skin Absorption*, CRC Press, Boca Raton, FL, 1990.

I. Kimber, D.A. Basketter, M. Butler, A. Gamer, J.L. Garrigue, G.F. Gerberick, C. Newsome, W. Sterling and H.W. Vohr, Classification of contact allergens according to potency: proposals, *Food Chem. Toxicol.*, 2003, **41**, 1799–1809.

S. Kuenzli and J.H. Saurat, Peroxisome proliferator-activated receptors in cutaneous biology, *Br. J. Dermatol.*, 2003, **149**, 229–236.

W. Levin, The 1988 Bernard B. Brodie Award lecture. Functional diversity of hepatic cytochromes P-450, *Drug Metab. Dispos.*, 1990, **18**, 824–830.

H.F. Merk, J. Abel, J.M. Baron and J. Krutmann, Molecular pathways in dermatotoxicology, *Toxicol. Appl. Pharmacol.*, 2004, **195**, 267–277.

H. Mukhtar and W.A. Khan, Cutaneous cytochrome P-450, *Drug Metab. Rev.*, 1989, **20**, 657–673.

H. Mukhtar, H.F. Merk and M. Athar, Skin chemical carcinogenesis, *Clin. Dermatol.*, 1989, **7**, 1–10.

B.J. Nickoloff and Y. Naidu, Perturbation of epidermal barrier function correlates with initiation of cytokine cascade in human skin, *J. Am. Acad. Dermatol.*, 1994, **30**, 535–546.

C. Nishigori, Y. Hattori and S. Toyokuni, Role of reactive oxygen species in skin carcinogenesis, *Antioxid. Redox. Signal*, 2004, **6**, 561–570.

V.D. Plueckhahn, B.A. Ballard, J.M. Banks, R.B. Collins and P.T. Flett, Hexachlorophene preparations in infant antiseptic skin care: benefits, risks, and the future, *Med. J. Aust.*, 1978, **2**, 555–560.

R.J. Pohl, R.M. Philpot and J.R. Fouts, Cytochrome P-450 content and mixed-function oxidase activity in microsomes isolated from mouse skin, *Drug Metab. Dispos.*, 1976, **4**, 442–450.

M.E. Sawaya and V.H. Price, Different levels of 5 alpha-reductase type I and II, aromatase, and androgen receptor in hair follicles of women and men with androgenetic alopecia, *J. Invest. Dermatol.*, 1997, **109**, 296–300.

E.E. Sisskin, T. Gray and J.C. Barrett, Correlation between sensitivity to tumor promotion and sustained epidermal hyperplasia of mice and rats treated with 12-*O*-tetra-decanoylphorbol-13-acetate, *Carcinogenesis*, 1982, **3**, 403–407.

R.M. Tyrrell and M. Pidoux, Singlet oxygen involvement in the inactivation of cultured human fibroblasts by UVA (334 nm, 365 nm) and near-visible (405 nm) radiations, *Photochem. Photobiol.*, 1989, **49**, 407–412.

J.G. van der Schroeff, L.M. Evers, A.J. Boot and J.L. Bos, Ras oncogene mutations in basal cell carcinomas and squamous cell carcinomas of human skin, *J. Invest. Dermatol.*, 1990, **94**, 423–425.

F. Vecchini, K. Mace, J. Magdalou, Y. Mahe, B.A. Bernard and B. Shroot, Constitutive and inducible expression of drug metabolizing enzymes in cultured human keratinocytes, *Br. J. Dermatol.*, 1995, **132**, 14–21.

R.G.M. Wang, J.B. Knaak and H.I. Maibach (eds), *Health Risk Assessment: Dermal and Inhalation Exposure and absorption of Toxicants*, CRC Press, Boca Raton, FL, 1993.

K. Weber-Matthiesen and W. Sterry, Organization of the monocyte/macrophage system of normal human skin, *J. Invest. Dermatol.*, 1990, **95**, 83–89.

M.L. Williams, G.K. Menon and K.P. Hanley, HMG-CoA reductase inhibitors perturb fatty acid metabolism and induce peroxisomes in keratinocytes, *J. Lipid Res.*, 1992, **33**, 193–208.

H. Zhai and H.I. Maibach (eds), *Dermatotoxicology*, CRC Press, Boca Raton, FL, 2004.

Chapter 14

Respiratory Toxicology

RAYMOND AGIUS AND ANIL ADISESH

14.1 INTRODUCTION

The respiratory tract is a very important organ of first contact for most environmental exposures. Some cause damage to the respiratory tract, others merely use it as a route to access other parts of the body. An adequate understanding of the toxicology of the respiratory tract requires some basic knowledge of the anatomy and physiology of this organ system and an outline is presented early in this chapter. An indication of the frequency of occupational lung disease is provided with reference to data from specialist reports in Britain. Reference to chemical interactions at a molecular level is made especially in relation to asthma. Practical aspects such as health surveillance are also outlined.

14.2 STRUCTURE AND FUNCTION

In common with other organisms, humans need a supply of oxygen for the metabolic processes of respiration and consequently a system for disposing of the carbon dioxide so produced. In unicellular and simple multicellular microorganisms, this can be achieved by diffusion alone. Complex organisms have evolved specialised tissues and structures to perform a respiratory function. In humans, the respiratory tract can be thought of in two parts, a gas-conducting and a gas-exchange system. This works in alliance with the cardiovascular system to perfuse tissues with oxygen and remove carbon dioxide.

The nose is the organ of the respiratory tract that is first exposed to inhaled agents (Figure 14.1). It is provided with a convoluted surface over the turbinates or conchae. Its main channel is to the nasopharynx and hence larynx (voice box), but it also drains tears from the eyes and communicates with the sinuses. Nasal breathing facilitates the sense of smell and allows the humidification of inhaled air as well as the removal of large (greater than 10 μm) particles before entrainment to the lungs.

The trachea (wind pipe) is a continuation of the larynx 2.5 cm in diameter, that divides into two main bronchi leading to the respective left and right lungs. There is a progressive (up to 25) division of the bronchi. The airways lacking in cartilage beyond the bronchi are the bronchioles. These lead into hollow spaces called alveoli, which have a diameter of about 0.1 mm each. There are approximately 300 million alveoli and their total surface area is about 140 m^2. The respiratory units (*i.e.* the alveoli and the smallest bronchioles called respiratory bronchioles) are responsible for the exchange of gases.

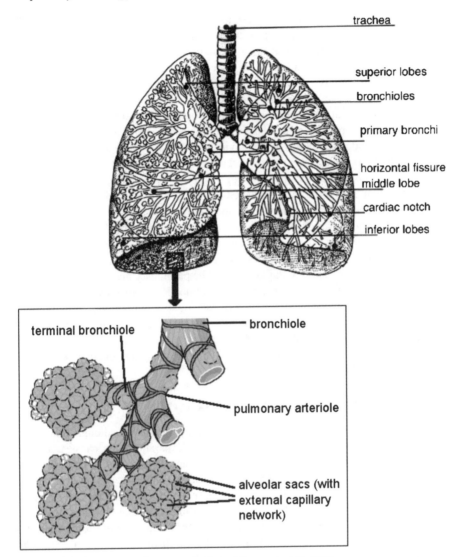

Figure 14.1 *Diagrammatic representation of the respiratory tract*

They are lined mainly by flat, extremely thin cells that permit easy diffusion of oxygen through them from the air in the alveolar spaces to the blood in the capillaries, and of carbon dioxide in the opposite direction. The ventilation of the lung varies such that the lower zones are better ventilated than the upper ones, thus affecting the deposition of particulates. It is estimated that 10,000 L of air are inhaled daily. The rate of breathing is stimulated by lowered blood pH, raised carbon dioxide or low oxygen partial pressure. Neural airway receptors can be stimulated by chemical mediators, *e.g.* histamine or capsaicin; other irritant receptors may react even to relatively inert dust (such as charcoal) causing coughing.

The conducting airways from the nose to respiratory bronchioles are lined by cells with cilia (small motile surface projections). Interspersed between these cells are mucus-secreting cells. This mucus along with a transudate from the respiratory epithelium forms a continuous fluid film 5 μm in depth. The cilia direct the mucus upwards to the larger airways by rhythmic wave-like movements thereby helping to clear deposited dusts, bacteria, *etc.* Particulates can be cleared from the large bronchi in 1 h but it may take several days from the respiratory bronchioles. At the alveolar level, the fluid consists of a lipoprotein surfactant monomolecular layer. The function of surfactant is not well understood but is thought to include the attenuation of tension within the alveoli, reduction of evaporative losses and protective effects against noxious particles, gases and vapours. If particles penetrate to the alveoli, then some may be cleared by transport within surfactant to the ciliated bronchioles, while others are absorbed. Absorption can be by direct penetration into or through an alveolar cell. Phagocytosis is the engulfment of a particle by a specialised cell, usually a macrophage. These mobile cells migrate to the lymphatic system and many insoluble particles such as silica or coal dust are deposited in the lymph nodes where they may reside indefinitely. Alveolar particle clearance may take 100 days or more. The pulmonary macrophages are part of the immune defences of the lung directed at protection from microorganisms. The inflammatory process aids the destruction of bacteria, *etc.* but if triggered by physicochemical means, *e.g.* by crystalline silica, they can provoke serious tissue damage.

The smooth lining of the outside of the lungs and the inside of the chest wall is called the pleura. It permits the lungs to slide within the chest wall but may be damaged by exposure, notably to asbestos.

14.3   EXPOSURE OF THE LUNG TO TOXICANTS

Most lung toxicants gain access to the organ by inhalation, though this is not universally the case, *e.g.* the bipyridium herbicide paraquat can cause fatal lung damage after it has been ingested. Inhaled harmful agents fall into two broad physical categories. The first category consists of those that are in the gaseous phase, *i.e.* gases and vapours (vapours are substances in the gaseous phase at a temperature below their boiling point). The second category consists of agents that are aerosols usually of fine solid (dust) particles though sometimes they can be fine droplets (mists). Mists are liquid droplets formed by the condensation of vapours (usually around appropriate nuclei), or the 'atomisation' of liquids. The term fume is usually applied to particulate aerosol generated by a thermal process such as combustion, or else vaporisation followed by condensation to fine solid particles.

The aerodynamic diameter of a particle is the diameter of a sphere of unit density that would settle at the same rate as the particle in question. When airborne particles come into contact with the wall of a conducting airway or a respiratory unit they do not become airborne again. This constitutes deposition and can be achieved in one of the following four ways:

- *Sedimentation.* Sedimentation is settlement by gravity and tends to occur in the larger airways from the nose downwards.
- *Inertial impaction.* This occurs when an airstream changes direction in the nose but also in other large airways.

- *Interception.* Interception applies mainly to irregular particles such as asbestos or other fibrous dusts, which by virtue of their shape can avoid sedimentation and inertial impaction. However, they are intercepted by collision with walls of bronchioles especially at bifurcations or if the fibres are curved.
- *Diffusion.* Diffusion is the behaviour of very small aerosol particles that are randomly bombarded by the molecules of air through Brownian movement. It significantly influences deposition beyond the terminal bronchioles.

Most particles larger than 7 mm in aerodynamic diameter are deposited in the nose or throat during breathing at rest. However there is a wide variation in the efficiency of this among apparently normal subjects. Moreover, conditions that favour mouth breathing *e.g.* high ventilation rates and obstructive disease of the nasal airways, will cause large particles to bypass this filter. Alveolar deposition is appreciable at particle aerodynamic diameters of between 1 and 7 mm (respirable particles) with the smaller particles being more likely to be deposited. There is evidence that very fine particles of less than say 0.1 μm can be particularly harmful if they traverse through the alveolar wall and into the circulation. During exertion, an increase in tidal volume (*i.e.* the volume of air inspired with each breath) and particularly in respiratory minute volume (*i.e.* the product of tidal volume and the number of breaths per minute) is the single most important determinant of the total load of particles in the alveoli and hence the total volume of particles deposited for a given aerosol.

In a sense, the respiratory tract tends to behave like a very complex elutriator with the particles of larger aerodynamic diameter depositing in the nose and larger airways, while smaller particles deposit in the smaller airways and alveoli; a proportion of the very smallest particles are not deposited at all. A similar analogy may be extended to the handling of gases and vapours by the lung. Gases that are very water-soluble such as ammonia will partition readily into the aqueous phase lining the mucous membranes and thus will achieve higher concentrations around the cells of the eyes and nose. Other less water-soluble gases or highly lipid-soluble vapours tend to be absorbed by the respiratory units of the lung.

14.4 THE FREQUENCY OF OCCUPATIONAL LUNG DAMAGE

Data on the frequency of occupational lung disease in the UK are available through The health and occupation reporting network (THOR), which incorporates the surveillance of work-related and occupational respiratory disease (SWORD) and the occupational physicians' reporting activity (OPRA) in which specialists in respiratory medicine and in occupational medicine, respectively, report cases to the Centre for Occupational and Environmental Health at the University of Manchester.

These schemes help to determine the scale and patterns of work-related respiratory disease in the UK, and help identify the agents thought to be responsible along with information on industry and occupation. The types of respiratory diseases reported include occupational asthma, benign and malignant pleural disease, mesothelioma, lung cancer and pneumoconiosis. The commonest reported occupational lung disease of recent causation in the UK is asthma and the commonest reported causal agents are the di-isocyanates (used in various industries such as in 'twin-pack' spray painting).

Table 14.1   *The commonest causes of occupational asthma as reported in the UK in 2003 (From THOR, 2005)*

Di-isocyanates (*e.g.* in twin-pack spray paints)
Laboratory animals
Flour
Paints (not otherwise classified)
Glutaraldehyde
Chrome compounds
Solder/colophony
Enzymes, *e.g.* amylase
Wood dusts

Other important asthma hazards include colophony fume (from soldering flux). The SWORD scheme was useful in picking up trends of concern such as a substantial increase in asthma associated with exposure to latex, and thus helped in raising awareness and reducing the risks of this problem. Unfortunately asbestos-related lung disease – both benign (such as pleural plaques) and malignant (mesothelioma) – have been shown by SWORD to continue to have a high incidence, the legacy of exposure before adequate control measures were put in place (Table 14.1).

14.5   ASTHMA AND OTHER TOXIC EFFECTS ON THE AIRWAYS

Sulfur oxides, nitrogen oxides, ozone, ammonia and chlorine are examples of so called irritant gases. Thus 'irritancy' is a relative term and, to paraphrase Paracelsus, "what matters is the dose". Many of them may be detected by their smell and irritant effect, but if evasive action is not taken in time, and if exposure is high enough, they can produce severe 'corrosive' damage throughout the lungs. A whiff of a smell of chlorine from 'dilute bleach' may be deemed an irritant but a higher concentration as in the trenches in World War I can cause a fatal chemical pneumonia (pneumonitis). These gases produce their harmful effects on the eyes, airways and even the alveoli in the respiratory units of the lungs.

Exposure to ammonia and chlorine may occur as a result of industrial accidents. High levels of nitrogen(IV) oxide can be encountered in agriculture (silo filling), during arc welding, as a result of shot firing in the mines, and in the chemical industry. It can achieve high levels in the vicinity of internal combustion exhausts. Ozone is usually a secondary pollutant. Sulfur(IV) oxide is produced by the combustion of sulfur-containing fossil fuels. Sulfur(IV) oxide, chlorine and ammonia are highly irritant and cause pain in the eyes, mouth and chest. In high concentrations, they can produce inflammation of the lining of the lungs and this causes breathlessness and may be fatal (see below for chronic effects).

Nitrogen(IV) oxide has less effect on the eyes, nose and mouth but can cause severe inflammation of the lungs. It is important to realise that although symptoms at first may be mild, serious breathing problems may follow later if the exposure is high enough.

Gases and vapours can produce harmful effects in other ways. Some can cause asphyxiation, deprivation of oxygen to the tissues, and after entering the body through the lungs they can cause damage to other tissues of the body (see also Section 14.9).

Asthma is a condition characterised by inflammation of the lining of the airways and intermittent spasm of the underlying smooth muscle. Comparatively more is known about the causes of asthma through work exposure ('occupational' asthma) than about other forms. It is often, but not invariably, the result of allergy to an inhaled dust or vapour in the workplace, and therefore there is usually a latent interval ranging from weeks to several years between first exposure and the onset of symptoms. Its symptoms include cough, wheeze, chest tightness and shortness of breath, which usually improve on days off work or longer holidays but the association with work may be difficult to establish in some cases, especially after long-standing exposure. Indeed in a substantial proportion of cases (depending on the duration of exposure, severity of the airway dysfunction and nature of the exposure) the asthma will persist indefinitely, even for life, after exposure has ceased.

An understanding of the chemical interactions between xenobiotic low molecular weight agents and native human molecules is important since it is probably the first step in the development of occupational asthma. In classical Type I allergy the exogenous substance behaves as a complete antigen but many low molecular weight asthmagens are not complete antigens but behave as 'haptens' or partial antigens. Their first chemical interaction is with a native human macromolecule, such as the protein albumin to form a complex, which the body then treats as 'foreign'. A number of very important occupational asthmagens such as di-isocyanates and resin acids (abietic acid and other related compounds) have revealed only limited and inconsistent evidence of being mediated by classical allergic reactions. There are other 'grey areas' where the chemical reactivity of low molecular weight agents is presumed to play an important albeit unclear role in the causation of occupational asthma. Thus, exposure to various low molecular weight agents has been reported as causing 'reactive airways dysfunction syndrome' (RADS) the mechanism of which is still unclear, but oxidation or other cell surface reactions are probably important in inflicting severe damage to the mucosal epithelium and setting off a cascade resulting in asthma.

Some transition metal compounds are notable asthmagens, acting by Type I IgE-mediated immediate hypersensitivity mechanisms. These tend to be in groups Vb, VIb and VIIIb of the periodic table. These asthmagens comprise platinum, nickel, cobalt, chromium and vanadium. The atoms of transition metals in the ground state tend to have partially filled d or f electron orbitals. They can therefore combine through a co-ordination bond with a wide range of ligands (ions or neutral molecules with an electron pair available to share with the metal). Moreover ligands can form chelate rings by holding the transition metal atom in a pincer like grip at more than one binding site. If the transition metal is very firmly bound to the ligand, it is probably less likely to behave in an antigenic way, since it is then not accessible to native human proteinaceous macromolecules. However, if the ligand can leave the metal and be replaced by, for example, a nucleophilic nitrogen-containing moiety of a human macromolecule, or a sulphur atom, then an antigenic complex could result. Thus, not all instances of the presence of atoms of a particular metal are equally likely to cause asthma, and chemical speciation of the metal is highly important.

Many asthmagens contain nucleophilic groups, with nitrogen atoms expressed in various ways: as aliphatic amines, *e.g.* diaminoethane or as aromatic amines such as 1,4-diamino benzene, or within heterocyclic groups as in 'piperazine'. Some examples are given in Figure 14.2.

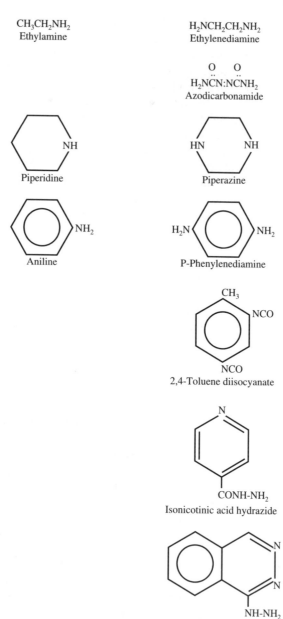

CH₃CH₂NH₂
Ethylamine

H₂NCH₂CH₂NH₂
Ethylenediamine

O   O
H₂NCN:NCNH₂
Azodicarbonamide

Piperidine

Piperazine

Aniline

P-Phenylenediamine

2,4-Toluene diisocyanate

Isonicotinic acid hydrazide

Hydralazine

Figure 14.2 *Nitrogen-containing low molecular mass chemicals in relation to asthma hazard. The left-hand column shows chemicals that appear not to present an asthma hazard in spite of extensive exposure experience in occupational settings. The right-hand column shows chemicals that have been reported as being occupational asthmagens. Note at least two reactive nitrogen containing functions in the asthmagens. In particular the three chemicals in the left-hand column are mono-functional homologues of their counterparts on the immediate right*

As previously mentioned potent as well as frequent nitrogen-containing low molecular weight asthmagens are the di-isocyanatess. It is notable that all these asthmagens have at least two nitrogen atoms in positions in which they could react with polypeptides or other human chemicals under physiologic conditions, *i.e.* they are least bi-functional. The so-called 'reactive dyes' are a heterogenous group and in some an unsaturated bond might be responsible for the conjugation of the hapten molecule to the free amino group in the lysine residues of albumin. In summary, molecules with two or more nucleophilic nitrogen atoms especially if $C=N$ or $N=N$ bonded are to be treated with caution.

As regards oxygen-containing organics, the category of dicarboxylic acid anhydrides (such as phthalic anhydride) has been recognised as including very important causal agents of occupational asthma. Aldehydes are another important category of asthmagens. Glutaraldehyde, responsible for many instances of occupational asthma especially among health care workers, is a dialdehyde. Formaldehyde also causes asthma and albeit arguably it is the simplest of mono-aldehydes, it behaves functionally as a dialdehyde since it has no alkyl groups but rather has two hydrogen atoms that can be readily displaced. Resin acids such as abietic acid in pine colophony and plicatic acid from cedars are another class of well-known causes of occupational asthma. Acrylic acids and some of their esters, epoxides (oxiranes), and some other unsaturated compounds are also potentially asthmagenic, especially if polyfunctional.

There are still considerable uncertainties as to the extent to which chemical exposure causes other diseases of the airways of the lung collectively termed 'chronic obstructive pulmonary disease' (COPD). These include chronic bronchitis (a condition characterised by chronic cough and sputum resulting from mucus hypersecretion) and emphysema (which is as much a disease of damage to the respiratory units of the lung as it is of collapse of the airways). Incontrovertibly, the best documented and most important environmental cause of COPD is tobacco smoke. However many substances (such as fine particulate matter arising from combustion and sulfur dioxide) can aggravate the symptoms of COPD and cause premature deaths from this condition, as occurred in the smogs that affected many big cities in the early 1950s, and probably still happens nowadays albeit to a lesser degree (as exposure has diminished) either from this condition or from cardiovascular damage. It has been estimated that to the order of 15% of the burden of COPD might be the result of chemical exposure of various kinds at work.

14.6 INTERSTITIAL LUNG DISEASE

The 'interstitium' of the lung can be considered to be the tissues between the specific lung structures. This includes the interepithelial spaces and the peribronchiolar, perivascular and perilymphatic tissues. The inhalation of noxious agents can cause injury to the cellular elements of the lung. One consequence is an inflammatory reaction, which results in the infiltration of the interstitium by cells and fluid (oedema). The proliferation of certain cell types particularly fibroblasts, and the laying down of increased quantities of collagen leads to a scarring or 'fibrosis' of the alveolar capillary units. The propensity of an inhalant to cause fibrosis is dependant on its ability to penetrate to the bronchioloalveolar level and persistence there, combined with its physicochemical properties.

14.6.1 Pneumoconiosis

Pneumoconiosis refers to the disease and tissue reaction caused by the inhalation of inorganic dusts and excludes asthma, emphysema, cancer and allergic alveolitis. It has been used as an all encompassing reference to dust disease of the lungs, but usage is best restricted to the aforementioned situation.

The pneumoconiosis due to crystalline silica is called 'silicosis'. The lung can respond with an acute illness as a result of high exposures of the order of say 2 mg m^{-3} of crystalline silica, such as quartz, measured in ambient air. Symptoms of cough and breathlessness may develop over a few months to a year after starting exposure and deterioration is progressive. More commonly there is a chronic illness with loss of lung function and characterised by collagen in the lung parenchyma, resulting from long-term exposure to levels of even 0.1 mg m^{-3} or less, *i.e.* lower even that current exposure standards.

Coal workers may develop coal workers' pneumoconiosis (CWP), silicosis or COPD due to the inhalation of coal dust. A 'simple pneumoconiosis' most commonly occurs, with fibroblast activity causing fibrotic change when coal dust laden macrophages die. This is accompanied by areas of loss of lung tissue. In itself, simple CWP does not cause symptoms, however it frequently coexists with COPD in smokers and non-smokers. Simple CWP can be complicated by 'progressive massive fibrosis'. This condition consists of large, several centimetre nodules forming in the upper part of the lungs accompanied by breathlessness and eventually respiratory failure.

Asbestos is a term encompassing a range of complex fibrous metal silicates that have long been widely recognised as a cause of occupational lung disease. Asbestosis is the pneumoconiosis caused by asbestos dust. The dust is a fibrous particulate with a length typically when inhaled to the order of approximately 50 μm and width 1–2 μm. The length of the fibres means that they cannot be engulfed whole by macrophages thereby limiting their clearance. The fibrotic process affects the bases of the lungs and patients present with gradual onset of breathlessness sometimes with a dry cough. The illness is progressive but with the lower exposures that have resulted from industrial hygiene measures most sufferers do not die of respiratory failure from asbestosis but instead succumb later to an associated lung cancer. Other effects linked to asbestos exposure can include thickening and/or calcification of the pleura. Asbestos can also cause a special cancer of the pleura called mesothelioma as well as bronchial cancer similar to that caused by smoking (see Section 14.7.4).

14.6.2 Berylliosis

Berylliosis is a rare occupational lung condition caused by the inhalation of dust containing as little as 2% beryllium. The lungs produce a type of inflammatory change known as a granulomatous reaction, which closely resembles the appearances of another spontaneously occurring and little understood condition, sarcoidosis. Berylliosis may be treated with steroids although complete reversal of lung changes may not occur.

14.6.3 Acute Alveolitis

Paraquat (1,1'-dimethyl-4,4'-bipyridilium dichloride) when ingested orally, usually accidentally, or with suicidal intent causes an interstitial lung disease. Paraquat may cause an acute alveolitis with pulmonary oedema (exudation of fluid into the airspaces and tissues), and this may cause death within hours to days, if not a second phase of toxicity ensues. The lung and kidney are target organs that accumulate the highest concentrations of paraquat. Death is usually from respiratory failure days to weeks after ingestion.

14.6.4 Extrinsic Allergic Alveolitis

Alveolitis (hypersensitivity pneumonitis) is the generic term applied to a plethora of diseases that share a common exposure to dusts of vegetable, animal or microbial origin, 'organic dusts'. It may not be readily appreciated that for instance grain dust varies from region to region and with climate. Largely dependent on humidity there is a progressive growth of various microorganisms, bacterial and fungal as well as a number of arthropods. This leads to the presence of bacterial endotoxins, fungal mycotoxins and proteases. Additionally, there may be the presence of insecticide or herbicide residues.

Some low molecular weight chemicals have also been recognised as causative of extrinsic allergic alveolitis. As the diseases were identified in relation to their specific, usually occupational exposures, some were given evocative names (Table 14.2).

This condition is usually associated with occupational or hobby exposure because a high exposure to the relevant allergen is required to induce sensitisation and then repeated exposure is usually necessary. Clinical diagnosis is based upon a compatible clinical history with a source of antigen being identified, an abnormal chest X-ray and lung function abnormalities consistent with interstitial lung disease. The presence of serum precipitins (immunogobulin G) to the inciting antigens may also be demonstrable.

Table 14.2 *Some diseases associated with occupational exposures*

Disease	Exposure	Antigenic agent
Farmer's lung	Mouldy hay, straw or grain	*Micropolyspora faemi, Thermoactinomyces vulgaris*
Bird fancier's lung	Feather 'Bloom', droppings	Avian proteins
Malt worker's lung	Mouldy barley	*Aspergillus clavatus*
Humidifier fever	Contaminated water	Endotoxin
Ventilation pneumonitis	Contaminated air	*Micropolyspora faemi, Thermoactinomyces vulgaris, etc.*
Di-isocyanate alveolitis	Chemical vapour, *e.g.* polyurethane foam manufacture	Toluene di-isocyanate, diphenylmethane di-isocyanate

14.6.5   Metal Fume Fever

This condition occurs when high concentrations of metal fumes, most commonly zinc oxide, are inhaled. From 4 to 12 h post inhalation there are symptoms of nausea, headaches, aching muscles and joints, weakness and cough. These are accompanied by shivering and fever. All of these subside within 24–48 h. Fumes of aluminium, copper, magnesium, antimony, cadmium, iron, nickel, silver, tin and manganese can all cause a similar illness.

14.6.6   Polymer Fume Fever

Pyrolysis products of polytetrafluoroethylene (PTFE) can cause an identical set of symptoms to metal oxide fumes.

14.6.7   Nuisance Dusts

The so-called 'nuisance dusts' are relatively inert and, by definition, cause no serious health effects, although they may be irritant to the upper airways. Examples include chalk, limestone, and titanium(IV) oxide. They may cause radiographic changes without disease. Dusts should be considered as nuisance dusts only when there is good evidence that they are inert and free from significant health effects, not when evidence for an effect is lacking because it has not been carefully sought. Some studies indicate that ultrafine particulate matter such as may be formed by combustion or condensation may be absorbed through the alveoli, be ingested by macrophages and, through cascade mechanisms, result in inflammation that may damage the lungs or other organs. Thus, while coarse titanium dioxide dust is usually regarded as non-toxic, ultra fine titanium dioxide has been shown to cause serious inflammation in animal experiments, and there is now evidence in humans that ultra fine particles of substances hitherto considered 'inert' can provoke significant inflammation and possibly long-term harm to the respiratory tract and/or cardiovascular system.

14.7   LUNG CANCER

Lung cancers may occur through environmental and occupational exposure to a wide range of toxic substances. The following are among the more significant examples.

14.7.1   Environmental Exposure

Environmental causes of lung cancer are thought to account for around 85% of cases, of these perhaps 5% are occupational. By far the most important cause of lung cancer is tobacco smoking not least because it is entirely avoidable. Active smokers have a death rate from lung cancer 14 times that of non-smokers. In a meta-analysis of passive smoking the excess risk of lung cancer was 24% among lifelong never smokers who lived with a smoker. The principal carcinogens of environmental tobacco smoke are polycyclic aromatic hydrocarbons (PAH), nitrosamines and aromatic amines. Radioactive elements such as polonium may also be contained in cigarette products.

The partial combustion of organic material leads to the formation of PAH among which benzo [a] pyrene ($C_{20}H_{12}$) is the main carcinogen. Exposure occurs in coal gas production, coke oven operators, aluminium refiners and steel founders. Motor vehicle exhaust, coal burning and the environmental effects of industry cause general population exposure. Working in a gas works may give exposures to benzo [a] pyrene of the order of 80 µg m^{-3} whereas urban air contains 0.002–2.0 µg m^{-3}.

14.7.2 Occupational Exposure

Chromium(VI) compounds in chromate manufacturing, chromium plating and lead chromate pigment manufacture have been reported as causing lung cancer. An excess risk of nasal cancers has also been noted with these exposures. Arsenic can be inhaled during mining operations, smelting and the manufacture or use of arsenical pesticides or fungicides. The lung cancer associated is usually a bronchial squamous cell carcinoma. Copper miners and smelters are the occupational groups best studied. Environmental exposure from being downwind of a smelting plant might be a significant risk if exposure is high. Drinking well water contaminated by arsenic may be associated with increased rates of lung and other cancers.

14.7.3 Radon

Exposure to radon and its daughters from radioactive decay is a hazard of mining in rock containing uranium or contaminated with radon. In 1531, Paracelsus recorded the effects of exposure on the Schneeberg and Joachimsthal miners who worked in an area now known to be contaminated with radon. Polonium (Po^{218} and Po^{214}) decay products of radon emit α particles of high energy that are then capable of damaging the DNA of the bronchial epithelium promoting carcinogenesis. Residents in areas of granitic rock such as Aberdeen may have higher environmental radon exposures, leading some to construct 'radon sinks' which when ventilated reduce concentrations. It has been estimated that in the UK residential radon exposure causes between 2000 and 3300 lung cancer deaths per annum, of which between 500 and 1300 are in non-smokers. Another radioactive gas thoron, which is an isotope of radon, $_{86}$Rn^{220} with a half life of 55.6 s, undergoes decay via Po^{212} and poses a hazard to workers involved in processing thorium.

14.7.4 Asbestos

Asbestos increases the risk of lung cancer (bronchial carcinoma) in persons who have been non-smokers but have had sufficient exposure to develop mild asbestosis, by a factor depending on the degree of exposure, typically say twofold but could be greater for very high exposures. However, in a smoker this risk has to be multiplied by the risk of tobacco smoking (say 10 or 15-fold in a very heavy smoker). Thus a smoker heavily exposed to asbestos may have a lung cancer risk 50 times that of a non-smoking, non-asbestos-exposed person. The tumour is usually a squamous cell carcinoma and up to 40% of persons dying of asbestosis have been found to have tumours present at post-mortem. Mesothelioma is a rare type of lung cancer, which is not a bronchial carcinoma, but arises from the pleural lining of the chest. Of tumours, 85% are attributable to asbestos especially crocidolite (blue asbestos), although it can occur spontaneously also. This tumour has the effect in advanced disease of encasing the lung and restricting its expansion leading to respiratory failure with 50% of cases dying in the year after diagnosis. Exposures do not have to be high and may only have been slight but many years previously. The 'latent period' of time since the relevant exposure may be over 35 years. Occupational exposures are the typical cause, however, some people have had environmental exposures from waste 'tailings' tips, or their parent's dusty work clothes.

14.7.5 Alkylating Agents

Alkylating agents such as bis(chloromethyl)ether (BCME, $ClCH_2OCH_2Cl$) may be potent lung carcinogens. Thus BCME occurred as an impurity of the production of chloromethyl methyl ether, (CMME, CH_3OCH_2Cl) for use as an alkylating agent in the manufacture of anion exchange resins. In this group of Philadelphia workers particularly high rates of small cell cancer, a less common lung tumour occurred. Alkylating agents are used in the treatment of cancers because of their ability to damage DNA that is rapidly reproducing. Consequently, unintended exposures to these agents can lead to cancers the site of origin dependant on target organ toxicity and route of entry. The chemical agent mustard gas is also an alkylating agent (a thio-ether and therefore a sulfur mustard as distinct from BCME). Those surviving its acute vesicant effects or working with the substance may, years later, develop lung cancer.

14.8 THE LUNG AS A PORTAL OF ENTRY AND ELIMINATION

So far the lungs have been considered as a target organ. However, they provide the main portal of entry for occupational toxicants and also a major way into the body for environmental exposures. The lungs are replete with enzymes capable of metabolising xenobiotics and provide a route of elimination.

Systemic toxicity may arise from lung uptake of particles, gases or vapours. Methylene chloride, various chloroethanes and chloroethylenes are examples causing systemic toxicity. Trichloroethylene is used industrially as a solvent and degreasing agent. It has also been used as an anaesthetic gas, but narcosis is one of its toxic effects. The effect of methylene chloride, although similar to that of vapours given off by other organic solvents such as trichloroethylene, presents the added hazard of being metabolised by the body to carbon monoxide. Initially these might cause a feeling of well being similar to that produced by alcohol. At higher concentrations they cause unconsciousness. Repeated exposure can lead to permanent brain damage.

Simple asphyxiants may interfere with the delivery of oxygen to the tissues by displacing oxygen from the air that is breathed in, *e.g.* methane and nitrogen. This happens usually in enclosed, poorly ventilated spaces.

Chemical asphyxiants on the other hand cause asphyxia by interfering with oxygen transport, *e.g.* carbon monoxide, hydrogen cyanide and hydrogen sulfide. This effect can be produced by reducing the oxygen-carrying capacity of the blood, *e.g.* carbon monoxide; or by preventing the cellular utilisation of oxygen, *e.g.* hydrogen sulfide. Carbon monoxide is present when there is incomplete combustion of carbon-containing fuels. It is odourless. Hydrogen sulfide occurs in mines, sewers and wherever there are chemical processes involving sulfur compounds such as in slurry pits. It may be detected by its smell at low concentrations, but this cannot be relied upon as a warning since it paralyses the sense of smell and can effectively become odourless. Hydrogen cyanide in the form of its salts (cyanides) is used in electroplating and other industries, and organic cyanides (nitriles) of varying degrees of toxicity are also used in industry. Absorption can also occur through the skin. At low concentrations, these gases cause the rapid onset of headache, dizziness, vomiting and confusion. At higher concentrations they are very rapidly lethal.

The deposition of particulates in the lungs allows for their absorption, resulting in toxic effects remote from the lung, *e.g.* lead and its salts. Similarly, gases such as arsine, and vapours like benzene can all utilise the inhaled route. Other sections of this book cover the organ-specific effects of inhaled toxicants.

We shall now consider the expiratory route of elimination. It is common to consider the expiration of increased levels of carbon dioxide as part of normal physiology. However, since the lung is efficient at gas exchange this also includes the elimination of gaseous toxicants or their metabolites. Several hundred hydrocarbons can be found in human breath, *e.g.* ethane and pentane as a result of lipid peroxidation during normal metabolism, propane and butane probably from faecal microbes. Other gases that can also be present above ambient environmental levels are carbon monoxide and nitrogen(II) oxide. These latter arise from the fact that specific enzyme systems exist, haem oxygenase and nitrogen(II) oxide synthase, which produce these gases particularly in response to inflammation. With knowledge of normal breath constituents and the lung elimination kinetics, it is possible to use exhaled breath as a medium for biological monitoring to assess exposure to dichloromethane (metabolised to carbon monoxide), benzene-1,1,1-trichloroethane and other volatile substances. This can be done by taking an end of exhalation breath sample, which is then representative of alveolar concentrations and analysing the breath sample for the relevant chemical. Breath samples can be collected as gas in bags or tubes, alternatively an absorptive medium such as Tenax® can be used with subsequent desorption and analysis.

14.9 CLINICAL EVALUATION

Physicians diagnose disease by taking a 'history' of events and eliciting symptoms, a patient's complaints and recognising physical signs, changes from normal appearances. These features are synthesised and combined with appropriately selected tests to arrive at, if possible, a single diagnosis or a number of likely diagnoses. Further investigations, which may be more invasive or involved, eliminate or confirm specific disorders. In toxicological diagnosis, the nature and circumstances of exposure must also be considered.

A change in the hue of skin or mucous membrane colour may suggest a problem with oxygenation. For example a bluish tinge 'cyanosis' being seen when the oxygen saturation of the blood is low (generally less than 7 kPa). The presence of methaemoglobin due to oxidation of the iron(II) to iron(III) by chemicals such as nitrates(III) gives a characteristic appearance. Conversely poisoning by cyanides or carbon monoxide can give a pinkish flush due to the formation of cyanmethaemoglobin and carboxy-haemoglobin, respectively. Patients experiencing hypoxia may be confused, irritable and restless.

The joints of the hands and feet especially may change in shape and be painful in response to the presence of a bronchial carcinoma, asbestosis or other lung fibrosis. Exposure to chromates or arsenic may have led to perforation or ulceration of the midline nasal cartilage.

Cough is a feature of many respiratory diseases including asthma, COPD, lung fibrosis and extrinsic allergic alveolitis. Expectorated sputum may be discoloured due to infection (green), blood (red with fresh blood or brown with old bleeding) or black with the contents of a pneumoconiotic coal miner's nodule.

Chest pain may be the presenting feature of mesothelioma or other lung tumours that have invaded the chest wall. Fever and weight loss may also be features of lung malignancy but are seen with lung fibrosis and extrinsic allergic alveolitis. Breathlessness is a predominant feature of lung fibrosis but also obstruction of the airways whether by a bronchial carcinoma or the reversible narrowing caused by asthma.

Further information is gleaned from observing the rate and rhythm of respiration, the chest shape, examining the structures associated with the respiratory system – nose, mouth and local lymph nodes, listening to the breath sounds at the mouth and with the stethoscope. Specific respiratory tests can be performed to measure lung function. A hand-held peak-flow meter measures the maximal flow that can be produced from a forced exhalation starting from full inspiration. This measurement when repeated several times a day (ideally 2 hourly) for 3 weeks both at work and at home is most helpful in diagnosing occupational asthma.

The measurement of lung volumes by spirometry is a basic investigation. This involves exhalation into an instrument that allows the recording of the volume of air that can be exhaled, vital capacity (VC) and the volume of air that can be exhaled in the first second of a forcible exhalation, forced expiratory volume (FEV1). Other measures can be derived from the information obtained at spirometry, however the aforementioned values are the most generally useful. These allow the recognition of restrictive lung disease (small lung volume, *i.e.* reduced VC), obstructive lung disease (reduced flow rate, *i.e.* low FEV1) or a mixed pattern. The measurement of the carbon monoxide transfer factor is a sensitive test of the efficiency of pulmonary gas exchange and may be impaired in lung fibrosis but would not be in asthma. It is measured by the change at exhalation in concentration of carbon monoxide from a known inhaled gas mixture that also contains helium as an inert tracer gas. From the same procedure an estimation of total lung capacity can be made using the information from the dilution of helium. There is a proportion of lung volume that cannot be expired, this volume increases in asthma and COPD, but is reduced in lung fibrosis.

Spirometry can also be performed with the inhalation of drugs that provoke airway narrowing in persons with 'reactive airways'. Histamine is one agent commonly used and dose response effects allow categorisation, which can be helpful in confirming an asthmatic tendency. The administration of drugs, such as salbutamol to relax the airways can achieve a similar effect a 'reversibility test'. In specialised centres provocation tests with sensitising agents suspected to be the cause of occupational asthma (or allergic alveolitis) can be performed but these are not without significant, possibly fatal risks.

The internal appearances of the lungs can be inspected with a thin flexible optical fibrescope, a bronchoscope passed with local anaesthesia through the nose. Obstructing lesions in the bronchi can be seen and whether there is inflammation or excessive mucous production. With this instrument, brushings of the airway lining can be taken and examined cytologically, washings of lung segments with physiological saline allow the analysis of cell types and other markers (bronchoalveolar lavage). A device may also be passed down the bronchoscope to take biopsy specimens of any lesions seen or passed deep into the lung tissue to assess whether lung fibrosis is present (transbronchial lung biopsy).

Various imaging techniques from chest X-ray, CT scan, radionuclide scintigraphy and sometimes ultrasound scan can be utilised in diagnosing respiratory diseases. Magnetic resonance imaging is not often used for chest diseases.

Supplementary investigations may be blood tests to look for anaemia, methaemoglobin or raised white blood cells indicating infection or inflammation. The measurement of blood oxygen and carbon dioxide tensions can help evaluate gas exchange. Immunological tests for elevated immunoglobulins such as IgE in occupational asthma or IgG in extrinsic allergic alveolitis.

All the available information must then be interpreted in the light of the individual patient's pre-existing medical condition, and likelihood of toxicant exposure versus other explanations of any pathophysiological findings.

14.10 CONCLUSIONS

The important lessons regarding respiratory toxicology are not so much a knowledge of the wide range of agents that can harm the lungs, their mode of action and their clinical effects, but rather the steps that should be taken to prevent harm caused by these agents, or at least to recognise possible problems at an early and hopefully remediable stage.

It is rarely possible to do away with a chemical hazard completely, if that chemical is needed for some industrial or other purpose. Nevertheless it is generally possible to replace the hazardous agent by one that is intrinsically less harmful, or to limit its exposure. Thus man-made mineral fibre of rock, ceramic or glass origin has been used as a substitute for asbestos. Glass fibre in particular is more soluble than asbestos, fractures to give shorter fibres (rather than thinner ones) and is therefore less harmful to the lungs, yet it can have adequate insulation properties for most purposes. In many applications it may be possible to substitute chemicals like di-isocyanates by others that are less harmful, or at least less volatile. However, care must be taken not to assume that a substituent has a lower risk simply because less ill health has been associated with exposure to it. The explanation for this might be that workers have not been exposed in large enough numbers or for long enough to become ill! There is a fast-growing industry producing substitutes for disinfectants and chemical sterilants. If low molecular chemicals have similar properties in attacking large macromolecules such as proteins in bacterial cell walls, or else in forming resins or other macromolecules, they may well have similar asthma-causing properties to the agents that they are intended to replace.

Last but not least, the suspicion and recognition of potential harm to the respiratory tract should not be left solely to the medical practitioner. Competent managers, having reviewed available information on toxic hazards and assessed and controlled likely exposures, should be aware of possible early symptoms. If the risk is significant, then a formal health surveillance strategy should be instituted. In its simplest form this could be a questionnaire regarding symptoms such as itching or running of the eyes or nose, cough, wheeze or shortness of breath, especially if it improves away from work. Thus toxicologic knowledge can be practically applied to the prevention and early recognition of adverse effects.

BIBLIOGRAPHY

R. Agius and A. Seaton, *Practical Occupational Medicine*, 2nd edn, Hodder, London, 2005.

B. Ballantyne, T. Marrs, T. Syverson (eds), *General and Applied Toxicology*, Macmillan Reference & Grove's Dictionaries, London, New York, 1999.

P.J. Baxter, P.H. Adams, T.-C. Aw, A. Cockroft and J.M. Harrington, *Hunter's Diseases of Occupations*, 9th edn, Arnold, London, 2000.

C.D. Klaasen (ed) *Casarett and Doull's Toxicology – The Basic Science of Poisons*, 6th edn, McGraw-Hill, New York, 2001.

THOR – The Health and Occupational Reporting Network, www.coeh.man.ac.uk/thor. The University of Manchester 2005.

Chapter 15

Hepatotoxicity

ALISON L. JONES

15.1 INTRODUCTION

Many potentially toxic substances enter the body via the gastrointestinal tract (gut). As the blood supply from the gastrointestinal tract, through the portal vein, drains into the liver, the liver comes into contact with the potentially toxic substances, and this exposure will often be at a higher concentration than that received by other tissues.

The liver is essential for the metabolic disposal of virtually all xenobiotics (foreign substances). This process is mostly achieved without injury to the liver itself or to other organs.

A few compounds such as carbon tetrachloride are toxic themselves and/or produce metabolites that cause liver injury in a dose-dependent fashion. However, most agents cause liver injury only under special circumstances when toxic substances accumulate. Factors contributing to the build-up of such toxic substances include genetic enzyme variants (metabolizing enzymes with altered function due to gene defects), which allow greater formation of the harmful metabolite, and induction (greater production) of an enzyme that produces more than the usual quantity of toxic substance. There may also be accumulation of toxic substances by interference with regular non-toxic metabolic pathways by substrate competition for enzyme, *e.g.* ethanol and trichloroethylene, or depletion of substrates used to metabolize the toxins or prevent toxic injury, *e.g.* glutathione.

15.2 THE ANATOMY OF THE LIVER

Two blood supplies serve the liver, the portal vein, which delivers 75% of the blood supply and the hepatic artery, which delivers the rest. The portal vein drains the gastrointestinal tract, spleen and pancreas and supplies nutrients and some oxygen. The hepatic artery supplies fully oxygenated blood to the liver.

The blood that enters the liver by the hepatic artery and portal vein flows through sinusoids (see Figure 15.1). Sinusoids are specialized capillaries with large pores, which allow large molecules to pass through into 'interstitial space' and into close contact with hepatocytes (liver cells); therefore foreign compounds are taken up very readily by hepatocytes. The liver is composed of hepatocytes arranged as plates: each plate is bounded by a sinusoid. The membranes of adjacent hepatocytes form the bile caniculi (tubes) into which bile is secreted. The bile caniculi form a network and feed

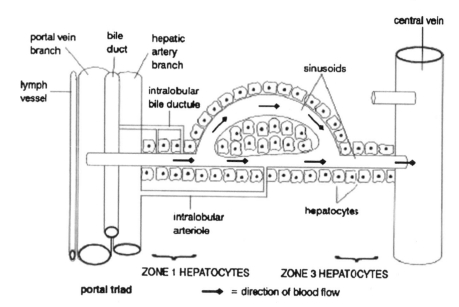

Figure 15.1   *The anatomy of the liver (This figure has been adapted with kind permission of the publisher from Figure 7.6 in J.A. Timbrell, Principles of Biochemical Toxicology. Taylor & Francis, London, 2nd edn, 1992)*

into ductules, which become bile ducts. Bile serves to excrete compounds from the body and aids in the digestion of food in the gastrointestinal tract.

Hepatocytes near the portal tract branches (zone 1) receive blood that is rich in oxygen and nutrients, but those near the hepatic vein branches (zone 3) receive blood that has lost much of its nutrients and oxygen. Therefore zone 3 is particularly sensitive to damage from toxic compounds. Zone 3 cells also have a higher level of some metabolic enzymes and higher lipid synthesis than zone 1, which may also explain why zone 3 tends to be the most damaged and why lipid accumulation is a common response to this damage (see Section 15.4.1).

However, 2-propen-1-ol causes zone 1 necrosis partly because this is the first area exposed to the compound in the blood and partly as a result of the presence of the enzyme alcohol dehydrogenase in zone 1, which produces reactive toxic metabolites.

15.3   MECHANISMS OF CELLULAR INJURY

Toxic substances can damage cells in target organs in a variety of ways. The eventual pattern of response may be reversible injury or an irreversible change leading to the death (necrosis) of the cell or perhaps to carcinogenesis. The toxic substance, or its metabolites, may cause cell damage by a variety of mechanisms, one or several of which may occur, and some of which are considered below.

15.3.1   Covalent Bonding

As far as free radicals, other reactive intermediates may be produced by metabolism. These can interact with proteins and other macromolecules binding covalently to them. Studies have demonstrated a correlation between the amount of binding and tissue damage, though this may merely reflect production of other damaging species. Binding to critical sites on proteins could, however, alter their function by, for example, inhibiting an enzyme or damaging a membrane, but binding could be to non-critical sites, and therefore be of no toxicological importance.

15.3.2 Lipid Peroxidation

Free radicals arise by cleavage of a covalent bond in a molecule, the addition of an electron or by the taking of a hydrogen atom from another radical. They have an unpaired electron centred on a carbon, nitrogen, sulphur or oxygen atom and hence are extremely reactive, electrophilic species, which can react with a variety of cellular components. Lipid peroxidation is caused by the attack of a free radical on unsaturated lipids, particularly polyunsaturated fatty acids found in cell membranes, the reaction being terminated by the production of lipid alcohols, aldehydes or malondialdehyde. Therefore there is a cascade of peroxidative reactions, which leads to the destruction of lipid unless stopped by a protective mechanism or a chemical reaction such as disproportionation, which gives rise to a non-radical product.

The structural integrity of membrane lipids will be adversely affected leading to alterations in fluidity or permeability of membranes, destabilization of lysosomes (intracellular organelles), and altered function of the endoplasmic reticulum and mitochondria. Such mechanisms are thought to be involved in liver damage caused by carbon tetrachloride and white phosphorus.

15.3.3 Thiol Group Changes

Glutathione is responsible for cellular protection. If depleted, a cell is made more vulnerable to toxic substances. Reactive intermediates of toxic substances can react with glutathione either by a direct chemical reaction or by a glutathione transferase-mediated reaction. If excessive, these reactions can deplete cellular glutathione and leave essential proteins vulnerable to attack by oxidation, cross-linking, formation of disulfides or covalent adducts.

15.3.4 Enzyme Inhibition

Sometimes inhibition of an enzyme may lead to cell death, for example cyanide inhibits cytochrome aa_3 leading to blockage of cellular respiration. This results in depletion of intracellular adenosine triphosphate (ATP) and other vital endogenous molecules. ATP is produced by mitochondria and is the main energy source within the cell.

15.3.5 Ischaemia

Reduction of oxygen or nutrients supplied to cells results in cell damage and eventual cell death if prolonged. Ischaemia may be a secondary event due to swelling of cells with reduction of blood flow. For example, phalloidin (a toxin from toxic mushrooms) causes centrilobular necrosis (see Section 15.4.2).

As a result of such mechanisms of damage as described above and the inability of the cell to compensate for changes in membrane structure and permeability, changes in subcellular skeleton and organelles may occur. This may result in ATP (energy) depletion, changes in Ca^{2+} concentration, damage to intracellular organelles and deoxyribonucleic acid (DNA) and stimulation of apoptosis (programmed cell death) may occur.

15.3.6 Depletion of ATP

Depletion of ATP may be caused by many toxic substances, usually by the uncoupling of mitochondrial oxidative phosphorylation or by DNA damage, which causes activation of poly (ADP-ribose) polymerase (where ADP = adenosine 5'-diphosphate). Depletion of ATP in the cell means that active transport in and out of the cell is altered or stopped and changes in electrolytes, particularly Ca^{2+}, lead to changes in biosynthesis within the cell such as protein synthesis, production of glucose and lipid synthesis.

A very important mechanism of cellular damage is alteration of the intracellular Ca^{2+} concentration. Changes in the intracellular distribution of this ion have been implicated in the cytotoxicity of many toxic substances including carbon tetrachloride. Interference with Ca^{2+} homeostasis may occur as a result of inhibition of Ca^{2+} adenosine triphosphatases (ATPases, which are responsible for the transport of Ca^{2+} across cell membranes), direct damage to the plasma cell membrane allowing leakage of Ca^{2+} or depletion of intracellular ATP.

15.3.7 Damage to Intracellular Organelles

Damage to intracellular organelles can result from the above mechanisms of injury, for example, carbon tetrachloride damages both smooth and rough endoplasmic reticulum leading to disruption of protein synthesis of the whole cell. Mitochondrial damage may occur, for example, after exposure to hydrazine, leading to functional changes and rupture of mitochondria. The mitochondria are crucial to the cell, and inhibition of their electron transport chain leads to rapid cell death.

15.3.8 DNA Damage

DNA damage may result from compounds such as alkylating agents, *e.g.* dimethyl sulphate can cause single-strand breaks in DNA resulting in the activation of poly (ADP-ribose) polymerase, which catalyses post-translational protein modification and is involved in polymerization reactions and DNA repair. Severe DNA damage may result from its activation and be sufficient to lead to cell death or carcinogenesis.

15.3.9 Apoptosis

Apoptosis is programmed cell death. Some foreign compounds may stimulate such cell death by, for example, the influx of calcium into a cell. In other cases cell death may be mediated by cytokines (*e.g.* interleukin 6); chemicals produced by activated white blood cells capable of mediating tissue injury.

15.4 PATTERNS OF RESPONSE TO INJURY IN THE LIVER

The resulting pattern of liver damage from the above mechanisms reflects the fact that the liver's response to injury is limited and includes fatty liver, necrosis (cell death), cholestasis, cirrhosis and carcinogenesis.

15.4.1 Fatty Liver

This is the accumulation of triglycerides (fats) in the liver cells. Fatty liver is a common response to toxicity, often occurring as a result of interference with protein synthesis in hepatocytes, such as after exposure to hydrazine (NH_2–NH_2). Normally, it is a reversible process that does not lead to cell death although it can occur in combination with liver cell death (necrosis) as is the case with carbon tetrachloride exposure. Repeated exposure to compounds that cause fatty liver, such as ethanol, may lead to cirrhosis.

15.4.2 Necrosis (Cell Death)

This may occur by direct cell injury, with disruption of intracellular function, or by indirect injury by immune-mediated membrane damage. As mentioned previously, 2-propen-1-ol causes periportal (zone 1) necrosis partly because alcohol dehydrogenase is present in zone 1 and partly because this is the first area exposed to the compound in the blood.

Conversely carbon tetrachloride and bromobenzene cause zone 3 (centrilobular) necrosis as a result of metabolic activation in that region. The hepatotoxicity of carbon tetrachloride has been extensively studied. It is a simple molecule, which upon exposure is distributed widely throughout the body, but despite this, its major toxicity is to the liver because it is dependent on metabolic activation by the P_{450} cytochrome system (a haem-like molecule with maximum absorbance at 450 nm: previously called mixed-function oxidase). Carbon tetrachloride is metabolically activated by dechlorination to yield a free radical as shown in Figure 15.2.

The carbon tetrachloride first binds to cytochrome P_{450} and then receives an electron from the reduced form of nicotinamide-adenine dinucleotide phosphate (NADPH) cytochrome P_{450} reductase. The enzyme–substrate complex then loses a chloride ion and the free radical intermediate is generated. This may then react with oxygen or take a hydrogen atom from a suitable donor to yield a secondary radical, or react covalently with lipids or protein.

Chronic exposure causes liver cirrhosis or tumours, particularly in the centrilobular area, which contains the greatest concentrations of P_{450} cytochrome. The production of reactive metabolites appears to start a cascade of damage, such as lipid peroxidation. Low-dose exposure may cause only fatty liver.

Mid-zonal (zone 2) necrosis is less common than the other two types of necrosis, but occurs with beryllium toxicity.

The explosive trinitrotoluene (TNT) can cause massive liver necrosis involving all zones. Ischaemia (impaired blood supply to the liver) may also contribute to necrosis, for example phalloidin, a toxic substance present in poisonous mushrooms, may cause swelling of the cells lining the sinusoids (Figure 15.1) and therefore reduce the oxygen and nutrients supplied to hepatocytes.

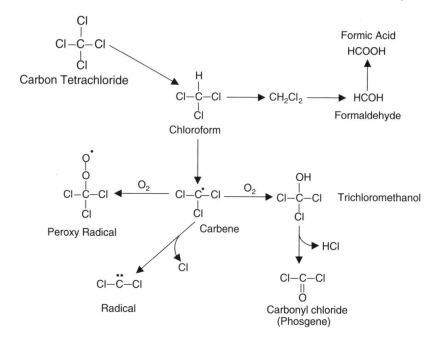

Figure 15.2 *The metabolic activation of carbon tetrachloride*

15.4.3 Cholestasis (Impaired Bile Flow)

Bile duct injury may result from exposure to a number of compounds, particularly those that are concentrated in bile, *e.g.* α-naphthyl isothiocyanate. The result of the damage will be cholestasis due to debris from necrotic cells blocking the ductules. Because of the close interrelation between bile ducts and hepatocytes this may also be accompanied by damage to hepatocytes by the build-up of bile, which in excess, damages cell membranes.

15.4.4 Cirrhosis

Cirrhosis is characterized by regenerative nodules (clumps of new cells) within fibrotic tissue (amorphous tissue) forming an irregular lobulated (defined) pattern. Any repetitive injury resulting in cell necrosis followed by repair mechanisms may lead to cirrhosis. This is because the liver has only a limited capacity to regenerate.

In addition, compounds that do not appear to cause acute necrosis such as ethionine may cause cirrhosis after chronic exposure.

15.4.5 Carcinogenesis

Liver tumours may be benign, *i.e.* grow *in situ*, or malignant, *i.e.* able to metastasize to other tissues. They may arise from any cell type within the liver, *e.g.* adenoma (aflatoxin B_1), hepatocellular carcinoma (dimethylnitrosamine) and haemangiosarcoma (vinyl chloride).

Among the various mycotoxins (toxins produced by moulds on nuts, oil seeds and grains) the aflatoxins have been the subject of intensive research because they are potent carcinogens. Aflatoxin B_1 is a very reactive compound. Its carcinogenicity is associated with its biotransformation to a highly reactive, electrophilic oxide, which forms covalent, adducts with DNA, ribonucleic acid (RNA) and protein (Figure 15.3). Damage to DNA is thought to induce tumour growth. There may be species differences due to differences in biotransformation and susceptibility to the initial biochemical lesion.

Dimethylnitrosamine is evenly distributed throughout the body, but exposure to single doses causes centrilobular hepatic necrosis indicating that metabolism is an important factor in its toxicity. One metabolite is a highly reactive alkylating agent that will methylate nucleic acids and proteins. The degree of methylation of DNA *in vivo* correlates with the risk of tumour induction in those tissues.

Vinyl chloride is the starting point in the manufacture of polyvinyl chloride. Chronic exposure leads to a 'vinyl chloride disease', which includes skin changes, changes to the bones of the hands and liver damage.

Haemangiosarcoma is a tumour of sinusoidal cells (not hepatocytes) that may also result from chronic exposure to vinyl chloride. This again appears to occur because the epoxide intermediate and fluoroacetaldehyde bind to DNA and proteins, respectively within the cell. Haemangiosarcoma has also been associated with arsenite exposure although the mechanism of this is still unclear: it should be stressed that the experience in humans with carcinogenicity has not been replicated in animal models.

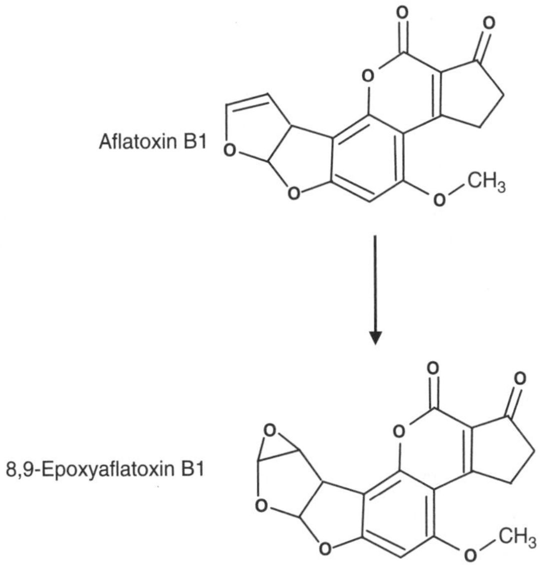

Figure 15.3 *Biotransformation of aflatoxin B_1*

15.4.6 Veno-Occlusive Disease

Rarely, a toxin may damage sinusoids and endothelial cells directly, for example monocrotaline, a plant alkaloid, which is metabolized to a reactive molecule causing damage and blockage of the venous return to the liver and secondary ischaemic death of hepatocytes.

15.4.7 Proliferation of Peroxisomes (Micro Bodies)

Peroxisomes are organelles found in many cell types but particularly in liver cells, and they take part in the metabolism of lipids predominantly. Exposure to certain substances leads to an increased number of peroxisomes and this may lead to an unwanted increase in the activities of various degradative enzymes.

15.5 DETECTION OF LIVER DAMAGE

Measurement of the plasma activities of the enzymes aspartate transaminase, alanine transaminase, alkaline phosphatase and γ-glutamyl transpeptidase (GGT) is a common way of detecting liver damage in live animals or humans. They are intracellular enzymes that are released when liver cell death occurs. Measurement of the plasma activity of alkaline phosphatase or GGT may reflect damage to biliary cells as they contain large quantities of these enzymes. Liver function may be estimated by measuring the hepatic clearance (uptake from blood) of a dye such as sulfobromophthalein, although this is outmoded now as a routine biochemical investigation of liver disease.

In pathology, damage to the liver can be detected by light or electron microscopy of liver sections. In animals or humans, liver biopsy may be undertaken to look at the pattern of liver injury. It is of limited value unless the potentially toxic substances are known, as the pattern of response to injury is diagnostically limiting within the liver.

15.6 CLINICAL PROBLEMS RESULTING FROM LIVER DAMAGE

Liver damage may be silent or the following may occur:

Jaundice. Excess quantities of bilirubin (a pigment produced by the liver from the degradation of haem in the red blood cells) is an indicator of hepatic cell death or dysfunction. The individual has yellow skin and eyes, which feel itchy.

Portal hypertension. This is due to cirrhosis, and is characterized by an enlarged spleen and diversion of blood flow from the liver to other areas because of resistance to flow through the scarred liver. In advanced cases individuals may vomit blood or have marked fluid retention causing swelling of the abdomen (ascites).

Clotting difficulties. The liver is an important producer of clotting proteins making up the clotting cascade. Therefore clotting difficulties may result from liver cell necrosis or cirrhosis and are seen as bleeding or bruising from the skin, gums or gastrointestinal tract.

Other features. In humans, anorexia, nausea or vomiting may be noted in the presence of liver damage.

In general, clinical features in humans do not correlate particularly well with the type of liver damage and thus biopsies may need to be undertaken to establish the pattern of liver injury. However, in general an individual with a fatty liver may not have symptoms and may be discovered only by increased plasma activities of aspartate transaminase or alanine transaminase (ALT). Necrosis may be asymptomatic if involving only a few liver cells, although more extensive damage may lead to marked jaundice, clotting difficulties and coma. Cirrhosis may be manifested by the features of portal hypertension. Tumours may present with jaundice and weight loss.

BIBLIOGRAPHY

A.L. Jones, in *Clinical Management of Poisoning and Drug Overdose*, 4th edn, Vol 11, M. Shannon, S. Borron and M. Burns (eds), Elsevier, USA, 2004.

D. Pessayre, D. Larrey and M. Biour, in *Oxford Textbook of Clinical Hepatology*, Vol 17, J. Bircher, J-P. Benhamou, N. McIntyre, M. Rizzetto and J. Rodes (eds), Oxford University Press, England, 1999, 1261.

M. Ruprah, T.K.G. Mant and R.J. Flanagan, Acute carbon tetrachloride poisoning in 17 patients, *Lancet*, 1985, **1**, 1027–1029.

J.A. Timbrell, *Principles of Biochemical Toxicology*, 3rd edn, Taylor & Francis, London, 1999.

Chapter 16

Nephrotoxicity

ROBERT F.M. HERBER

16.1 INTRODUCTION

Nephrotoxicity was defined in 1991 by the World Health Organization as a renal disease or dysfunction that arises as a direct or indirect result of exposure to medicines and industrial or environmental chemicals.

16.2 PHYSIOLOGY

Human kidneys are paired, bean-shaped organs. Each kidney weighs 120–170 g, and in the adult measures approximately $11 \times 6 \times 2.5$ cm. The total kidney mass correlates with the body surface area. The kidney consists of the cortex and the medulla (Figure 16.1). The human kidney is a multilobar organ containing 4–18 pyramids of medullary substance and is situated so that their bases are adjacent to the cortex. In humans, 180 L of fluid are filtered into the tubular lumen every 24 h of which 178 L must be returned to the blood. Each human kidney contains approximately one million functional units called nephrons (Figure 16.2). Each nephron is made up of a renal corpuscle, the glomerulus, and of a complex tubular portion, which drains into a unifying tubular system, the collecting duct system. The tubular portion consists of a proximal and distal tubule, collectively referred to as the renal tubule. A connecting tubule lies between the nephron and the collecting ducts.

The renal corpuscle (first segment of the nephron) is the site at which an ultrafiltrate of the blood is produced. The filtrate moves from the capillary lumen into Bowman's space. This flow is influenced by the following factors: renal blood flow; the colloidal osmotic pressure (part of the osmotic pressure due to the presence of large protein molecules) and the hydrostatic pressure in the capillaries and in Bowman's space, the size, shape, and charge of the serum molecules; and the various morphologic components of the wall separating the capillary lumen from Bowman's space. The filtrate contains only barely detectable quantities of serum proteins. The filtration barrier increasingly restricts the passage of larger molecules, with very little filtration of the molecules larger than albumin. In the renal corpuscle, a series of narrow slits between pediculated cells provide a long extracellular path for filtration of water and solutes. The collecting ducts are the final regulators of fluid and electrolyte balance, playing important roles in the handling of Na^I, K^I, chloride, Cl^-, acids, and bases.

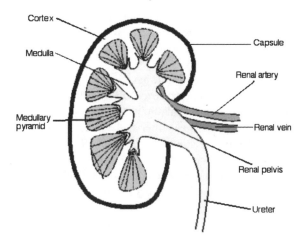

Figure 16.1   *Diagram of the kidney*

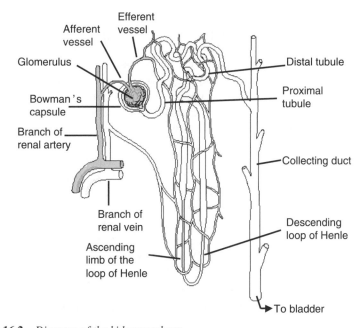

Figure 16.2   *Diagram of the kidney nephron*

The capability of the kidneys in achieving their homeostatic function, *i.e.* the maintenance of the internal environment within tolerable limits, is optimized by an intricate microvascular system that adjusts vascular resistance to maintain an appropriate intrarenal haemodynamic environment. Essentially, all the renal blood flow traverses through the glomerular capillaries, where about 20% of the serum is filtered. The volume and composition of the body fluids are regulated through independent but coordinated homeostatic mechanisms that control solute and water excretion.

Both the volume (50–60% of the body weight) and the osmolality (osmolality measures the concentration of particles in solution) of body water are maintained within a narrow range. Water intake from all sources (liquids, water in solid foods, and metabolic water of oxidation) is typically 1.5–3.0 L day^{-1}. The chief causes of water loss are evaporation, sweating, faeces, and urine. Sodium is the primary extracellular cation and, as such, is of critical importance to the maintenance of extracellular volume. The kidney is the dominant organ regulating the excretion of sodium ions. Each day, approximately 24,000 mmol L^{-1} Na^I is filtered, the kidneys normally absorbing over 99%, which they do with remarkable precison, 99.0–99.6%.

The most abundant cation in intracellular fluid is K^I, which is critical for a variety of cell functions. Several overlapping regulatory mechanisms serve the important functions where potassium is involved by maintaining the total body potassium content (50–55 mmol kg^{-1} body weight) and by partitioning potassium between extracellular and intracellular fluid. Variations in potassium intake are compensated for by adjustments in K^I excretion over a period of hours, or sometimes days. This is referred to as the external potassium balance. Most of the adaptive variation in excretion is accomplished by the kidney. Regulation of external potassium balance is primarily a function of the kidney. Even though the dietary potassium intake is approximately equal to the sodium intake (80–120 mmol day^{-1}), the concentration of potassium in serum, and therefore the rate at which potassium is filtered by glomeruli, is only 3% of that of sodium. Because many processes within the body are dependent on the pH of the ambient fluid, it is important to maintain intracellular and extracellular pH within narrow limits. The extrarenal generation of acid by metabolism is referred to as the net acid production.

To maintain acid–base balance, the kidney must excrete an amount of acid, referred to as the net acid excretion, equal to the net acid production. Animal diets tend to be rich in meat containing sulfur and phosphorus, and nucleic acids, causing them to be acid. Vegetarian diets tend to contain more organic anions, causing them to be more alkaline. In a typical adult with an glomerular filtration rate (GFR) of 180 L day^{-1} and a serum hydrogen carbonate concentration of 24 mmol L^{-1}, the glomerulus will filter 4320 mmol L^{-1} hydrogen carbonate day^{-1}. This is an enormous amount of hydrogen carbonate, and loss of any significant fraction of this filtered hydrogen carbonate in the urine will lead to severe metabolic acidoses. Thus, the first function of the renal tubules in acid–base regulation is the reclamation of filtered hydrogen carbonate. As hydrogen-secretory mechanisms in the kidney cannot generate steep pH gradients, buffering is achieved mainly by an NH$_3$/NH$_4^I$ buffer system.

The kidneys play a principal role in the conservation of many substances essential to the body namely, the secretion of several physiologically active compounds, and the maintenance of acid–base balance by producing ammonia. In doing this there is a high oxygen consumption by the kidney, which is only exceeded by that of the heart.

The kidney transports a large number of organic and non-organic solutes, and water. With a normal diet, urea is the most abundant urinary solute, and its concentration accounts for the largest fraction of 'solute-free' water re-absorbed by the kidney, about 75% of the total, see Table 16.1, or in other words, most of the 'osmotic work' achieved by the kidney. Thus, the kidney transforms large quantities of 'urea-poor'

Table 16.1 *Concentration of different solutes in plasma and urine, their daily excretion, and their relative concentration in urine with respect to serum*

Solute	Serum concentration (mmol L^{-1})	Urinary concentration (mmol L^{-1})	Excretion (mmol day^{-1})	Urinary/serum ration
All solutes	296	590	825	1.99
Na^I	142	93	133	0.65
K^I	4.1	51	70	12.4
Urea	6.2	270	380	43.5
NH$_4^I$	0.03	20	28	670

serum into a small volume of 'urea-rich' urine, with only modest changes in sodium concentration. Two hormones involved in the control of urea excretion are corticosterone and glucagon. Factors enhancing urea excretion are water diuresis, and a meat meal or protein diet.

16.3 HORMONE EFFECTS

Glomerular and tubular functions of the kidney are regulated by an intricate network of circulating and locally produced hormones, including proteins, peptides, lipids, nucleosides, steroids, and amino-derived molecules. Not only is the kidney the target of hormone action, but it also synthesizes, modifies, and secretes hormones that have an influence on non-renal functions (Table 16.2).

Furthermore, the kidney plays an important role in the clearance and inactivation of peptide and protein hormones. All peptide hormones are excreted by the kidney, which on average removes between 16 and 40% of peptide hormones entering the renal circulation. In many cases, the liver also contributes significantly to peptide hormone metabolism. A few hormones, however, undergo negligible hepatic extraction and the kidneys are their predominant site of degradation, *e.g.* calcitonin, and C-peptide, the inactive fragment of proinsulin.

16.4 HANDLING TOXICANTS

All animal species produce endogenous organic compounds and are exposed to environmental xenobiotics of natural and man-made origin, including drugs. Many of these compounds are weak acids or bases, ionized at physiological pH or become ionized after biotransformation. A large number of xenobiotics are first hydrolyzed by microsomal enzymes in the liver, and to a lesser extent, the kidney; this is followed by glucuronidation, acetylation, glutathione, glycine, or sulfate conjugation. The kidney is endowed with transport mechanisms allowing rapid excretion of these ionized compounds, a property that is useful in dealing with toxic products, but undesirable in handling therapeutic drugs. Some compounds are almost completely removed during a single pass through the kidney. Metabolites, usually more polar than the parent molecules, are in principle

Table 16.2   *Examples of hormones affecting, and produced by, the kidneys*

Hormone(s)	Effect or function
Vasopressin (AVP) or antidiuretic hormone	Maintains serum osmolality by water excretion Regulates blood pressure and response to stress
Atrial natiuretic peptide (ANP)	Increases urinary Na^I output and urine volume Decreases systemic blood pressure
Urodilatin	Action similar to ANP Stimulates hyperfiltration, diuresis, and natiuresis
Corticosteroids	Modulate renal function Glucocorticoids effect GFR, tubule production of NH_4^I and ion transport Mineralcorticoids regulate Na^I, K^I, and hydrogen ion excretion
Catecholamines, primarily adrenaline, noradrelaline, and dopamine	Regulate renal blood flow, GFR, renin secretion, and tubular transport
Kinins (includes kallikrein/kinin system)	Vasodilator peptides Modulate renal excretion of salt and water Control blood pressure
Endothelium derived relaxing factor (EDRF), nitric oxide (NO)	May be one and the same thing Produced by renal endothelial cells Regulate renal haemodynamics and glomerular function
Cytochrome P450	Wide range of metabolites produced by kidneys Effect tubular re-absorption of water and ions Moderate renin/angiotensin axis Vasoactive
Endothelin (ET)	Potent, long-lasting vasocostrictor produced by kidneys
Fatty acids derivatives, eicosanoids, prostaglandins, lipoxygenase products, leukotrienes	Produced in the kidneys Wide range on hormonal actions
Adenosine	Produced in the kidneys Wide range of regulatory functions
Parathyroid hormone (PTH)	Produced by the parathyroid glands Enhances kidney re-absorption of calcium
Erythropoietin	Produced in the kidneys Stimulates bone marrow production of blood cells

excreted more efficiently, but there are exceptions to this rule. As many of these compounds are extensively bound to serum proteins, their excretion by the kidney is often achieved more by tubular secretion, than by filtration. Typically, those drugs listed in Tables 16.3 and 16.4 are excreted by the human kidney. Many of these have different nephrotoxic effects.

Table 16.3 *Selected anionic drugs excreted by the human kidney*

Acetazolamide	Pantothenate
Cephalosporins	Penicillins
Chlorothiazide	Phenylbutazone
Diatrizoate	Probenecid
Enalaprilat	Salicylate
Furosemide	Sulfinpyrazone
Iodipamide	Sulfonamides
Methotrexate	Zomepirac
Niconitate	

Table 16.4 *Selected cationic drugs excreted by the human kidney*

Adrenaline (Epinephrine)	Neostigmine
Amiloride	Noradrenaline (Norepinephrine)
Atropine	Procainamide
Cimetidine	Procaine
Dopamine	Quinine
Histamine	Thiamine
Isoproterenol	Tolazoline
Levamisole	Triamterene
Morphine	

16.5   TOXIC NEPHROPATHIES

As the kidney concentrates and excretes metabolic waste, chemicals, and drugs, it is easily exposed to toxic concentrations of these substances. The term 'toxic nephropathies' refers to renal disorders produced by a broad range of drugs, diagnostic agents, and chemicals. Nephrotoxic substances may cause injury at a number of sites along the nephron and produce characteristic clinical syndromes. Table 16.5 gives some examples of groups of compounds that can produce nephrotoxic effects.

16.5.1   Proximal Tubule Injury

Although proximal tubule injury is a common manifestation of nephrotoxic agents such as aminoglycosides, *cis*-platin, and metals, proximal tubular dysfunction without tubular necrosis also may be an important aspect of nephrotoxicity. Histopathologic studies reveal early non-lethal dysfunction, such as mitochondrial swelling, blebbing of the endoplasmic reticulum, and sloughing of portions of the plasma membrane. With more severe injury, tubular cell death, intratubular plugging with cellular debris, and interstitial oedema develop leading to a reduction in renal blood flow and a marked fall in GFR.

16.5.2   Renal Medullary Injury

Nephrotoxicity may be persistent without histologic evidence of proximal tubular necrosis. Thus, other segments can be the focus of the injury of some toxins. With a

Table 16.5 *Compounds with potential nephrotoxic effects*

Penicillins and cephalosporins: vancomycin, sulfonamides, amphotericin, acyclovir, ganciclovir, pentamidine, foscarnet
Radiocontrast agents
Drugs used in transplantation: cyclosporine, FK-506
Antineoplastic drugs: *cis*-platin, carboplatin, cyclophosphamide, streptozotocin, semustine, carmustine, ifosfamide
Antibiotics: mitomycin C, mithramycin
Antimetabolites: methotrexate, cytosine arabinoside, 6-thioguanine, 5-fluorouracil
Other agents: interleukin-2
Metals: lead, cadmium, mercury, uranium, others
Antirheumatic agents: penicillamine, gold, NSAIDs

mild nephrotoxic insult, impaired urinary concentration, due to medullary impairment, is often the first and sometimes the only apparent injury. Cells of the medullar tubule are at risk from exposure to polyene antibiotics, cyclosporine, and radiocontrast agents.

16.5.3 Intratubular Obstruction

Agents that have a low solubility in tubular urine when given in high doses may precipitate within the nephron and obstruct the flow of urine, leading to a reduction of GFR.

16.5.4 Distal Tubule Dysfunction

Agents that interfere with the renin–angiotensin–aldosterone axis produce hyperkalemia. These abnormalities include impaired production of renin, reduced production of aldosterone, and tubular insensitivity to the action of aldosterone.

16.5.5 Oxygen-Free Radical Production

Evidence for the participation of free radicals in nephrotoxic injury includes aminoglycoside antibiotics, cyclosporine, and radiocontrast agents.

16.6 METAL TOXICOLOGY

Generally, the absorption of excessive amounts of metals leads to a condition resembling the Fanconi syndrome with broad inhibition of proximal tubular solute transport. Because of their high reactivity with proteins and other ligands, metals do not exist *in vivo* as free ions, but renal uptake, metabolism, and excretion of metals depend on the renal handling of the metal–ligand complex. Many ligands are naturally found in biological systems; their number includes low molecular mass molecules like glutathione and citrate, as well as larger compounds, such as polypeptides and proteins. In addition, exogenous chelating compounds may be introduced via the food chain, or through therapeutic prescribing. Metals enter proximal tubular cells by endocytosis following the binding of the metal or a metalloprotein complex to the brush-border membrane. Entry is followed by intracellular release of the metal from the membrane or protein–metal complex by lysosomal degradation. Low doses of certain metals cause leaking of glucose and amino acids (renal tubular acidosis) associated with increased diuresis. Then renal tubular necrosis occurs and this may lead to renal shutdown, raised blood urea, and finally death. Necrosis is thought to be due to a combination of ischaemia, caused by vasoconstriction, and direct cytotoxic action.

16.6.1 Cadmium

Exposure to the non-essential element cadmium takes place both in the general environment and in occupation. In the general environment, the oral intake via food or water is the most important intake route. Occupational uptake is either directly via the lungs, or indirectly via hand–mouth contact (orally).

Dysfunction of the proximal renal tubules is characteristic of chronic but not acute Cd^{II} exposure in both humans and experimental animals. Acute doses of cadmium accumulate mainly in the liver, but chronic exposure leads to accumulation in the kidney. Cd–protein complexes accumulated by hepatic cells induce the synthesis of metallothionein, a low molecular mass, cysteine-rich, protein that binds cadmium avidly. This Cd–metallothionein complex is thought to be released, very gradually, from the liver and taken up by endocytosis into renal tubule cells, where free cadmium ions can be generated owing to lysosomal degradation of the complex. The critical cadmium concentration, when kidney damage occurs, is about $100 \, mg \, kg^{-1}$ of renal cortex. Cadmium in the kidney also induces the synthesis of metallothionein in the proximal tubules, and dysfunction probably reflects saturation of the available binding sites. An early effect of chronic exposure is the excretion in urine of the low molecular mass proteins such as β_2-microglobulin, α_1-microglobulin, and retinol binding protein, which are all indicators of effects on the tubules. Also, there is increased activity in urine of some tubular enzymes, *e.g.* N-acetyl-β-D-galactosidase (NAG). These early or subclinical effects are found to be adverse. Effects on the glomerulus, *e.g.* enhanced concentrations of albumin in urine, and IgA are also reported. Prolonged exposure at higher concentrations may lead to a diminished glomerular filtration ratio and a decreased kidney function, see Figure 16.3.

It has been found in the general population that environmental exposure to cadmium leads to osteoporosis and renal dysfunction, and that tubular damage is particularly related to osteoporosis. Finally, the International Agency for Research on Cancer (IARC) defines cadmium as carcinogenic to humans.

16.6.2 Chromium

Cr^{III} compounds are essential to humans, while Cr^{VI} compounds are non-essential. There is occupational exposure to Cr^{VI} in industrial processes such as welding, plasma, and industrial-laser processing of chromium-containing stainless steels. Cr^{VI} compounds are characterized by the IARC as (lung) carcinogenic to humans. Cr^{VI} is deposited in the lungs and minor amounts are reduced to Cr^{III} and occur in the bloodstream. Other sources of exposure are chrome smelting, tanning of leather, the concrete industry, and galvanizing in the chrome-plating industry. There are some indications that glomerular toxicity occurs following occupational exposure to chromium.

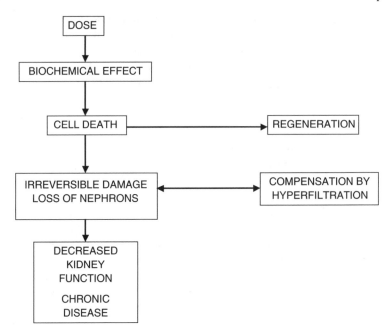

Figure 16.3 *Effect of exposure to xenobiotics on the kidney. The severity of effect increases from top to bottom*

16.6.3 Gold

Dietary gold intake is estimated as less than 7 µg day^{-1}, and the absorption of gold compounds from the gastrointestinal tract is low. About 70% of injected gold from drugs (chrysotherapy) passes from the kidney into the urine. From recent studies, proteinuria and nephrotic syndrome occur at treatment levels of 2–6 mg day^{-1} and above in about 17% of patients with rheumatoid arthritis treated with gold salts, *e.g.* auranofin, probably due to the presence of gold in the kidneys.

16.6.4 Lead

Exposure to the non-essential element lead occurs through environmental and occupational exposure. In general, environmental exposure is due to oral intake via food or water. Occupational uptake is either direct via the lungs, or indirect via hand–mouth contact. Lead as Pb^{II} is the most abundant nephrotoxic metal. Nephrotoxicity, however, is observed only after relatively high occupational exposure giving a blood concentration of 600 µg L^{-1} or more. Lead binds to high-affinity lead-binding proteins that are present in the kidney tubule cells in high concentrations. These proteins carry lead to the nucleus where *de novo* synthesis of a unique acidic protein results in metal precipitation to form lead intranuclear inclusion bodies.

Mitochondria are extremely sensitive to lead and, following chronic *in vivo* administration, mitochondrial swelling occurs. The mitochondria show impaired oxidative phosphorylation and thus the tubule cells are less able to re-absorb or secrete solutes and metabolites. Acute effects result a Fanconi-like syndrome in the proximal tubules, giving rise to aminoaciduria, glucosuria, and hyperphosphaturia.

An early effect of exposure to lead on the kidney is the presence of NAG in the urine and an enhanced concentration of the tubular proteins, retinol binding protein, and α_1-microglobulin. In a study of lead-smelter workers performed in the USA, an increased mortality rate was found due to chronic renal failure.

Inorganic lead is classified by IARC as probably carcinogenic to humans (Group 2A).

16.6.5 Mercury

The non-essential element mercury exists environmentally in inorganic and organic forms. Environmental and occupational exposure to inorganic mercury, *i.e.* Hg(II) is relatively common. Amalgam fillings (elemental Hg) are common sources of environmental exposure, moreover accidental exposure to elemental mercury due to broken thermometers and barometers is not unusual. Elemental mercury is oxidized to Hg(II) in air and within the body, and is distributed as such to the target organ, the kidney. Elemental mercury as mercury vapour, is oxidized in the blood within minutes. The highest mercury concentrations after exposure to inorganic mercury are found in the kidney. Early effects on the kidney are the enhanced activities in urine of the tubular enzymes NAG, γ-glutamyl transferase, β-galactosidase, and tubular antigens. In the past, cases were described of mercury(II) chloride causing cytotoxicity of the proximal tubules leading to oliguria, anuria, and tubular necrosis. Often occupational exposure to elemental mercury vapour leads to proteinuria, and less commonly to nephrotic syndrome. Sometimes a skin-lightening cream used by dark-skinned humans can cause this.

Although in the past, some indications of kidney toxicity from organic mercury were described, either the compound causing the problem has been banned from use (phenyl mercury), or the described effects were not confirmed.

Mercury and its inorganic compounds are not classifiable as to carinogenicity to humans (Group 3).

16.6.6 Platinum

Exposure to the non-essential element platinum occurs in the platinum-refining industry, and more severely, in the production and usage of organic platinum compounds, used as anti-cancer drugs, *i.e.* the antineoplastic drugs, *cis*-platin, and carboplatin. Tissue concentrations of platinum following therapy can be as high as 10 mg kg^{-1}. The therapeutic application of *cis*-platin is hampered by its high nephrotoxicity. Carboplatin is less nephrotoxic than *cis*-platin.

Nephrotoxic side-effects of *cis*-platin include increased plasma creatinine concentration, decreased glomerular filtration, enhanced enzyme activity in the urine and increased concentrations of alanine aminotransferase, of β-galactosidase, and of NAG, retinol binding protein, and albumin. In the kidneys of patients treated with *cis*-platin, areas of focal tubular necrosis, dilatation of the tubules, and cast formation are sometimes seen.

Cis-platin is classified by IARC as probably carcinogenic to humans (Group 2A).

16.6.7   Uranium

The non-essential element uranium's occurrence in the Earth's crust may lead to significant direct exposures above WHO guidelines, for example through drinking water from local wells filtered through granite. Dietary intake is the main contributor to environmental exposure-effective dose. With naturally occurring uranium, chemotoxicity is the dominant toxic effect, not radiotoxicity, causing renal failure.

The renal toxicity of uranium in man is caused by the precipitation of U(VI) in the proximal tubules in the process of clearance. The resulting tissue damage leads to kidney failure and the emergence of enhanced concentrations of proteins, glucose, and creatinine in the urine. Acute intoxication may lead to irreversible damage and to death due to renal dysfunction.

16.6.8   Other Inorganic Compounds

There are other inorganic compounds that can cause nephrotoxicity and result in increased urinary protein output and/or tubular enzyme activity, for example, arsenic(V) (retinol binding protein) and SiO_2 (albumin and α_1-microglobulin).

16.7   ORGANIC SOLVENTS

The nephrotoxicity of some halogenated hydrocarbons such as chloroform depends on the generation of phosgene by cytochrome P450 action. Phosgene is trapped by two molecules of glutathione to form diglutathione dithiocarbamate and undergoes further metabolism to 2-oxothiazolidine 4-carboxylic acid, which is excreted. Other haloalkenes may undergo metabolic activation by other routes.

Prolonged exposure to volatile hydrocarbons may lead to effects on the kidney (glomerulonephritis). Early renal effects of exposure to the following compounds have been described: tetrachloroethene, styrene, toluene, and mixtures of solvents. The effects reported are small and sometimes not related to the dose. Other variables such as shift work or heavy physical work may be responsible for the effect.

Balkan endemic nephropathy (BEN), a disease restricted to Balkan countries (Bulgaria, Rumania, Serbia, and Croatia), is characterized by a progressive shrinking of the kidneys and, in some cases, tumours in the proximal regions of the urinary tract. The disease may be related to drinking water polluted with organic compounds. It is not clear if these compounds are aliphatic or aromatic solvents or compounds like humic acids. Other studies point at mycotoxins or a virus as possible cause.

16.8   MYCOTOXINS

Mycotoxins may have profound effects on the fundamental biochemistry and physiology of the kidney. For example, ochratoxin A, a secondary metabolite of aspergillus and penicillium species, may be present in contaminated grain products,

Table 16.6 *Risk factors for renal cell cancer*

Aflatoxin B
Analgesics
Asbestos
Diethylnitrosamine
Dimethylnitrosamine
Diuretics
Oven operation
Printing press operation
Streptozotocin
Tobacco products

nuts, and in meat from animals eating contaminated feeds. In pigs, it produces porcine nephropathy, a disease characterized by degeneration and atrophy of proximal tubules, interstitial fibrosis, and hyalinization of glomeruli.

16.9 PESTICIDES

The herbicides diquat and paraquat are nephrotoxic. Acute kidney insufficiencies have been reported, but there are no studies on chronic exposure in humans. Nematicide 1,3-dichloropropene is suspected of being nephrotoxic in humans, but possibly the *trans* compound only, is responsible for the nephrotoxicity.

16.10 ANTINEOPLASTIC DRUGS

Antineoplastic drugs are a group of powerful cytotoxic drugs used in treating some cancers. Many have nephrotoxic effects that may be compounded by the use of other drugs. For example, methotrexate toxicity is increased by concomitant administration of non-steroidal anti-flammatory drugs (NSAID), ibuprofen, salicylates, and flurbiprofen. Currently, as NSAIDs are widely used, this kind of renal toxicity is encountered more frequently. *cis*-Platin is a cytotoxic drug and is nephrotoxic (see Section 16.6.6). Like other nephrotoxic metal ions such as mercury and chromate(VI), it is taken up by tubular cells as a sulfydryl conjugate.

In primates, uric acid (2,6,8-trihydroxypurine) derives exclusively from the degradation of purines. Uric acid has powerful free radical scavenger properties and may thus exert cell protective effects. Human kidneys absorb a large proportion of the filtered urate, potentially affording some protection.

16.11 RENAL CARCINOGENS

Renal cell cancer accounts for about 75% of all kidney cancer in adults. A number of risk factors for renal cell cancer have been identified including those listed in Table 16.6. All of the factors are associated with DNA damage and mutagenicity.

BIBLIOGRAPHY

M.W. Anders, W.K.D. Henschler and H.O.S. Silbernagl (eds), *Renal Disposition and Nephrotoxicity of Xenobiotics*, Academic Press, San Diego, 1993.

B. Ballantyne, T. Marrs and T. Syverson (eds), *General and Applied Toxicology*, Macmillan Reference & Grove's Dictionaries, London & New York, 1999, 2199 (text), 2154 (indexes).

A.M. Davison, J. Cameron Stewart, J.P. Grünfeld, D.N.S. Kerr, E. Ritz and C.G. Winearls, *Oxford Textbook of Clinical Nephrology*, 2nd edn, Oxford University Press, Oxford, 1998.

R.L. Jamison and R. Wilkinson (eds), *Nephrology*, Chapman and Hall, London, 1997.

C.D. Klaasen (ed), *Casarett and Doull's Toxicology – The Basic Science of Poisons*, McGraw-Hill, New York, 2001, 1236.

E. Merian, M. Anke, M. Ihnat and M. Stoeppler (eds), *Elements and their Compounds in the Environment*, 2nd edn, Wiley-VCH, Germany, 2004.

World Health Organization, Principles and Methods for the Assessment of Nephrotoxicity Associated with Exposure to Chemicals, Environmental Health Criteria 119, *International Programme on Chemical Safety*, World Health Organization, Geneva, 1991.

Chapter 17

Neurotoxicity

JOHN H. DUFFUS

17.1 INTRODUCTION

Neurotoxicity is the term applied to a toxic effect on any aspect of the central or peripheral nervous system. Effects may be functional (behavioural or neurological abnormalities), biochemical, physiological, or morphological. It has been suggested that over one-third of chemicals may be neurotoxic in some sense. Behavioural effects are considered later in Chapter 18.

17.2 THE NERVOUS SYSTEM

The central nervous system (CNS) is the main mass of nerve tissue lying between the effector and receptor organs, co-ordinating the nervous impulses between receptors and effectors. The CNS is present in vertebrates as a dorsal tube, which is modified anteriorly to form the brain and posteriorly to form the spinal cord; these organs are enclosed in the skull and backbone respectively. In higher organisms, the CNS takes on an additional activity in the form of memory, which is the storage of past experiences.

The peripheral nervous system (Figure 17.1) connects with the CNS and is usually subdivided into two parts, somatic and autonomic. The somatic nerves supply voluntary skeletal muscles, giving control of movement, and also supply skin sensory organs. The acetylcholine receptors in the somatic system are activated by nicotine and similar compounds.

The autonomic nervous system controls the involuntary activities of the body. There are two main parts to the autonomic nervous system: sympathetic nervous system (SNS) and parasympathetic nervous system (PNS).

17.2.1 The Sympathetic Nervous System

The main role of SNS is in enabling 'fight or flight' responses and in adaptation to stress. It is characterized by complete activation once it is triggered: virtually all structures under SNS control are affected when the SNS 'fires'. In more detail, the sympathetic system inhibits peristalsis, stimulates contraction of the sphincters of bladder and anus, inhibits bladder contraction, stimulates the heart pacemaker and speeds up the heart, stimulates arterial constriction, inhibits contraction of the bronchioles and permits their dilation, inhibits contraction of iris muscle, and permits dilation of the pupils.

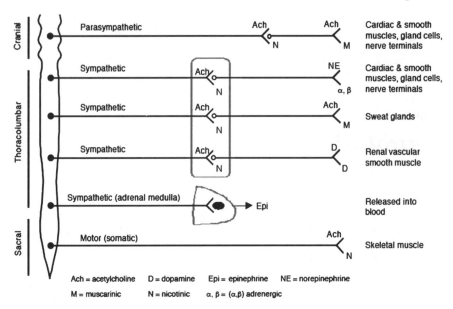

Figure 17.1 *The peripheral nervous system showing the location of nicotinic and muscarinic receptors*

In the SNS, complexes of synapses form ganglia alongside the vertebrae of the spinal cord. The preganglionic fibres from the CNS are therefore short and mostly adrenergic, synaptic transmission being mediated by noradrenaline (norepinephrine) except for the nerves activating the sweat glands. The acetylcholine receptors activating the sweat glands and pilo-erector muscles are muscarinic, *e.g.* activated by muscarine and muscarine-like compounds, but all the other acetylcholine receptors are nicotinic, *e.g.* activated by nicotine and similar compounds.

17.2.2 The Parasympathetic Nervous System

The PNS is mainly concerned with 'housekeeping' functions of the body, *e.g.* digestion and maintenance of various organ systems or their structures. When the PNS is activated, the activity of one or more structures that it controls can change without changing activity of all the structures to which it is connected. In detail, the parasympathetic system stimulates intestinal peristalsis, inhibits contraction of the sphincters of bladder and anus, stimulates bladder contraction, inhibits the heart pacemaker and slows down the heart, inhibits arterial constriction and permits dilation, stimulates contraction of the bronchioles, and stimulates contraction of iris muscle and thus of the pupil.

In the parasympathetic system, ganglia are embedded in the wall of the effector. Thus, the preganglionic fibres are long and the postganglionic fibres are short; these fibres are cholinergic, synaptic transmission being mediated by acetylcholine. The preganglionic acetylcholine receptors here are nicotinic and the postganglionic receptors are muscarinic.

It will be seen that the sympathetic and parasympathetic systems may innervate the same end organs, but the effects produced by the two systems generally oppose one another. The principal cells of nervous tissue are the neurons (Figure 17.2). Each neurone has a cytoplasm rich in granular endoplasmic reticulum, with granules of Nissl substance, carrying out active protein synthesis. Nissl is a chromophilic substance in the form of granules found in cell bodies and dendrites of neurons but not axons. Each neurone also has several short dendrites and an axon, which can be very long. The axon contains no granular endoplasmic reticulum but it contains neurofilaments and microtubules, which drive bidirectional movement of molecules and organelles. At the synaptic ending, the axon may endocytose neurotoxicants, viruses, and neurotransmitters, and allow them to enter the neurone to be transported to the cell body, the perikaryon.

In addition to the neurons, there are the glial cells, which include macroglia (astrocytes and oligodendrocytes) and microglia. Astrocytes are star-shaped cells, the processes of which are often in close relation to endothelial cells and also to the meningeal cells, which cover the CNS; a trophic, *i.e.* nutrient, and axonal guiding role have been attributed to them. They may also contribute to the blood–brain barrier. Microglial cells show macrophage activity. The neuronal axon is often in very close relation to oligodendroglial cells in the CNS or Schwann cells in the peripheral nervous system. Schwann cells are responsible for myelination. The myelin sheath produced by them speeds up the transmission of electrical signals along peripheral nerves.

17.3   THE BLOOD–BRAIN BARRIER

The blood–brain barrier is formed by the endothelial cells of brain capillaries, which differ from other endothelial cells. These endothelial cells are bound to each other by tight junctions, which prevent the entry of proteins and other water-soluble substances of low molecular mass. Some areas of the nervous system lack a blood–brain barrier; these areas are the choroid plexus, neurohypophysis, eminentia media, pineal gland, area postrema, and the subfornical organ (see the Glossary). In the brain, compartments filled with cerebrospinal fluid (CSF) communicate with the extracellular space. The CSF is secreted mainly by the choroid plexus. Among the cavities filled by the CSF are the ventricles, which are lined with ependymal cells (a variety of macroglia), and the subarachnoid space, which separates the pia mater from the arachnoid.

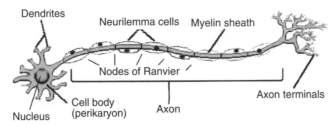

Figure 17.2 *Diagrammatic representation of a neuron*

Sensory and autonomic ganglia are not protected by the blood–brain barrier and so their neurons are particularly exposed to neurotoxicants. Peripheral nerves have a special protective barrier isolating them called the perineurlum. The perineurium is a connective tissue sheath; the inner layers of perineurial cells are connected by tight junctions, which prevent free movement of fluid and solutes. There is also a blood–nerve barrier provided by capillary and endothelial cells bound by tight junctions.

17.4   SPECIAL FEATURES OF THE NERVOUS SYSTEM

Cells of the nervous system have high energy requirements: they have glucose as their only substrate and are unable to use ketones. Thus, they are highly dependent on a good blood supply of both glucose and oxygen for oxidative phosphorylation. Although, many of the cells have blood–brain barrier protection, the nervous system extends throughout the body with a complex geometry and so there are many possible vulnerable areas. Further, the way in which the nerve impulse is transmitted chemically across extracellular space is particularly at risk from interference by toxicants.

Neurons are unusual in that they require an environment rich in lipids to be maintained in order to function properly. Thus, drugs that lower blood fat and particularly cholesterol, may adversely affect nerve function.

The inability of neurons to regenerate makes it important to prevent any neuronal death, as excessive loss of neurons over a lifetime may lead to dementia in old age.

17.5   TOXICITY TO THE NERVOUS SYSTEM

The main toxic injuries to which the nervous system is liable are neuronopathy, axonopathy, myelinopathy, carcinogenesis, developmental injury, and complex damage leading to behavioural abnormalities. Behavioural toxicology is discussed in Chapter 18.

17.5.1   Neuropathy

Neuropathy is defined as any functional disturbance and/or pathological change in the peripheral nervous system. Inflammatory damage may come within this definition, but it is more commonly called neuritis.

Neurons are highly susceptible to damage from lack of oxygen or glucose. Thus, they are highly sensitive to carbon monoxide, which prevents transport of oxygen by the formation of carboxyhaemoglobin. Cyanide and azide have similar effects, *i.e.* by blocking the use of oxygen by inhibition of cytochrome oxidase. Lack of oxygen may lead to death of neurons and permanent damage to the brain and nervous system.

There are various kinds of neuropathy. They are listed below together with some of the causative agents:

- Neuronal chromatolysis is the disappearance of Nissl substance from the cell body. It may be caused by methylmercury acting on dorsal root ganglion neurons, alkyl lead acting on brain stem neurons, and podophyllotoxin acting on spinal ganglion neurons. Ultimately the entire neurone dies.

- Neuronal atrophy and degeneration, acute toxic peripheral neuropathy, are caused by acrylamide (acute toxic peripheral neuropathy) and by trimethyltin.
- Neuronal swelling is caused by trimethyltin in the brain stem neurons and in the anterior horn cells of the spinal cord.
- Neurofibrillary degeneration is caused by eating cycad nuts (Parkinsonism – dementia of Guam), aluminium (in the cerebral cortex and hippocampus neurons), alkyl lead, and podophyllotoxin.
- Dendropathy, damage to the dendritic processes, is caused by ethyl alcohol, lead ions and mercury.
- Alteration of cell body function may be caused by alanosine, glutamate, kainic acid, and related chemicals, in very large doses. These changes affect areas of the CNS, which are not protected by the blood–brain barrier. The cell bodies are affected but the axons are not. The effects are both neurotoxic and neuro-excitatory. The toxicity may be mediated by nitrogen(IV) oxide.

17.5.2 Axonopathy

Proximal axonopathy is caused by β,β-iminodipropionitrile (IDPN), n-hexane, methyl n-butyl ketone; distal axonopathy by acrylamide; and general axonopathy by carbon disulfide, isoniazid, lead ions, leptophos, polychlorinated and brominated biphenyls, taxol, trichloroethylene, and triorthocresyl phosphate (TOCP). It should be noted that the symptoms, which could result from a single dose of TOCP, may not be apparent for 10–14 days after exposure.

IDPN causes proximal axonopathy, damage to the axon adjacent to the cell body. This follows from incorrect phosphorylation of the neurofilaments, which run along the axon, causing them to accumulate in the proximal region and the distal region to waste away.

TOCP and some other organophosphorus compounds, such as EPN (*O*-ethyl *O*-4-nitrophenyl phenylphosphonothioate) and leptophos, cause derangement of neurofibrillary structures, resulting in atrophy of the distal region of the axon, the part farthest away from the cell body. Thallium ions have a similar effect as a result of causing mitochondrial swelling and degeneration.

17.5.3 Myelinopathy

This is the destruction of myelin sheaths around the axons, and is caused by triethyltin, lead ions, acetylethyltetramethyl tetralin (AETT), cuprizone, cyanate, dichloroacetate, ethidium bromide, hexachlorophene, and isoniazid. Some drugs, such as triparanol, which was used to lower blood cholesterol but is now withdrawn, disrupt myelin as a result of interfering with normal cholesterol metabolism, a requirement for myelin sheath synthesis.

17.5.4 Toxic Degeneration of the Synapse

The following toxic actions may occur at the synapse:

- Inhibition of the action potential
- Blocking of ion channels, altering electrical transmission, and intracellular signalling

- Interference with synthesis of neurotransmitter
- Interference with storage, release, or uptake of neurotransmitter
- Interference with binding of neurotransmitter to a target cell receptor
- Interference with enzymic breakdown of excess or unused transmitter

Among the substances that produce such effects are anatoxins, batrachotoxin, botulinum toxin, cocaine, x-dendrotoxin, lead ions, nicotine, organophosphorus insecticides, pyrethroid insecticides, saxitoxin, scorpion toxins, tetanus toxin, and tetrodotoxin.

17.6   MECHANISMS OF PRODUCTION OF NEURONAL LESIONS

In general lipid-soluble compounds, as reflected by their octanol/water partition coefficient, enter the brain easily. This is why methylmercury, for example, is more neurotoxic than mercury(I) or (II) ions.

Mechanisms of chemical damage to the nervous system can be of any kind known to toxicology and so it is possible to list only some of the more common examples:

- *Methylmercury*. Methylmercury causes the disappearance of Nissl substance, see Section 17.5.1, through an action on protein synthesis. It also acts on –SH groups in membranes, altering permeability and transport systems. In the cerebral cortex this causes degeneration and necrosis of neurons, especially those dealing with vision.
- *Carbon monoxide*. Carbon monoxide binds reversibly, but more avidly than oxygen, to the iron in haemoglobin, myoglobin, and the cytochromes so that the oxygen-carrying capacity is reduced. Oxygen reaches the tissues less easily and its use in oxidative phosphorylation is inhibited. This inhibits neurone metabolism causing peripheral neuropathy, and by depriving the brain of oxygen causes loss of consciousness. Since dead neurons are not replaced, the effects of carbon monoxide can be cumulative although carbon monoxide is readily lost to the air once exposure ceases.
- *2,5-Hexanedione, carbon disulfide, and acrylamide*. These compounds inhibit some glycolytic enzymes directly by reacting with proteins to form pyrrole adducts.
- *Arsenic*. In the form of arsenic(III), arsenic binds to lipoic acid, which is involved in pyruvate decarboxylation. This causes symptoms similar to thiamine deficiency.
- *Thallium*. Thallium ion binds to mitochondrial membranes and causes similar symptoms to arsenic, resembling thiamine deficiency.
- *Aluminium*. Aluminium ion causes abnormal accumulation of neurofilaments in the perinuclear region of neuronal cell processes so that other cell organelles are displaced peripherally. Aluminium ion intoxication has been observed in patients undergoing renal dialysis with aluminium ion-containing dialysates and results in progressive dementia.
- *Tetrodotoxin*. Tetrodotoxin blocks sodium channels in the nodes of Ranvier causing paralysis by inhibition of peripheral nerve function. The nodes of Ranvier are small spaces within the myelin sheath, see Figure 17.2.
- *Botulinum toxin*. Botulinum toxin blocks the release of acetylcholine at the synaptic ending of the neurone.

- *Black widow spider venom.* Black widow spider venom causes continuous release of acetylcholine.
- *Snake venom bungarotoxin.* Snake venom bungarotoxin binds to acetylcholine postsynaptic receptors, blocking neurotransmission and causing paralysis.
- *Organophosphorus compounds.* These compounds cause distal axonopathies to occur in both the CNS and the peripheral nervous system following reduction in the number of microtubules in the cytoskeleton. This was first observed in neuropathies due to chronic exposure to organophosphorus compounds. Organophosphorus compounds also cause acute damage to the nervous system by phosphorylating and inhibiting the neurone membrane-bound protein, acetylcholine esterase, essential for the breakdown of acetylcholine, and therefore for the maintenance of normal neurotransmission.
- *Diphtheria toxin.* Diphtheria toxin causes lesions in Schwann cells and in oligodendrocytes by inhibiting the synthesis of the proteolipid of myelin and of the basic protein of the Schwann cell, and thus slowing conduction in the affected axons.
- *Triethyltin.* Triethyltin affects the myelin sheath, causing wide-spread oedema in the white matter of the brain. Hexachlorophene neurotoxicity is similar in effect. Care has to be taken in interpreting these effects when they are seen in experimental animals as they may be species specific. For example, AETT is used in perfumes and does not normally cause toxicity in humans; however, it causes vacuolation of myelin in experimental animals.

Some toxic substances show a degree of specificity for the part of the nervous system on which they act. For example, tetanus toxin specifically binds to the neuronal surface of the spinal cord motor neurone and may be used to distinguish neurons from non-neuronal cells. It inhibits the release of γ-aminobutyric acid and glycine, causing muscle spasms and rigidity. Some toxicants attack the blood–brain barrier itself, *e.g.* mercury. The critical organ for inorganic mercury is the kidney, see Section 16.6.5, but the main target for alkyl mercury is the nervous system and part of its effect is due to damage to the blood–brain barrier.

The mechanism of production of nervous system tumours is presumably very similar to that of tumours of other organs (see Chapter 9). Some of the chemicals known to induce such tumours in rats are N-methyl-N-nitrosourea, N-ethyl-nitrosourea, azoxyethane, 1,2-diethylhydrazine, 1-phenyl-3,3-diethyltriazine, acrylonitrile, and ethylene oxide. Whether these substances have similar effects in humans is a question still to be answered by epidemiological studies.

BIBLIOGRAPHY

B. Ballantyne, Y. Marrs and T. Syverson (eds), *General and Applied Toxicology*, 2nd edn, Macmillan, London & New York, 1999.

C.D. Klaasen (ed), *Casarett and Doull's Toxicology – The Basic Science of Poisons*, 6th edn, McGraw-Hill, New York, 2001.

J. Turton and J. Hooson (eds), *Target Organ Pathology*, Taylor & Francis, London, 1998.

Chapter 18

Behavioural Toxicology

GERHARD WINNEKE

18.1 INTRODUCTION

The study of behaviour and its reversible or irreversible modification by chemicals or therapeutic agents in toxic doses is a relatively young approach within neurotoxicology. Generally, neurotoxicology draws upon a broad spectrum of neuroscience methods, such as neuroanatomy, neurochemistry, neurophysiology, neurogenetics or neuropharmacology to study adverse effects of chemicals on the structure and function of the mature or developing nervous system, and behaviour is within this spectrum. Adding the prefix 'neuro' to 'behaviour' may be used to emphasise this link with other neuroscience approaches.

From a historical perspective, behavioural toxicology has evolved from empirical psychology, which has itself been defined as the objective study of behaviour, and it has strong methodological roots in the animal studies carried out in experimental and comparative psychology. Despite its closeness to other neuroscience approaches, behavioural toxicology has clear distinctive features of its own. Behavioural changes are dependent on underlying biochemical or physiological changes at various morphological sites within the body, and behavioural analysis deals with the integrated output of these bodily changes, which may not necessarily be restricted to the nervous system.

There is a tendency to use the term behavioural toxicology for animal models and the term neuropsychological toxicology for the behavioural consequences of human exposure to chemicals. This distinction will not be made in this review. Instead, selected human and animal models and typical studies will be treated under the heading of behavioural toxicology.

It has also become customary to discuss studies on postnatal behavioural deficit following pre- or perinatal exposure to chemicals under the heading of behavioural teratology. Although this subspecialty within the general field of behavioural toxicology is very important and has received close attention in environmental toxicology, particularly with respect to the higher vulnerability of the brain of young children, the distinction between adverse effects of chemicals on the adult and those on the developing brain will not be considered here. This is because it is usually less important for the selection of behavioural models. Exceptions to this rule occur where specific developmental end points may be selected. Such end points include developmental milestones or neurodevelopmental tests. Pertinent examples will be given in the following text.

18.2 ANIMAL APPROACH TO BEHAVIOURAL TOXICOLOGY

It is beyond the scope of this review to describe in detail the principles and methods underlying behavioural toxicology; this has been done elsewhere by McMillan (Niesink *et al.*, 1999). Much of the terminology and methodology in this field has been developed in the behaviourist tradition of experimental and comparative psychology. In this tradition, behaviour, defined as the movement of an organism or its parts in a temporal/spatial context, can be either unconditioned (unlearned) or conditioned (learned). Unconditioned behaviour may either be respondent, *i.e.* elicited by known, observable stimuli or spontaneous, *i.e.* emitted in the absence of any observable eliciting stimulus. Examples of the former are reflexes or goal-directed movements; examples of the latter are muscular or spontaneous locomotor activities. Conditioned or learned behaviour includes classical or type-S conditioning in the Pavlovian sense, whereby new stimulus–response associations are formed by repeated pairing of an unconditioned behaviour with a neutral stimulus (NS) to become a conditioned stimulus (CS), and instrumental or operant conditioning, whereby behavioural changes are brought about by the consequences of a behavioural response, namely reward, punishment or negative reinforcement.

Animal models used in behavioural toxicology are available in each of the four classes of behaviour mentioned above, namely (1) unconditioned/respondent, (2) unconditioned/emitted, (3) classically conditioned and (4) instrumentally conditioned. For each of these classes, selected models will be described briefly and explained by illustrative examples. Chemically induced deficits of motor and sensory functions are typically assessed by means of models in classes (1) and (2), whereas many models in classes (3) and (4) have been adapted for the assessment of chemically induced cognitive deficit, *i.e.* learning and memory dysfunction.

The distinction between learning and memory is somewhat artificial and more a matter of emphasis rather than one of substance. If learning is defined as an experience-based behavioural modification or adaptation to environmental change, memory means the retention of such adaptation over a period of time. Two types of memory are studied in animals, working ('short-term') and reference ('long-term') memory. Learning requires intact memory to occur. Conversely, memory dysfunction cannot be tested in the absence of a successfully learned task. It follows that, in practice, neither function can be studied in isolation. It is also true that in some behavioural procedures in classes (3) and (4) above, the emphasis is mainly on learning or acquisition of behavioural adaptations, whereas in others the retention aspect is more prominent. This will be considered in subsequent sections.

Although animal paradigms for use in behavioural toxicology have been adapted for different species, including pigeons, rodents and non-human primates, the animal part of this brief overview will, with few exceptions, concentrate on rodent paradigms. The preferred animal model in behavioural toxicology is the rat. It should also be stressed that much detail of the methodology cannot be given here. The interested reader should consult a more practical text such as Sahgal (1993) for more detailed information.

18.2.1 Unconditioned Emitted Behaviour

Motor activity (MA) is a typical example in this behavioural category. MA is often regarded as an 'apical' test of neurotoxicity, because it is influenced by sensory, motor and even associative processes in the nervous system. In this capacity, MA tests are typical components of screening batteries.

Many different devices are available for investigating MA, ranging from simple open-field arenas to complex photocell-controlled alleyways. The functional comparability of different devices has always been a matter of concern. In order to clarify this issue, interlaboratory comparisons have been carried out using stimulants, depressants and other chemicals, as well as a number of different photocell-to-actometer devices in circular, quadratic or figure-8-form. These studies produced reasonably consistent outcomes. A large number of chemicals consistently gave rise to hyper- or hypoactivity as measured in different activity devices. In some instances, for example following prenatal exposure to ortho- and non-ortho-substituted polychlorinated biphenyls (PCBs 47 and 77), hyperactivity that was not present in rats as young adults became increasingly pronounced with ageing whatever technique was used.

In addition to MA, sweet preference (SP) is another example of unconditioned behaviour. SP can be used to characterise modifications of sex-specific behaviour in rodents following exposure to chemicals with endocrine-disrupting properties. This behavioural paradigm recognises the fact that female rats tend to express preference for sweetened relative to neutral tap water to a much stronger degree than males. It has been shown, for example, that male offspring from dams treated with low doses of nicotine during pregnancy exhibit female-like elevated SP postnatally, and that this apparent feminisation relates to a prenatal decrease in brain aromatase activity. Aromatase is a cytochrome P450 enzyme (CYP19) that plays a key role in determining the 'male direction' of brain development by converting testosterone to 17 β-estradiol.

Using the SP model in an environmental toxicology context, feminisation of male rats following prenatal exposure to an environmental PCB mixture was found to be associated with reduced prenatal brain aromatase activity (Hany *et al.*, 1999). Male rats from PCB-treated dams exhibited a long-lasting postnatal SP compared to those from untreated dams. This resembled the spontaneous SP behaviour of female offspring. The SP behaviour of female offspring was not affected by comparable PCB treatment.

Other variables in this behavioural category include measures of emotionality (defecation, urination) in the open-field test, or sleep–wakefulness patterns measured by ultrasonic vocalisations. Consummatory and other reproductive activities may also be included in this category. The 'functional observation battery (FOB)' developed for screening purposes contains some of these variables and will be described in more detail in the following section.

18.2.2 Unconditioned Respondent Behaviour

Examples of unconditioned respondent behaviour are the acoustic startle reflex (ASR) and its amplification, the prepulse inhibition technique or reflex modulation (RM). ASR techniques are widely used in behavioural toxicology for testing the excitability of the central nervous system (CNS) and/or muscle weakness. RM is used for the measurement of hearing thresholds in rodents. In ASR testing, the animal is placed on a strain gauge-coupled platform in a sound-attenuated chamber and a short burst of white noise is presented. The amplitude of the accompanying motor reflex is measured. Since this measure is ambiguous and may contain excitability, motor and sensory components, more specific information about auditory functioning may be sought through RM. In this procedure, the eliciting stimulus (S2) is preceded by brief prepulse stimuli of different intensities from faint to loud (S1) in random order. The amplitude reduction for the S2 response associated with the prepulse stimulus is recorded as a function of the S1 intensity. This allows for auditory threshold testing in rats and a complete audiogram may be established in about 3 h.

Using this model Goldey and colleagues demonstrated the role of hypothyroid dysfunction induced by developmental exposure to Arochlor 1254, a technical mixture of PCBs in producing hearing loss (ototoxicity). Having first shown that a goitrogen, namely propylthiouracil (PTU), produced dose-dependent hearing loss between 1 and 40 kHz, they then showed that Arochlor 1254 had the same effect for the lower frequencies. Finally, in order to support the hypothesis that thyroid hormone depletion caused this effect, they showed that treatment with thyroxine (T4) to compensate for the PCB-induced reduction in this hormone prevented some of the hearing loss (Goldey and Crofton, 1998).

Other variables in this behavioural category include developmental milestones, such as placing, homing, the righting reflex, negative geotaxis and various measures of forelimb or hindlimb grip strength or of motor co-ordination as measured by performance on a rotating drum (ROTAROD test). In addition, social interactions in the presence of a partner of the same or different sex (approach, avoidance, aggression) can be assessed by systematic observation.

Some of the above-mentioned behavioural signs are also part of the FOB developed and validated for general neurotoxicity screening of chemicals in rats. It is based both on cageside observations and on simple manipulations, and its main structure as described by Kulig (Niesink *et al.*, 1999) is given in Table 18.1.

The data collected from the FOB are mostly not quantitative but they do permit binary or rank-order decisions to be made. The information obtained by using the FOB in a WHO-supported interlaboratory validation exercise has demonstrated its usefulness as a screening device for detection of neurotoxic potential of chemicals. Owing to the subjective nature of this information and the difficulties of standardisation among observers and laboratories, more quantitative and objective neurobehavioural tests are needed for in-depth characterisation of the exact type of chemical-induced nervous system toxicity observed. Some examples of such tests, including methods for the assessment of learning and memory, are briefly described elsewhere in this chapter.

18.2.3 Classically Conditioned (Learned) Behaviour

This type of associative learning or type-S conditioning, which was first studied and described by Pavlov, is based on the principle of temporal contiguity. An unconditioned stimulus (US), *e.g.* food in the mouth, elicits a reflex-type unconditioned response (UR), *e.g.* salivation, whereas an NS, *e.g.* sound, is ineffective in this respect. By systematically pairing NS and US, the NS gradually acquires the ability to elicit the response, which was initially associated with the US alone, and is, thus, turned into a CS. The response to the CS is now termed a conditioned response (CR).

Paradigms based on principles of classical conditioning are widely used in the basic neurobiology of learning and memory, and also in behavioural toxicology. One early example is the classical conditioning of the nictitating membrane extension

Table 18.1 *Functional observation battery (Kulig, in Niesink et al., 1999)*

Category	Sign
Neuromuscular	Observation and ranking of gait abnormalities
	Landing foot splay measurement
	Ranking of righting reflex
	Forelimp grip strength measurement
	Hindlimb grip strength measurement
Arousal	Observation of posture in the home cage
	Ranking of ease of removal from the home cage
	Ranking of ease of handling
	Observation of vocalisations
	Observation of piloerection
CNS hyperexcitability	Observation and ranking of clonic movements
	Observation and ranking of tonic movements
Physiological	Measurement of body weight
	Measurement of body temperature
Sensorimotor reactivity	Ranking of rat's reaction to:
	− approach of a blunt object
	− being touched with an object
	− a sound stimulus
	− a tail pinch
Bizarre behaviour	Observations of bizarre behaviours such as walking backward, self-mutilation, stereotyped motor behaviour
Autonomic	Ranking of salivation
	Ranking of lacrimation
	Observation of response of pupil to a light stimulus
	Observation and ranking of palpebral closure
	Observation and ranking of degree of urination
	Observation and ranking of degree of defecation

reflex in the rabbit. The US in this model was a para-orbital shock and the CS a tone. Following the systemic injection of aluminium lactate, acquisition of the CR was impaired at the two highest dose levels (200, 400 μmol Kg^{-1} bw).

The more generalised classical eyelid conditioning (CEC) model is receiving increased attention in basic neurobiology. It is a robust behavioural paradigm that can be elicited across a broad spectrum of species including man. The relevant neural circuitry is well characterised (Steinmetz, 2000). Toxicological applications have so far concentrated on the effects of ethanol. It has been shown by the Steinmetz group that neonatal exposure of rats to ethanol by three daily intubations on PD 4–9 results in pronounced disruption of CEC acquisition in adult animals aged 3–9 months. This is paralleled by disrupted activity in a deep cerebellar structure, the nucleus interpositus, known to be involved in CEC. The US was a peri-ocular stimulation eliciting an eyelid response and the CS a tone.

Other examples include conditioned taste or flavour aversion (CTA) and Pavlovian fear conditioning (PFC). A typical feature of both models is that a single CS–US pairing is usually sufficient for stable conditioning to occur ('one-trial-learning'). A classical approach for CTA is the pairing of lithium chloride, an emetic agent, with novel saccharin containing water. The typical outcome is that the subsequent intake of sweetened water is reduced relative to preconditioning intake. Applications in behavioural toxicology include 2,4,5-trichlorophenoxyacetic acid, ethanol, acrylamide and chlordimeform. Prenatal treatment of rats with a brominated flame retardant (PBDE99) has been shown to affect CTA induced by 1,25 dihydroxyvitamin D_3 in a dose-dependent manner. PFC usually takes the form of cued and/or contextual fear conditioning. For both variants the US is usually a foot shock, the CS either a tone (cued FC) or the test environment (contextual FC) and the typical response is behavioural freezing (immobility). Both contextual and cued learning are amygdala dependent, whereas the contextual variant is also hippocampus dependent. PFC has received broad attention for studying basic neurobiological features of retention involving the amygdala and the hippocampus but has not been used extensively in a toxicology context. One example is the modification of PFC following prenatal exposure of rats to a brominated flame retardant (PBDE99).

18.2.4 Instrumentally Conditioned (Learned) Behaviour

The principle of instrumental conditioning or instrumental learning, as formulated by Thorndike in his 'Law of Effect', is that behaviour is controlled by its consequences: "Of several responses made to the same situation, those which are accompanied or closely followed by satisfaction to the animal will ... be more firmly connected with the situation, so that, when it recurs, they will be more likely to recur; those which are accompanied or closely followed by discomfort to the animal will ... have their connections with that situation weakened, so that, when it recurs, they will be less likely to occur. The greater the satisfaction or discomfort, the greater the strengthening or weakening of the bond".

Most of the paradigms used in behavioural toxicology for the assessment of cognitive deficit, *i.e.* disruption of learning and memory, belong to this class of instrumental learning. A variety of different procedures is available, and the choice among these depends on the question being asked. The different "cognitive" models used in behavioural toxicology are mostly based on Thorndike's "satisfaction" principle. These are either based on learning through reward (positive reinforcement), or on learning through removal of punishment (negative reinforcement).

18.3 MODELS BASED ON NEGATIVE REINFORCEMENT

18.3.1 Passive Avoidance Conditioning

This is a simple retention test (long-term memory) based on one-trial learning. In the step-through model the animal receives a foot shock as soon as it enters a darker compartment from a brightly lit one through a guillotine door connected to a runway.

In the step-down procedure the animal receives a foot shock as soon as it steps down from a pedestal on to a grid-floor. Animals are often retested at varying intervals afterwards, *e.g.* 55 min, 3 h, 24 h, and latencies to re-enter or to step-down are measured. Good retention is characterised by prolonged latencies upon retesting.

Many toxicological applications of this simple paradigm exist. For example, in trimethyltin-treated rats, dose-dependent reduction of re-entering latencies was found 24 h after initial training. Following prenatal exposure to the organophosphate pesticide, dichlorvos, disruption of passive avoidance performance has recently been reported. Also, perinatal hypothyroidism induced by methimazole in drinking water has been found to result in impaired passive avoidance performance in adult animals (using the step-through model). This observation suggests that there is a long-lasting cognitive deficit produced by this condition.

Although passive avoidance tests are quick and economical, performance in this task is influenced by many non-associative factors, *e.g.* pain sensitivity, reactivity and MA. The interpretation of outcome may, therefore, require additional information from other behavioural models.

18.3.2 Active Avoidance Learning

Active avoidance learning (AAL) is a frequently used paradigm in behavioural toxicology. It has a stronger emphasis on acquisition than on retention, and creates active avoidance conditioning, either by one-way step-through or pole-jump avoidance, or by active two-way avoidance.

In one-way or pole-jump avoidance, one-directional movements result in foot shock avoidance. Active pole-jump avoidance conditioning with warning stimuli varying in pitch has been used to demonstrate high-frequency hearing loss in rats exposed to toluene at weanling or adult age, with weanling rats exhibiting higher vulnerability than adults.

Bidirectional movements or shuttling between the two compartments of a shuttle box are necessary for avoiding the signalled foot shock in two-way active avoidance conditioning. This is a frequently used model in behavioural toxicology. The two-way active avoidance conditioning is used to assess components of learning and memory as affected by neurotoxic chemicals. In an example of this using a cross-fostering approach, Wistar-rats were given a commercial mixture of PCBs such as Chlophen A 30 in the diet pre- and postnatally as well as permanently. The two-way active avoidance conditioning at postnatal day 120 was impaired in offspring with an *in utero* exposure, but not in those with only postnatal exposure. These findings support evidence from neuroepidemiological work indicating the particular importance of *in utero* exposure for neurodevelopmental deficit. Similar studies have shown a deficit of AAL after neonatal exposure to chlordecone, or after prenatal exposure to low levels of carbon monoxide.

Some caution is needed when interpreting results from two-way active avoidance conditioning in terms of cognitive factors. Non-associative factors, such as changes in MA or pain sensitivity, may contribute to the observed outcome. Counting inter-trial activity within the ongoing conditioning procedure typically controls for MA. Measuring pain thresholds using standard methods such as the hot-plate test or the tail-flinch response can control for alteration of pain sensitivity caused by the experimental manipulation. The memory or retention aspect of this task may be emphasised by comparing acquisition curves in consecutive sessions (see above).

18.3.3 Morris Watermaze

This test is considered as a model for the assessment of spatial learning and retention, and has also been shown to be sensitive to hippocampal dysfunction. It is a swim test and escape from water is the negative reinforcement principle. Typical protocols are given in Stewart and Morris (Sahgal, 1993). The acquisition element is very simple: animals must learn the spatial position of a hidden platform from varying starting positions over several days of training. If the learning criterion is reached, retention is tested after removal of the platform as a percentage of time spent in the correct quadrant. This is the procedure for the assessment of reference (long-term) memory. If the position of the platform is changed from day to day with daily retention testing, working (short-term) memory can be tested as well.

Specific pharmacological interference with normal hippocampal function, such as application of *N*-methyl-*D*-aspartate receptor (NMDA) antagonist AP5, has been found to disrupt spatial learning and retention in the Morris watermaze, as well as disrupting electrophysiological hippocampal long-term potentiation (LTP) according to the cellular model of learning and memory.

Use of the Morris watermaze has shown that prenatal exposure of mice to cocaine gives rise to long-term alterations in both reference and working memory of the offspring when they become adults.

Use of the Morris test in rats exposed prenatally to inorganic lead has shown that early postnatal environmental enrichment reverses the long-term deficit in spatial learning induced by such exposure. The recovery in spatial learning is associated with reduction in the deficit of NMDA subunit 1 mRNA and induction of brain-derived neurotrophic factor mRNA in the hippocampus.

18.4 MODELS BASED ON POSITIVE REINFORCEMENT

18.4.1 Radial-Arm Maze (RAM)

This is the positive reinforcement analogue of the Morris watermaze for the assessment of spatial learning and of both working and reference memory. It was developed by Olton as an octagonal central arena from which eight arms radiate outwards. It is easily learned by normal rats or mice, and a variety of other species as well, can be automated, and is sensitive to hippocampal lesions. Detailed protocols are described by Rawlins and Deacon (Sahgal, 1993). Food-deprived rats are allowed to choose freely from among the eight arms, but only one food pellet is available at the end of each. The rat is free to choose arms continuously until all novel arms have been entered and food taken. Generally, rats soon learn to avoid re-entering arms that have already been visited.

The RAM can be used to differentiate between working or short-term memory and reference or long-term memory in the following manner. Always baiting the same four arms and leaving the remaining four unbaited results in the rat learning to avoid reliably the set of unbaited arms (reference memory), as well as avoiding the baited arms already successfully visited during that particular session (working memory).

Early developmental lead exposure was found to impair acquisition and retention of RAM performance. Re-acquisition of RAM performance was impaired in rats treated with trimethyltin.

Other significant studies with the RAM have shown that subcutaneous injection of the glutamate NMDA receptor antagonist, ketamine, at postnatal day 10 to mice results in a pronounced deficit in RAM learning and increased apoptosis. Another important study has shown that environmental enrichment in aged Alzheimer's transgenic mice improves performance in the RAM and in the Morris watermaze test 6 months later but without decreasing brain beta-amyloid deposition.

18.4.2   Delayed Alternation

In this task, the animal is required to alternate between two different responses to receive positive reinforcement. A delay element between choices is introduced to cover the short-term memory aspect. Testing is mostly done in Skinner boxes with two levers or in t-mazes (right/left turn). Performance on delayed alternation tasks depends on the integrity of the prefrontal cortex.

Some of the results obtained using this method are of note. Marked performance deficit in a delayed alternation task has been shown in 5–6-year-old rhesus monkeys exposed to inorganic lead only during the first year of life. Disruption of delayed t-maze spatial alternation has been found in female (but not male) rats at postnatal day 90 after gestational and lactational exposure to specific ortho-substituted PCB congeners. On the other hand, prenatal exposure of rats to four doses of cocaine had no effect on various automated delayed alternation tasks in the adult offspring, although increasing delay intervals between 0 and 80 s produced increasing performance deficit.

18.4.3   Delayed Matching or Non-Matching to Sample

This test is the animal equivalent of recognition memory in man. Although initially developed for use in monkeys, adaptations for rats do exist. In both tasks the animal is first presented a sample stimulus, which is rewarded, and then, after some delay, it is given a choice between the sample and a novel stimulus. In the matching variant (delayed matching to sample – DMS), the animal is rewarded for choosing the original, and in the non-matching model (delayed non-matching to sample – DNMS), it is for selecting the novel stimulus. Since the information to be remembered in each trial is independent from the next one, both tasks are mainly tests of working (short-term) memory rather than of reference (long-term) memory. The delayed non-matching-to-sample ('oddity problem') is the easier task of both because it takes advantage of the animal's spontaneous novelty preference. The delay element emphasises the short-term memory component. It is typically done in Y-mazes with visual stimuli placed at the end of each of the three arms, and equipped with guillotine doors to produce varied intervals of delay. Applications appear to be primarily in behavioural pharmacology.

18.4.4 Repeated Chain Acquisition (RCA)

Once a task has been learned according to chosen criteria, performance changes in this task can be tested. However, there are problems if the time course of agent-induced learning impairment is the focus of interest. RCA is an elegant method of overcoming this difficulty, because new learning is required in each in a series of repeated learning sessions. If, for example, three levers are available, the animal may be required to press these levers in a different sequence on each successive session, or to respond to a different sequence of coloured cue lights on each session, in order to be rewarded.

In the rat, RCA learning was found to be disrupted following a single injection of trimethyltin. Impairment was observed to last for at least 5 weeks post injection. The RCA model has also been used to study the role of excitatory amino acid receptors in the neurobiology of learning. In one such study, various competitive and non-competitive NMDA-receptor antagonists and glycine/NMDA-site antagonists were studied for effects on RCA learning (using a new chain of activity each day) or performance (using the same chain of activity each day). The NMDA-receptor antagonists impaired RCA learning at doses not affecting performance. The glycine-site antagonist had no comparable effect.

18.4.5 Schedule-Controlled Operant Behaviour

Several models based on the variants of reinforcement schedules as described by Ferster and Skinner have been applied successfully in behavioural toxicology (Laties and Wood, 1986). In these behavioural models, pressing a lever for reward in a Skinner box is modified by the specific behaviour–reinforcement contingencies. Thus, a food pellet may be given after a fixed or variable time-interval has elapsed since the last response or, else, after a fixed or variable number of behavioural responses has been emitted. Each of these contingencies produces characteristic response patterns, the modifications of which have been studied following exposure to neurotoxic chemicals, such as inorganic lead or hexachlorobenzene (HCB).

Schedule-controlled operant behaviour in the Skinner box is often used in drug discrimination learning (DDL). In DDL, animals learn to discriminate the stimulus properties of drugs. This is a powerful tool for testing the involvement of specific transmitters in chemically induced neurotoxicity. Receptor agonists and antagonists or saline are injected and conclusions on the involvement of specific receptor families are derived from subsequent differential responses to two levers, namely the 'drug' lever and the 'saline' lever. This *in vivo* paradigm is complementary to receptor binding studies *in vitro*. Using DDL the differential involvement of dopamine D_1 and D_2 receptors in postweaning and lactational lead exposure of rats was demonstrated. Similarly, DDL has been used to support the hypothesis that thyroid hormones are involved in PCB-induced neurodevelopmental toxicity in rats.

18.5 SUMMARY AND CONCLUSIONS FOR ANIMAL MODELS

A variety of animal models is available to cover motor, sensory and cognitive functions, and a selection of these has been discussed here. Applications in neurotoxicology have shown their value in screening for the neurotoxic potential of chemicals and in helping to establish cause–effect and dose–response relationships. These studies have contributed to better risk assessment in relation to both occupational and environmental exposures. They have also added to our understanding of the mechanisms underlying neurotoxicity and helped to identify critical periods of exposure.

There is an ongoing debate about the sensitivity and specificity of the current experimental systems. Are the current behavioural toxicology end points capable of detecting adverse effects of chemicals at lower doses than traditional toxicology end points such as mortality, pre- and postweaning body weight or food and water consumption? Do the behavioural end points contribute anything significantly new to hazard identification and characterisation? A recent comprehensive report compared the results from behavioural toxicity studies with those from general toxicology parameters across a large number of studies and chemicals for parental (F_0) and filial (F_1) measures (Middaugh *et al.*, 2003). For the F_0 measures it was concluded "that hazard identification was improved by the inclusion of multiple tests and neurobehavioural assessments". As for the F_1 behavioural parameters, it was concluded that, although they mostly did not detect agent effects as readily as other more general measures, "they provide information about agent effects on specialised functions of developing offspring not provided by other standard measures of toxicity" (p.260).

It should also be pointed out that the developing field of behavioural toxicology is still characterised by a certain eclecticism rather than by an established set of unifying principles capable of integrating the findings produced by a large and varied set of testing paradigms. However, owing to ever-closer interaction with basic neurobiology, knowledge about neural structures underlying different paradigms has markedly improved in the recent years. Examples are the elaboration of the neural circuitry underlying Pavlovian eye blink conditioning, the distinction between hippocampus- and amygdala-dependent aspects of fear conditioning and the involvement of the prefrontal cortex in delayed alternation tasks. This knowledge allows for more specific hypotheses to be tested in behavioural toxicology. It should also be emphasised that most animal models are mere structural analogues of neuropsychological functions in man. Efforts are currently being intensified to develop behavioural models that allow for a better functional extrapolation to effects observed at the human level.

18.6   THE HUMAN APPROACH TO BEHAVIOURAL TOXICOLOGY

Both experimental and field studies on the behavioural effects of occupational and environmental exposure will be considered here.

18.6.1   Experimental Human Exposure

Experimental exposure studies in human subjects are mainly conducted to identify the presence or absence of acute neurobehavioural effects at low levels of exposure that may serve as indicators of subclinical effects, or to check for an impairment of human performance capabilities by short-term exposures to neurotoxic chemicals, which may increase the risk of unsafe job performance. Recently, there has been an increasing focus on the objective study of odour and irritation aspects of volatile chemicals in order to improve indoor air quality or sensory aspects of air quality at the workplace.

Table 18.2 *Selection of domains, behavioural tests and functions typically covered in experimental human exposure studies*

Domain	Test	Function
Psychomotor performance	Simple reaction time	Visuomotor speed
	Choice reaction time	Visuomotor speed
	Tapping	Motor speed
	Pursuit tracking	Coordination
	Steadiness, aiming	Tremor, coordination
Attention	Digit span	Attention span
	Mackworth clock	Sustained attention
	Auditory vigilance	Sustained attention
Visual perception	Tachistoscopic vision	Perceptual speed
	Critical flicker fusion	Temporal resolution
Cognition	Benton test	Visual memory
	Paired associate	Verbal learning
	Learning and recall	Verbal memory
Affect	Mood scale	Mood

Human experimental studies have been carried out using a variety of tests, but these have largely concentrated on the functional domains and associated tests shown in Table 18.2. Although not exhaustive, this list can be regarded as representative of the type of tests and functions covered in most of the human chamber studies published up to about 1990. More recent studies have used computerised tests such as the Neurobehavioural Evaluation System (NES) (see below and Table 18.3) which, from a functional point of view, do not differ substantially from the tests listed in Table 18.2.

Among the chemicals studied, organic solvents have received particular attention, namely chlorinated hydrocarbons (*e.g.* methylene chloride, trichloroethylene, tetrachloroethene, 1,1,1-trichloroethane), aromatic hydrocarbons (*e.g.* toluene, styrene, xylene), ketones (acetone, methyl ethyl ketone), carbon monoxide and anaesthetics. Exposure levels have typically been at or below threshold limit values (TLVs) for the individual compounds.

Although results differ somewhat both between as well as within chemicals, depending on the level and/or the duration of exposure, a typical finding for a group of solvents, particularly some halogenated hydrocarbons, is impaired vigilance and generalised slowing of perceptual and psychomotor functions suggestive of CNS-depression or prenarcotic state. In addition to psychomotor impairment, vigilance decrement and elevated flicker fusion thresholds have been found in several experimental studies of the effects of solvents on humans.

Experimental studies of effects such as these have in some instances contributed to the confirmation or modification of existing TLVs or short-term exposure limits (STELs) for occupational exposure, although there is as yet no conclusive evidence that acute neurotoxic exposure has contributed to accident-prone workplace performance.

18.7   FIELD STUDIES: OCCUPATIONAL EXPOSURE

A variety of different behavioural tests and test batteries have been and are being used in studying neurotoxic effects associated with workplace exposure to chemicals. A rather typical collection of domains, tests and functions covered in many occupational field studies is given by the computerised NES (Table 18.3).

An up-to-date critical review of computer-based neurobehavioural testing in occupational and environmental human toxicology, also covering the Swedish Performance Evaluation System (SPES), as well as some second-generation computer-based test systems such as the third version of the NES (NES3) and the 'Behavioural Assessment and Research System (BARS)' has recently been published (Anger, 2003).

Groups of chemicals of primary concern for behavioural toxicity in workplace settings include solvents, metals, pesticides and organohalogen compounds. Most behavioural studies have concentrated on solvents and metals, however, and brief examples will be given to illustrate the main approaches and typical findings.

18.7.1   Organic Solvents

Solvents represent a large, chemically heterogeneous group of chemicals that are liquids between 0 and 250 °C and are widely used for the extraction, solution or suspension of water-insoluble materials. They may be grouped into broad chemical

Table 18.3   *Tests and functional domains covered in the computerised neurobehavioural evaluation system*

Domain	Test	Function
Psychomotor performance	Symbol digit	Coding speed
	Hand–eye coordination	Coordination
	Simple reaction time	Visuomotor speed
	Continuous performance test	Vigilance (sustained attention)
	Finger tapping	Motor Speed
Perceptual ability	Pattern comparison	Perceptual speed
Memory and learning	Digit span	Short-term memory/attention
	Paired associate learning	Associative learning
	Paired associate recall	Intermediate memory
	Visual retention	Visual memory
	Pattern memory	Visual memory
	Memory scanning	Memory processing
	Serial digit learning	Learning/memory
Cognitive	Vocabulary	Verbal comprehension
	Horizontal addition	Numerical ability
	Switching attention	Mental flexibility
Affect	Profile of mood states	Emotional state/mood

classes, namely aliphatic, aromatic and halogenated hydrocarbons; alcohols, ketones, esters and mixtures. Owing to the wide variety of applications, occupational exposure is frequent, although the recreational abuse of solvent sniffing in children and adolescents, *e.g.* of toluene or of glue thinners, for achieving desired emotional/ mental states, should not go unnoticed.

Because of the lipophilic nature of organic solvents, the lipid-rich nervous system is a major target for them. Narcotic action is the predominant biological effect in the CNS. Functional and structural effects ranging from neurophysiological changes to severe polyneuropathies have been reported to occur in the peripheral nervous system. A comprehensive overview covering both laboratory and workplace studies on neurobehavioural effects of solvents has been written by Dick (1995).

Much of the CNS neurotoxicity of organic solvents may be explained in terms of narcotic action. Prenarcotic reversible effects, such as psychomotor slowing or vigilance decrement, have been documented in experimental human exposure studies (see above). As yet it is not clear, however, whether repeated prenarcotic exposure over many years may eventually result in irreversible brain damage. It has, however, been shown for some compounds, such as trichloroethylene, styrene, carbon disulfide and solvent mixtures, that chronic low-level exposure is associated with perceptual, cognitive and motor retardation which, from the very design of the different studies, cannot be explained as acute reversible effects.

18.7.2 Metals

Lead, mercury and manganese have received particular attention in occupational exposure studies using behavioural end points. Both cross-sectional and prospective studies have been and are being used in studying behavioural effects of neurotoxic metals in humans. In cross-sectional studies, the preferred approach in occupational exposure studies, behavioural functions are measured in exposed subjects at one point in time and typically related to concurrent exposure, either relative to unexposed but otherwise comparable subjects or relative to one continuous exposure variable using regression models. In prospective studies, associations are sought between changes in behavioural functions and the history of exposure. This approach offers an opportunity to identify critical periods of exposure and to determine if observed effects are reversible or not. Furthermore, problems of causality are more easily handled in prospective studies, partly because they avoid the difficult problem in cross-sectional approaches of estimating the premorbid state.

18.7.2.1 Lead. It is probably the most studied toxic element. This metal occurs both in organic and inorganic form. Inorganic lead is mostly present as lead salts, which have widely different water solubilities. Although the organometallic lead species, owing to their lipid solubility, are highly neurotoxic in acute exposure situations, chronic low-level exposure to inorganic lead constitutes the more serious occupational and public health risk.

Lead serves no known physiological function. Its toxicity can be largely explained by its inactivation of various enzyme systems, caused by binding to sulfhydryl groups (SH–) or by competitive interactions with other metal ions. Consequently, a broad spectrum of biomedical effects has been found to be associated with lead, but the more critical are those related to haem biosynthesis, erythopoiesis and nervous system function. The effects on the nervous system have received considerable attention in both occupational and environmental exposure settings using behavioural performance as the main outcome variable.

Cognitive and sensorimotor functions have often been studied in occupationally exposed subjects. Cognitive functions are typically covered by standard intelligence tests, such as the 'Wechsler Adult Intelligence Scale' (WAIS) or parts of such assays. Tests for the assessment of sensorimotor functions include different reaction-time models and tasks involving sensorimotor speed and coordination, as well as task involving perceptual speed and sustained attention. A recent meta-analysis of neurobehavioural studies in adult workers concluded that adverse effects were associated with blood-lead levels (PbB) down to about 400 μg/L (Meyer-Baron and Seeber, 2000). This conclusion led the German MAK-Commission (TLV expert body) to lower the acceptable biological tolerance level (BAT = Biologischer Arbeitsstoff-Toleranzwert) for inorganic lead in occupational exposure from the previous 700 to 400 μg/L.

18.7.2.2 Mercury. Like lead, mercury belongs to those metals known to, and used by man since ancient times. Mercury (Hg) has three oxidative states, namely metallic mercury, Hg(0); mercurous, Hg(I) and mercuric, Hg(II), each of which has its own characteristics of target organ toxicity. Exposure to metallic mercury vapours is most commonly associated with occupational exposure and with mercury amalgam fillings. Mercury may form stable alkylmercury compounds. A particularly important member of this group is methylmercury.

The classic symptoms of exposure to metallic mercury vapour are caused by effects on the CNS. The kidney is the target for Hg(I) and Hg(II) salts. The differences in toxicity between inorganic mercury compounds are largely determined by their physical properties and redox potentials. The rest of this section will be concerned exclusively with the behavioural toxicity of elemental mercury in occupational exposure settings.

Well-designed studies of the behavioural toxicity of elemental mercury have been carried out using computerised psychometric tests with matched controls. In addition, other assessments such as mental arithmetic performance, perceptual speed, attention in its various forms, simple and choice reaction time and finger tapping were carried out. All studies came to the same overall conclusion that exposure to mercury at the workplace produces both behavioural performance changes and altered mood states. These effects occur at low levels of exposure. Because of the cross-sectional character of all of the available studies there are some inconsistencies, but these may be due to inadequate matching.

18.7.2.3 Manganese. It is an essential trace element which, in cases of excessive exposure, induces signs and symptoms of CNS involvement similar to those of Parkinson's disease. Most of our current knowledge of the neurotoxicity of manganese comes from neurological and behavioural studies in occupationally exposed workers. In contrast to lead or mercury exposure, there are no biochemical markers available to relate the absorbed dose of manganese to neurotoxicity. Instead, manganese concentrations in dust must be used to quantify workplace exposure.

Steel-smelting workers with moderate dust exposure (190–1390 mg m^{-3} total dust) were compared with matched controls. No group differences were observed for most neurophysiological, psychiatric and cognitive measures. However, the steel-smelting workers exhibited slower motor impulse change (diadochokinesis), prolonged reaction times and impaired tapping- and digit-span performance (short-term memory). In another study of workers from a dry alkaline battery factory exposed to manganese dioxide dust (948 mg Mn m^{-3}), highly significant group differences were observed when compared with matched controls in sensorimotor performance, namely impaired eye–hand coordination, prolonged reaction times and impaired steadiness (tremor). As a consequence, the authors proposed a drastic revision of the currently accepted TLV for manganese, as well as regular surveillance of exposed workers, using simple sensorimotor tests for early detection of deficit.

18.8 FIELD STUDIES: ENVIRONMENTAL EXPOSURE

Behavioural research dealing with neurotoxic effects of environmental pollutants has emphasised the developmental perspective. Infants and children have been the subjects of such studies and the focus of interest has been on inorganic lead, organic mercury and on PCB. An important methodological issue in such studies is proper adjustment for relevant confounders. Confounders are variables exhibiting associations with both exposure and outcome. If not considered adequately, their presence may give rise to spurious associations between the exposure variable and the target variable. Particularly important variables, as elaborated in lead cohort studies, are maternal intelligence quotient, to account for genetic background, and quality of the home environment, to account for differences in environmental stimulation during development.

18.8.1 Lead

Acute symptomatic lead poisoning is often associated with encephalopathy which, if the victim survives, may give rise to resultant neurological and psychological damage. This is particularly true for children. Such clinical findings have led to the hypothesis that long-term low-level exposure during infancy and childhood may also result in subclinical neurobehavioural deficit in asymptomatic children.

Much of the relevant literature has recently been reviewed with an emphasis on longitudinal cohort studies (Koller *et al.*, 2004). A variety of psychological tests has been used to assess the degree of CNS involvement, and both lead concentrations in blood (PbB) and tooth lead levels have served as internal markers of current or past environmental exposure. The dependent variables used in these studies included psychometric intelligence, assessed in many studies by means of the Wechsler IQ scales. Other intelligence tests (*e.g.* McCarthy Scales of Children's Ability, Stanford Binet Intelligence Scale) have been used as well. Additional functional domain tests have been added to psychometric intelligence tests in several such studies. These included tests of perceptual-motor functions, of gross or fine motor coordination, of sensorimotor speed and of attention. Aspects of educational attainment have also been studied.

A particularly convincing body of evidence was collected in several prospective cohort studies, which started to examine children shortly after birth, and followed them both in terms of internal exposure and developmental outcome until school or adolescent age. The strength of this set of studies, apart from their prospective design, results from the initial agreement reached on the main features of the core protocol covering independent, dependent and main confounding variables. Neurobehavioural tests in these studies included the 'Bayley Scales of Infant Development (BSID)' until age 2, the 'McCarthy Scales of Children's Ability' beyond age 3 and the Wechsler or the Stanford Binet Intelligence Scales at or beyond age 5 to adolescence.

In reviewing this set of studies, the following conclusions have been drawn:

1. lead-related intellectual deficit has consistently been demonstrated,
2. integrated early postnatal lead exposure is more important than prenatal exposure, and
3. no clear cut effect threshold can be found.

It was further concluded that "current environmental lead exposure accounts for a very small amount of the variance in cognitive abilities (1–4%), whereas covariates such as social and parenting factors account for 40% or more (Koller *et al.*, 2004, p 993). Although the adverse effect recorded was small, it may be argued that even small lead-related IQ shifts can have substantial consequences for the extremes of a population and may, therefore, be of public health significance.

18.8.2 Methylmercury

Although exposure to metallic mercury is relevant in occupational settings, organic mercury complexes, particularly methylmercury compounds (MeHg), are of greater concern than metallic mercury from a neuropaediatric perspective. Methylation of mercury by micro-organisms occurs in the marine environment and biomagnification in the aquatic food chain may give rise to elevated MeHg exposure in fish-eating populations.

Much of the current evidence on neurodevelopmental and behavioural effects is summarised in a recent monograph from the National Research Council (2000). Initial knowledge about the neurological and neurobehavioural consequences of high MeHg exposure was gained from studies of two episodes of mass poisoning in the Minamata bay area (Japan) in 1950 and in Iraq 20 years later. MeHg exposure in the Minamata tragedy was through contaminated fish and in Iraq through eating bread made from seed grain treated with a fungicide containing mercury. In both incidents several thousand victims were hospitalised, neurological signs and symptoms were predominant and effects were generally more pronounced in the children born to poisoned mothers.

Relatively few studies have been conducted on infants and children exposed to low-level environmental exposure to MeHg, but one has been performed in New Zealand, one in the Seychelles Islands and two in the Faroe Islands. Each of these study populations was characterised by a high proportion of marine food in their diet. Both reports from the New Zealand study are based on relatively small groups of cases of 4-and 6-year-old children matched to control children. A cross-sectional approach was used. Cases were defined on the basis of markedly elevated MeHg levels in the hair of their mothers during pregnancy. Significant MeHg-related neurobehavioural deficit was found in both age groups. The Seychelles study is a prospective cohort approach based on 779 mother–infant pairs. Neurodevelopmental and neurobehavioural examinations were performed at several ages up to 66 months. Prenatal MeHg exposure was estimated from maternal hair samples collected at birth. At no age, significant exposure-related neurodevelopmental or neurobehavioural deficit was observed.

The two prospective Faroe Islands cohort studies were based on 182 newborns at 2 weeks of age and 917 children at 7 years of age, respectively. MeHg in maternal hair during pregnancy and in umbilical cord serum served as the exposure markers. In the smaller study, a significant inverse association between Hg in umbilical cord blood and neurodevelopment was observed. In the larger cohort, with children at school age, significant inverse associations were found between Hg in cord blood and outcome in a number of neurobehavioural tests. In general, children's performance was more closely associated with Hg in cord blood than with Hg in either maternal or child hair.

The two large cohort studies on neurodevelopmental and neurobehavioural effects of methylmercury in fish-eating populations have produced different results. The reasons for this are unclear because they do not suffer from serious methodological flaws. A recent risk assessment effort presenting a benchmark dose estimate was based on the Faroe Islands studies alone (National Research Council, 2000).

18.8.3 Polychlorinated Biphenyls

PCBs are mixtures of congeners that differ in the number and position of chlorine atoms on the two rings. From a toxicological point of view, some of them are dioxin-like and some are not. Although no longer used or produced in most countries, PCBs are still of environmental concern in the context of waste disposal and because of their long persistence in environmental media. Lacking biodegradability they accumulate in the food chain and finally reach the human biosystem with elevated levels in different tissues, notably human milk.

In a comprehensive review, with an emphasis on more recent studies from the last decade, neurobehavioural effects of PCBs as assessed by neurodevelopmental and neuropsychological tests have been discussed by Schantz *et al.* (2003). Dutch and German studies, a study from the Faroe Islands, two studies from the United States and one in the Inuit population from Canada were covered, and compared with the older studies published in the 1980s. A common feature of most of these studies was that some or many PCB congeners were detected in maternal or cord blood, and in samples of maternal milk collected shortly after birth. Developmental tests such as the Fagan Test of Infant Intelligence (FTII) or the BSID and standard intelligence tests were applied at various ages. Almost all of the studies have considered important confounders such as maternal intelligence and the quality of the home environment. An example directly comparing the positive developmental impact of the home environment with the negative impact of early PCB exposure, taken from Walkwiak *et al.* (2001) is given in Figure 18.1.

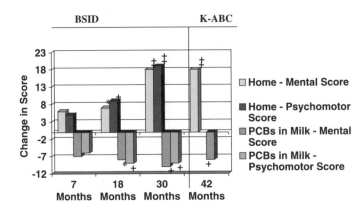

Figure 18.1 *Adjusted effects sizes for mental and psychomotor development of the Bayley scales (BSID) or the Kaufman-test (K-ABC) in relation to PCBs in milk (downward columns indicating deficit) and the quality of the home environment (upward columns indicating improvement). *<0.10; † <0.05; ‡<0.001 (after Walkowiak et al., 2001)*

Schantz *et al.* (2003) came to the following conclusion: "As the data from ongoing PCB studies are published, the weight of evidence for PCB effects on neurodevelopment is growing. In particular, studies ... have now all reported negative associations between prenatal PCB exposure and measures of cognitive functioning in infancy and childhood." It should be pointed out, however, that questions remain about the reversibility of the neurodevelopment deficit, the mechanistic basis of the deficit and the possibility of differential effects of dioxin-like and non-dioxin-like PCB congeners.

18.9 CONCLUSIONS

The application of neurobehavioural tests covering cognitive and sensorimotor functions in experimental and epidemiological studies in children and adults has helped to document the risk of neurobehavioural toxicity associated with occupational or environmental exposure to low levels of solvents, various chemical species of metals and PCBs. In many instances, such observations have been used to establish or revise exposure limits, particularly in occupational settings.

It should, however, be pointed out that frequently such observations have not been sufficiently consistent to be fully acceptable within the scientific community and/or to regulatory bodies. This is particularly true if, as in the case of lead or PCBs in the environment, the critical neurobehavioural effects are small and embedded in a complex background of confounding. In such cases questions are still raised as to the true cause–effect contingencies. In these situations, behavioural studies in experimental animal models have helped substantially to corroborate field observations, to contribute to elaborating the underlying neurobiological mechanisms and to identifying critical periods of exposure.

BIBLIOGRAPHY

W.K. Anger, Neurobehavioral tests and systems to assess neurotoxic exposures in the workplace and community, *Occup. Environ. Med.*, 2003, **60**, 531–538.

R.B. Dick, 'Neurobehavioral assessment of occupationally relevant solvents and chemicals in humans', in *Handbook of Neurotoxicology*, L.W. Chang and R.S. Dyer (eds), Marcel Dekker, Inc., New York, Basel, Hong Kong, 1995, 217–322.

E.S. Goldey and K.M. Crofton, Thyroxine replacement attenuates hypothyroxinemia, hearing loss, and motor deficits following developmental exposure to Arochlor 1254 in rats, *Toxicol. Sci.*, 1998, **45**, 94–105.

J. Hany, H. Lilienthal, A. Sarasin, A. Roth-Härer, A. Fastabend, L. Dunemann, W. Lichtensteiger and G. Winneke, Developmental exposure of rats to a reconstituted PCB mixture or Arochlor 1254: effects on organ weights, aromatase activity, sex hormone levels, and sweet preference behavior, *Toxicol. Appl. Pharmacol.*, 1999, **158**, 231–243.

K. Koller, T. Brown, A. Spurgeon and L. Levy, Recent developments in low-level lead exposure and intellectual impairment, *Environ. Health Perspect.*, 2004, **112**, 987–994.

V.G. Laties and R.W. Wood, 'Schedule-controlled behavior in behavioral toxicology', in *Neurobehavioral Toxicology*, Z. Annau (ed), Johns Hopkins University Press, Baltimore, 1986, 69–93.

M. Meyer-Baron and A. Seeber, A meta-analysis for neurobehavioural results due to occupational lead exposure with blood lead concentrations <70 µg/100 ml, *Arch. Toxicol.*, 2000, **73**, 510–518.

L.D. Middaugh, D. Dow-Edwards, A.A. Li, J.D. Sandler, J. Seed, L.P. Sheets, D.L. Shuey, W. Slikker, W.P. Weisenburger, L.D. Wise and M.R. Selwyn, Neurobehavioral assessment: a survey of use and value in safety assessment studies, *Toxicol. Sci.*, 2003, **76**, 250–261.

National Research Council (NRC), *Toxicological Effects of Methylmercury*, National Academy of Sciences, Washington, DC, 2000.

R.J.M. Niesink, R.M.A. Jaspers, L.M.W. Kornet, J.M. van Ree and H.A. Tilson (eds), *Introduction to Behavioral Toxicology. Food and Environment*, CRC Press, Boca Raton, London, New York, Washington, DC, 1999.

A. Sahgal (ed), *Behavioural Neuroscience. A Practical Approach*, Vol I, IRL Press, Oxford, 1993 (Recommended chapters: A. Sahgal, Passive avoidance procedures, 49–56; J.N.P. Rawlins and R.M.J. Deacon, Further developments of maze procedures, 95–106; C.A. Stewart and R.G.M. Morris, The watermaze, 107–122.)

S.L. Schantz, J.J. Widholm and D.C. Rice, Effects of PCB exposure on neuropsychological function in children, *Environ. Health Perspect.*, 2003, **111**, 357–376.

J.E. Steinmetz, Brain substrates of classical eyeblink conditioning: a highly localized but also distributed system, *Behav. Brain Res.*, 2000, **110**, 13–24.

J. Walkowiak, J.-A. Wiener, A. Fastabend, B. Heinzow, U. Krämer, E. Schmidt, H.-J. Steingrüber, S. Wundram and G. Winneke, Environmental exposure to polychlorinated biphenyls and quality of the home environment: effects on psychodevelopment in early childhood, *Lancet*, 2001, **358**, 1602–1607.

Chapter 19

Pathways and Behaviour of Chemicals in the Environment

MONIKA HERRCHEN

19.1 INTRODUCTION

The environmental chemicals considered here are those substances that enter the environment as a result of human activity. Such chemicals can be differentiated as inorganic and organic substances or differentiation can be made between natural compounds and xenobiotics. Xenobiotics are man-made substances and their structure and biological properties differ from those of natural origin. Some of them, such as many flavouring agents and synthetic pyrethroid insecticides, are structurally comparable to natural compounds while others are quite different. Because of this, simply being 'xenobiotic' is no indication of the intrinsic environmental hazard of a chemical.

Chemical compounds may enter the environment directly or indirectly either intentionally as in the application of biocides and plant protection products, or unintentionally through accidents, spills and waste streams. Any entry of a xenobiotic to the environment modifies the environmental quality whatever the chemicals' properties. The environment consists of the lithosphere, hydrosphere and atmosphere including the organisms present (the ecosphere). Although the organisms' environment can, in principle, be defined as an infinite system, the influence of the environment on organisms can be defined by a limited number of local systems.

The 'environment' and its quality can be described analytically. The term 'environmental quality', however, also includes the requirement of an evaluation of quality. The direct impact of the environmental compartments, soil, water and air, on humans, as well as on their food and living conditions must be considered when evaluating environmental quality. In addition, in current environmental policy, the maintenance and improvement of environmental quality is a target independent of any consideration of environmental hazard to humans. The term 'sustainable development' can be seen as summarising this policy.

Changes in composition and concentrations of environmental chemicals in soil, water and air, as well as changes in their spatial distribution, result in changes in environmental quality. Furthermore, the compartments' influence on the bioavailability of the compounds must be considered. The bioavailable fraction of a compound is the fraction that can be taken up by the organism, can be internally transported, is available for metabolism, and might cause adverse effects. Different exposure routes result in differences in bioavailability. In other words, the bioavailable fraction depends on whether exposure is dermal, through the lungs or through the gut.

19.2 CONCEPTS FOR ENVIRONMENTAL EXPOSURE ASSESSMENT

19.2.1 Medium-Oriented Approach

Different goals for exposure assessment are appropriate for different concepts of environmental protection Bahadir (1987) and Parlar and Angerhöfer (1991). In order to design and implement technical protection measures, such as filtering and recirculation of waste streams and to derive quality standards for environmental compartments, a medium-oriented approach is needed. In particular, the primary receiving compartment and the transport medium, respectively, must be targeted. Thus, quality standards must be set for inorganic and organic priority substances in surface waters (a receiving compartment), trigger values are required for contaminants in sewage sludge (a transport medium) as applied to the agricultural soil and limit values are needed for concentrations of substances in the atmosphere (a receiving compartment).

The chief advantage of a medium-orientated approach is that objectively measured data are used for the derivation and control of emission and immission limit values. However, state-of-the-art analytical methods differ in their effectiveness for different substance groups, and thus, detection limits, limits of quantification and precision vary. For several substance groups the quality standards and trigger values are far below the currently available analytical limits of quantification, whereas for other substance groups we can detect ultra-traces that have no adverse effects.

A major disadvantage of the medium-oriented approach is the lack of consideration of exchange processes at the boundary layer between two compartments. Thus, each compartment is considered separately. Furthermore, in many cases different regulatory agencies are responsible for each compartment. Thus, interaction between the different competent regulatory agencies in this regard is an important issue in ensuring that regulation is effective. One way of facilitating interaction is to consider pathways in a defined sequence. Once the sequence is defined and incorporated in appropriate legislation, one authority can start the process of risk assessment and management and then another authority can take over the process and it can be continued to completion in a concerted way.

19.2.2 Sectoral Approach

The medium-oriented approach is designed for risk assessment and management, and the sectoral approach extends this to include risk/benefit analysis. Benefits arise from specific uses of substances, for example, plant-protection products benefit agriculture, additives give plastics special properties, *etc.* Benefits must be compared with risks to the environment and to the consumer. Such an approach facilitates assessment and evaluation of appropriate substitutes when unacceptable risks are identified. A serious limitation of the sectoral approach is the fact that a chemical substance can be used in different industrial sectors, making an integrated assessment of all emissions and impacts difficult.

19.2.3 Substance-Oriented Approach

The substance-oriented approach focuses on use, exposure and effects of a single substance or group of closely related substances. Intended and unintended concentrations and effects of the substance(s) and possible resultant changes in environmental quality are assessed and evaluated. Such an approach can be applied only to well-defined substances and products. Knowledge of major and minor components, and of by-products and impurities is necessary since they may all have influence on the quality of an ecosystem.

For existing substances, *i.e.* substances already in use, a long-term mass balance scan be established. The mass balance should allow for production volume, use pattern, transport within and between the environmental compartments, degradation and possible accumulation processes. Compiling such a balance makes possible an assessment of environmental capacity for degradation and elimination of the substance and its derivatives. This is necessary for the prediction of potential changes in environmental quality.

The substance-oriented approach is favoured in much of the legislation related to specific substance groups, such as pharmaceuticals, plant protection products and industrial chemicals.

19.2.4 Integrated Exposure Assessment

A complete harmonisation of approaches to exposure assessment is probably unattainable. However, an integrated exposure assessment should be carried out as far as it is possible, since this must lead to the best possible risk assessment. Realistic description of the exposures of consumers and environment requires stratification of input data in relation to the full life cycle of any substance of concern. Where exposure is through different media or a given chemical is used in different sectors, important exposure pathways may be overlooked if media and sectors are not considered together. An integrated approach must be used, which considers all possible exposure routes and has the potential for in-depth analyses to account for specific problems (*e.g.* substances with unusual properties or intended uses, new industrial sectors, known consumer vulnerabilities and particularly dangerous exposure paths).

19.2.5 Harmonisation of Monitoring Programmes

Approaches to exposure assessment have a profound effect on monitoring programmes. Monitoring programmes mostly follow sectoral approaches, considering selected targets and environmental pathways only. Differing objectives lead to differences in the quality of data required and in its analysis. Differing regulatory authorities may have different statutory requirements. Some examples may help to clarify the situation.

The monitoring of food for food safety and consumer protection accounts for the most probable direct routes of food contamination only. Routinely, vegetables, corn and other plant products are analysed only for plant-protection product residues, certain metals and some toxins; food of animal origin will be checked for selected contaminants such as chlorinated hydrocarbons (including PCBs), muscarinic compounds and selected metals. Occasionally, further selected substances of current concern are included in the analyses.

Soil samples may be collected and analysed for any or all of the following reasons:

1. To monitor the ongoing state of the environment and to relate changes in air and soil quality; this requires a continuing programme of sampling and analysis.
2. To establish an environmental specimen bank, which allows for retrospective analyses, and for observation of time trends to improve decision-making on future actions and emerging issues.
3. To set up a database for reference sites and background values, for example, to establish reference values needed to assess change in soil function and ecological diversity in contaminated sites.
4. To identify sites requiring remediation (for example, under Superfund legislation in the USA) and to follow the remediation process as it takes place.
5. To check whether emission reduction measures are being applied successfully by frequent, regular surveillances in the vicinity of point sources of concern.

19.2.6 Parameters of Environmental Exposure Assessment

The substance-oriented approach is applied to predict local, regional and global environmental changes caused by a substance throughout its life cycle from entry to the natural environment, through distribution, transformation and dispersion, to degradation or persistence. Several parameters have been identified as necessary to permit a local, regional or global exposure assessment. These parameters are:

– Production
– Use pattern
– Distribution in the environment
– Physicochemical properties and environmental reactivity.

19.2.6.1 Production. If one knows the regional or global use of a certain substance, ideally with the total tonnage since the start of production and use, regional and global fluxes and loadings can be predicted and assessed fairly easily. In the case of organic xenobiotics, the production volume and maximum loadings are directly correlated. In the case of inorganic substances and naturally occurring organic compounds, the natural background levels must also be taken into account. However, these levels are often difficult to determine, and agreed estimated values must be used.

Not all high-production volume chemicals are of environmental concern. Persistence and ecotoxicity have to be taken into account in their risk assessment. Inert products, such as concrete, cement or steel, may be of indirect environmental relevance since substances necessary for their production or use may cause harm. Consequently, not only end products must be assessed but also substances such as auxiliary products, by-products and intermediates. Pure intermediates that are readily biodegradable are of no environmental concern, but a persistent intermediate that is dumped must be included in any environmental risk assessment.

19.2.6.2   Use Pattern.   Substances can enter the environment at all stages of their life cycle. Production, processing, transport, storage, industrial and private use of end products, recycling and disposal, by either incineration or dumping, must all be reviewed. The probability of a release into the environment varies along the life cycle and can generally be divided into intended releases during use (such as the application of plant-protection products or the use of cosmetics) and unintended emissions (such as volatilisation of plasticisers from user products or the formation of dioxins during incineration).

The terms 'use' and 'use pattern' are defined as the quantitative description of the range of applications of an individual chemical during its normal life. Depending on their technological properties and applications, substances can be allocated to product groups. However, substances within such a product group mostly differ with respect to their structure and properties, and thus have different behaviours in the environment. Consequently, it is more sensible to jointly assess substances of comparable environmental behaviour and subsequently subdivide them according to their end use and application range, *i.e.* the use pattern.

Besides differentiation of substances according to intended and unintended emissions, differentiation between closed use and wide-dispersive use is also important in environmental exposure assessment. In the case of closed use, an emission into the environment is – apart from accidents – unlikely to occur, and reuse or systematic controlled safe destruction is possible. A wide-dispersive use usually results in the emission of substances into the environment with the potential for subsequent effects.

19.2.6.3   Distribution.   The degree of distribution and dispersion depends on the stability of the substance and on the rate and distance of transport in the environment. Physical, chemical and biological mechanisms influencing the transport of a chemical are complex and not easy to quantify. Figure 19.1 summarises the most important exposure pathways of chemicals in the environment to be considered in environmental and human health-risk assessment (modified from Baker *et al.*, 2003).

Dispersion usually starts with a substance's distribution between the phases of the receiving environmental compartment. If a substance is applied directly to the soil, it distributes between the solid phase, the pore water, the gaseous phase in the soil capillary system and also – depending on the volatility of the compound and mode of application – the atmosphere. For short-range transport in soil, sorption and desorption processes are predominant. Both can be described by the sorption coefficient. The adsorption depends on temperature, chemical structure and soil properties (*e.g.* sand, clay, silt content, organic carbon content, pH-value and cation exchange capacity). Desorption into salt solutions and water is rarely complete since a portion of the compound can be irreversibly bound to the soil (bound and non-extractable residues, NER). The difference between adsorption- and desorption-isotherms is called hysteresis. The NER are partly covalently bound to precursors of humic substances and partly irreversibly incorporated into cavities formed by macromolecules or clay minerals. Adsorption and ion exchange determine the

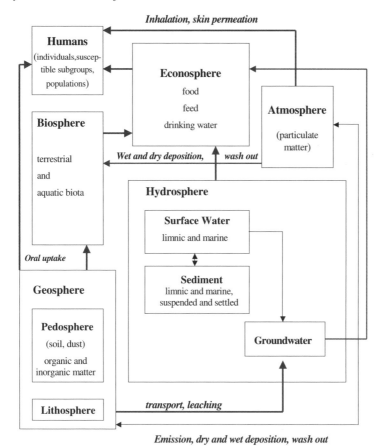

Figure 19.1 *Important exposure pathways and processes including bioavailability to be considered in environmental and human health risk assessment, modified from Baker et al. (2003)*

volatility of a compound and also plant-uptake. Besides these two above-mentioned processes the transport is also influenced by convection and diffusion. Thus, mobility depends on soil density, permeability and flow rate, as well as on the relation of the direction of movement processes of the aqueous phase.

Chemicals can enter groundwater and surface water by leaching, *i.e.* by vertical transport into deeper soil layers and groundwater, and by surface run-off, *i.e.* horizontal transport on the soil surface. Wind erosion and volatilisation may transfer chemicals into the atmosphere. The volatilisation rate of a chemical depends on its concentration in the soil, temperature, wind speed and water content of the soil. The vapour pressure of substances that are sorbed to dry soils is reduced in the lower concentration range (ppm range) and significantly increases with increasing soil concentrations to reach a value that is identical to that of the pure substance. For humid soils the volatilisation rate is much higher because of the partial desorption of the substance by the pore water.

Atmospheric transport is predominantly horizontal. Consequently, most air pollutants can be found in the troposphere (<3 km). Turbulence in the troposphere is influenced by the surface of the earth below. Vertical transport into the upper atmosphere by diffusion is much slower and can take several years (as compared to several months for similar horizontal movement). As a result of these processes, global distribution and dilution of pollutants occur.

Most chemicals in the atmosphere are in the form of particulates. Particulates in the air reach land or water by dry deposition, rainout and washout. Table 19.1 summarises the estimated mean flux of trace elements from the atmosphere to the surface of the sea. For substances in the oceans, transport may occur from deeper layers to the surface by advection or by diffusion or in association with microorganisms as they move, often with a circadian rhythm, between surface and deeper layers.

In general, organisms are an important factor in the environmental transport of chemicals. The extent of transport depends on the trophic level of the accumulating organisms and their mobility. Accumulation, in or on environmental components, counters dispersion and results in an inhomogeneous distribution of the substance. In aquatic systems, for example, the concentration of metals and hydrophobic organic compounds is much higher in suspended and settled sediments than in surrounding water. Similarly, in the atmosphere, the concentration of chemicals in particulates is higher than their concentration in the gaseous phase.

19.2.6.4 Accumulation. Accumulation is the enrichment of a xenobiotic in an environmental compartment. The ratio between concentration in the organism and that in the surrounding environmental compartment, or in the organism's food, is called the accumulation factor. When the accumulation factor is higher than 1, bioaccumulation is occurring.

Table 19.1 *Estimated mean fluxes [ng cm^{-2} year^{-1}] of trace elements from atmosphere to the marine surface. (Data taken from Parlar and Angerhöfer, 1991)*

Trace element	North Sea	South Atlantic Bay	Bermuda	Tropic North Pacific
Al	30,000	2900	3900	1900
Cr	210	—	9	6
Mn	920	60	45	18
Fe	25,500	5900	3000	1300
Co	39	—	1.2	0.6
Cu	1300	220	30	2
Zn	8950	750	75	13
As	280	45	3	—
Se	22	—	3	4.5
Ag	—	—	—	0.2
Cd	43	9	4.5	0.5
Hg	—	24	—	—
Pb	2650	660	100	7

For elements, differentiation between essential and non-essential elements is made. Essential elements are required by the organism for its optimal functioning. The following metals have been identified by Senesi *et al.* (1999) as being essential:
Essential for higher plants: Cu, Fe, Mn, Mo, Ni, Zn
Essential for animals: V, Cr, Mn, Fe, Co, Ni, Cu, Zn, As, Se, Mo

Some of the enzymes and processes that may involve essential elements are listed in the following table:

Element	Function
Co	Prosthetic group of cytochromes and haemocyanin
Cr	Insulin-increase, glucose-tolerance factor
Cu	Oxidizing enzymes
Fe	O_2-transport
Mn	Metabolism of mucopolysaccharides, superoxide-dismutase, arginase, pyruvate carboxylase, malate enzyme
Mo	Xanthine oxidase, sulfoxidase
Ni	Iron resorption, cofactor of urease
Se	Gluthatione peroxidase, type-I-iodide-thyronine de-iodase and phospholipid hydroperoxide glutathione-peroxidase
Sn	Gastrin
V	Control of cholesterol synthesis; nitrite reductase
Zn	Essential for many enzymes in the energy metabolism; transcription; cofactor of carbonic anhydrase

Certain metals such as cadmium, mercury, lead and silver are generally considered non-essential, as they are not needed for tissue metabolism and growth.

Given the number and importance of essential elements to life, organisms have developed homeostatic control mechanisms for these chemicals. Internal body concentrations are regulated to a level at which optimal functioning of an organism can be achieved. Non-essential elements are also regulated. Mechanisms vary between organisms and, within an organism, between elements. In general, there are two types of mechanisms:

1. Those that actively regulate the internal body concentration through the regulation of uptake and excretion rates
2. Those that store metal compounds in a detoxified form in the body

For example, copper and zinc are actively regulated in fish, decapods and some bivalves, while they are stored in granules in barnacles, isopods and most of the bivalves. Cadmium and lead concentrations are regulated both by storage and active regulation in fish and decapods.

As a result of evolution, species or populations have become conditioned to the background availability of essential elements in their natural environment. In environments that have high concentrations relative to a 'normal environment', organisms are not only more tolerant to high-environmental levels, but may even depend on relatively high-background concentrations for optimal functioning. Many examples are known of metal-resistant plant species or populations. Similarly, where environmental-element concentrations exist that are substantially below 'normal', the native biota will be conditioned to these naturally low concentrations.

Anthropogenic substances are accumulated by organisms that do not possess specific mechanisms either to avoid absorption of the substances or to excrete them. The accumulation of organic compounds is not an unlimited process but, depending on exposure, saturation of uptake may be achieved. Since accumulation may be a long-term process and environmental concentrations are not constant, accumulation factors for xenobiotics are difficult to determine under normally occurring conditions. However, standardised tests may be carried out to permit comparison of uptake of substances of concern. Table 19.2 shows accumulation factors for algae (*Chlorella fusca*) for selected organic compounds.

Ecotoxicological effects caused by accumulation of chemicals are difficult to predict. There may be no effects but accumulated substances form a 'reservoir' that may be mobilised upon stress and subsequently cause adverse effects. Furthermore, accumulated substances that are not toxic for the accumulating organism may be toxic to its predators, *for example uptake and accumulation in aquatic organisms.*

The following processes of accumulation of chemicals must be differentiated:

Bioconcentration. Accumulation of a chemical compound by uptake from the surrounding medium without consideration of uptake from food.

Biomagnification. Accumulation of a chemical compound by uptake from food without consideration of uptake through the body surface.

Bioaccumulation. Accumulation by both paths, from the surrounding medium or from food.

Ecological magnification. Increase of substance concentration through a food chain or food web by transfer from a lower to a higher trophic level.

In natural aquatic systems, bioconcentration and biomagnification are parallel processes with bioconcentration usually being the more important. The relative importance of both paths is influenced by the growth rate.

The uptake of chemicals by an organism from water can be described by a typical saturation curve. The accumulation rate depends on the concentration both in the surrounding medium and in the organism, and on the uptake and elimination rates:

$$\frac{dc_A}{dt} = k_1 \times c_W - k_2 \times c_A$$

Table 19.2 *Accumulation factors of several organic xenobiotics for algae (Chlorella fusca), 24 h after application. (Data taken from Parlar and Angerhöfer, 1991)*

Compound	Accumulation factor (24 h)
Hexachlorobenzene	120,000
Sencor	290
Pentachlorophenol	5200–7400
2, 4, 6-Trichlorophenol	580
Monolinuron	60
4-Chloroaniline	1200
Pentachlorobenzene	14,000
Aldrin	29,000
2, 5, 4'-Trichlorobiphenyl	110,000

where c_W is the concentration in the aqueous phase ($ng \times g^{-1}$), c_A the concentration in the organism and k_1, k_2 the rate constants for uptake and elimination.

The saturation concentration in the organism is given by:

$$c_{As} = \frac{k_1}{k_2} c_w$$

The ratio k_1/k_2 is called bioconcentration factor (BCF). Since accumulation of organic substances can be considered to be a transfer from a hydrophilic phase to a hydrophobic phase, the BCF correlates well with the *n*-octanol/water partition coefficient. The more hydrophobic the compound, the higher is the BCF. For extremely high *n*-octanol/water partition coefficient ($\log P_{ow} > 6$), the correlation is poor because the substance tends to remain in the lipid layer, which becomes saturated.

19.2.6.5 Persistence and degradation. There are various definitions of 'persistence'. The following are some of them:

1. Persistence of organic compounds is defined by standardised OECD tests for substances that are not readily or inherently degradable.
2. Persistence in an environmental compartment is a measure of the time that a substance is present in the compartment before it is removed physically or chemically modified.
3. Persistence in an environmental compartment is the final elimination rate, the rate of mineralisation (conversion to carbon dioxide and water), including mineralisation of derivatives of xenobiotics of concern.

Persistence in the context of inorganic substances has proved more difficult to define than in relation to organic compounds. If the definitions applied to organics were applied to inorganics, all inorganic compounds would be characterised as being persistent since they are never converted to carbon dioxide and water. Hence, definitions are being developed based on considerations of change in chemical speciation from bioavailable potentially toxic species to non-bioavailable or low toxicity species. Such a definition should prove to be generally applicable to all substances.

A distinction should be made between intended and unintended persistence. Intended persistence of a compound is a prerequisite for its technological use. Consequently, stabilisers are added to many technical products in order to increase their stability, for example, lubricants, plastics, varnishes and coatings. In all cases, the product has to be stable during its entire use-phase (which might be long-term).

Unintended persistence reflects stability that lasts longer than the use-phase. In particular, for plant-protection products the term 'unintended persistence' has been characterised as a property occurring when a chemical exists over a long-term in any analytically detectable form. An 'optimal compound' would be one, which exists only during the use-phase and mineralises thereafter. However, since the optimum is unattainable, modern chemical plant-protection products are designed to have a half-life of about 30 days. Such a half-life ensures that at the beginning of the next plant growth period after application the substance is no longer present. In other words, the use-phase and the chemical's properties are adapted to each other.

Table 19.3   *Degree of mineralisation [%] 5 days after aerobic and anaerobic incubation in suspended sediment (15 µg applied). (Data taken from IUPAC, 1971)*

Substance	Aerobic	Anaerobic
Urea	66.3	70.1
Methanol	53.4	46.3
Aniline	17.4	n.d.
Dodecylbenzenesulfonate	13.8	26.0
Diethylhexylphthalate	5.6	2.9
Phenantren	3.0	4.2
Trichloroethylene	1.8	2.5
p-Chloroaniline	1.5	n.d.
Lindane	0.4	0.4
DDT	0.1	0.3
Anthracene	0.1	0.3
2,4-D	0.1	0.2

Note. n.d., not determined.

Mineralisation of organic compounds is affected by physical conditions. Table 19.3 shows the biological mineralisation of several organic compounds measured in soil suspension under aerobic and anaerobic conditions. It can be seen how variable mineralisation is, ranging from 0.1 to 70%, 5 days after the start of incubation.

19.2.6.6   Transformation of chemicals in the environment.   The transformation of an environmental chemical can be described as a modification of its chemical structure caused by environmental influences. Transformation caused by radiation, light or oxygen is referred to as abiotic transformation. Modification of chemical structure by organisms or by enzymes is called biotic transformation. The term 'biotic transformation' is identical with the term 'metabolism of environmental chemicals'.

Abiotic transformation reaction pathways, transformation rates and products depend on the surrounding environment (soil, air or water), on energy sources and on other reactive chemicals present including catalysts. The most common abiotic transformations involve hydrolysis, oxidation and reduction.

Hydrolytic reactions introduce a water molecule into a molecule. Examples are the formation of alcohols from alkyl halogens, conversion of epoxides to diols, and the formation of alcohol plus acid from an ester under alkaline conditions.

Oxidative processes are reactions involving oxygen or reactive oxidizing agents (*e.g.* free radicals), the latter being formed in the environment photochemically or enzymatically. The relative contribution of different oxidising agents depends on their reactivity and their concentration. In an aqueous system, peroxide-radicals and activated oxygen are more important than hydroxyl-radicals. Oxidative processes may be subdivided into chemical and photochemical processes. A substance can be oxidized photochemically following absorption of light energy. Light activation may lead immediately to oxidation (direct photo-oxidation) or it may lead to the formation of radicals that react with oxygen (indirect radical photo-oxidation).

Reductive processes occur under anaerobic conditions, most commonly in sediment and the deeper soil layers (saturated zone). They can be characterised as being at the borderline between abiotic and biotic reactions. Redox-systems (such as protein-bound porphyrins and Fe(II)/Fe(III) systems) are produced by degradation of biological material and subsequently can transfer electrons from a reducing organic substrate to xenobiotics organic compounds.

19.2.6.7 Bioavailability. In general, bioavailability, the ability to be taken up by organisms, is an essential prerequisite to the production of adverse effects. Not all physical and chemical species of an element can be absorbed and cause toxicity to organisms. Uptake depends both on the chemical species of the element and on the organism at risk.

In the aquatic environment, if we consider metals, the free ion is often the form which is absorbed. When metals or metal compounds enter the water, their solubility and so their ability to form free ions depends on the physicochemical characteristics of the water body. Metals can be found as free (hydrated) ions and as complexed ions in water and pore water in soil and sediment, or adsorbed/bound to organic and inorganic matter in sediment and soil. Sparingly, soluble metal compounds may also be taken up under appropriate conditions. Thus, in order to assess the bioavailability of any element, all the chemical species of that element present in any relevant environmental compartment must be subdivided into those that are bioavailable and those that are non-bioavailable. This requires consideration of the following environmental parameters:

pH, redox potential, hardness, alkalinity, ionic strength, organic carbon content, temperature, inorganic ligands, presence of oxides of Fe, Mn, Al and Si, sulfides, organic chelating agents (*e.g.* humic substances), concentrations of other metal ions, methylating agents, cation exchange capacity, *etc.*

Consideration of the availability of free metal ions has helped to explain observations of metal behaviour in many natural systems. For example, the decreasing availability of Cd in waters of increasing salinity is caused by the reduction of free Cd^{2+} concentrations, as chloride complexes are formed.

In different soils, large variations in the ecotoxicity of given metal compounds can be found. Clear correlations with soil properties such as pH, ion exchange capacity and organic carbon content can be found indicating the close connection between the available portion of the metal compound and its effects. For example, cadmium toxicity (expressed as reduction of arylsulfatase activity) is highest in sand ($ED_{50} = 1.08$ mmol kg^{-1}) and silty loam and lowest in sandy peat ($ED_{50} = 9.04$ mmol kg^{-1}).

Inconsistencies in organism responses lead to further variations in ecotoxicity, in addition to those resulting from differing bioavailability. For example, inconsistencies in effects on plant growth and on soil respiration with respect to soil cadmium load have been observed. These inconsistencies may be attributed to microbial soil properties.

19.3 HUMAN AND VETERINARY MEDICINES IN THE ENVIRONMENT

19.3.1 Potential Exposure Routes

This review has described general aspects of pathways and behaviour of xenobiotics in the environment without usually referring to a specific substance group. Some of the parameters and processes described above will be presented here with the focus on human and veterinary pharmaceuticals. An overview of potential exposure routes and thus of compartments that may be affected is given by Velagaleti (1997). This overview is illustrated in Figures 19.2–19.4. The distinction is made between use, disposal, and accidental spills. Emissions from production sites should not significantly contribute to the impact on the environment owing to modern production techniques and effective risk management.

19.3.2 Human Pharmaceuticals

The main source of pharmaceuticals and their metabolites in the environment is human excretion. Depending on the application form and the physicochemical properties of the pharmaceuticals, they are excreted in urine or faeces or expired air. Generally excretion in expired air is not considered to be significant.

In industrialized countries most excreted pharmaceuticals should pass through a wastewater treatment plant before entering surface waters or, through sewage sludge, dumping sites or agricultural soils. Compounds in sewage sludge may enter surface or ground water through runoff or leaching.

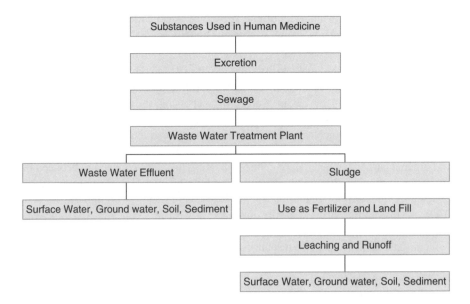

Figure 19.2 *Potential exposure routes for human pharmaceuticals in normal use*

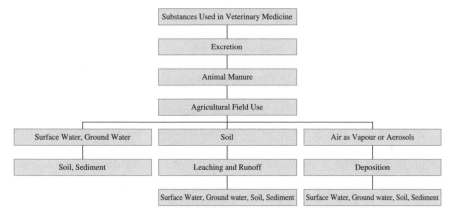

Figure 19.3 *Potential exposure routes for veterinary pharmaceuticals in normal use*

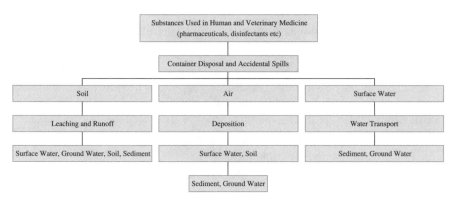

Figure 19.4 *Potential exposure routes for human and veterinary pharmaceuticals following container disposal and accidental spills*

Much of the release of pharmaceuticals into the natural environment is due to inappropriate disposal by consumers in domestic waste or into toilets. According to expert opinion reported in the daily newspaper "Frankfurter Rundschau" on April 20th, 1995, about half of all prescribed medicinal products are applied wrongly or not at all. Thus, pharmaceutical products are included in domestic waste and dumped on municipal dumping sites. The pharmaceuticals can then enter the environment through leachate, evaporation or erosion as long as they are not encapsulated. In general, two types of emission must be evaluated:

1. Point sources such as production sites. Ideally, the amount released should be negligible with correct application of good manufacturing practice and good pharmaceutical practice.
2. Diffuse release by the consumer, by pharmacies and by hospitals. Amounts released can be high (inappropriate disposal) or low (excretion), and, in the case of excretion, may include metabolites.

For pharmaceuticals that are prescribed by general practitioners, Steger–Hartmann found a good correlation with those medicines that are applied exclusively to in-patient therapy.

In wastewater treatment plants pharmaceuticals may be assigned to one of three groups:

1. Those that are completely mineralised such as acetylsalicylic acid
2. Those that are partly metabolised such as penicillin
3. Those that are persistent such as cyclophosphamide

The extent of distribution of pharmaceuticals through wastewater treatment plants depends on the local environmental conditions. Downstream of the wastewater treatment plant efflux, the concentration of extracellular enzymes may increase and thus the probability of degradation of complex organic molecules also increases. However, even after treatment the emission of non-metabolised pharmaceuticals may occur. In particular, it must be remembered that, during the course of a year, the ratio of effluent to receiving water varies. Generic scenarios suggest a dilution factor of 10 whereas extreme values may be in the range from 1:3 to 1:30.

19.3.3   Veterinary Products: Aquaculture and Agriculture

In contrast to human pharmaceuticals, veterinary products used in aquaculture are intentionally applied to the environment. Mostly they are directly applied to the aquatic system and they may easily enter adjacent surface waters if no measures are taken. However, such measures may be complex and costly,

In fish farming there are three important modes of application of pharmaceuticals:

1. External therapeutic application (primarily disinfectants)
2. Oral application (mostly antibiotics)
3. Parenteral and local application (*e.g.* injections for precious single fish).

"Medication in aquaculture almost always means mass application" (Chapman and Scott, 1992). Estimations show that, when no filtering system is used, about one-third of applied medicinal product is taken up by the treated organisms whereas two-thirds enter the adjacent water and sediment phase. Systems allowing for "fish farming in complete compatibility with the environment" (Schlotfeldt, 1992) are effective, but are quite expensive.

In agriculture, veterinary medicinal products can enter soil and surface waters via faeces or sewage sludge. Up to 75% of any applied antibiotic is excreted. However, liquid manure is usually stored prior to application to agricultural soils and during storage, the amount of the medicinal product is reduced by degradation. Quantities and concentrations of pharmaceuticals in soil have, at the time of writing, not been analysed in detail but have to be predicted (see below). Modern products and application forms such as 'slow-release-bolus' or 'interval-bolus' can reduce the amount entering the environment.

19.3.4   Prediction of Environmental Concentrations

(Note: The following contribution is taken from "GDCh-Advisory Committee on Existing Chemicals" (BUA). 'Risk assessment of substances in the soil', S. Hirzel,

Wissenschaftliche Verlagsgesellschaft, Frankfurt, Main, 2003, Report No. 230, ISBN 3-7776-1241-3. The text has been edited to conform to the style of this book.) As a result of the use of waste from livestock farming as manure, emissions of pollutants into the soil are to be expected. Such emissions may occur through the use of veterinary medicines, cleaning and disinfecting agents, feed or feed additives and also as a result of the animals' metabolism. If the emission rates of individual compounds in liquid manure and the characteristic amounts of manure for individual animals are known, the concentration of a particular substance in the liquid manure can be derived. These values can also be obtained by analysis of random samples of liquid manure.

Knowing concentrations in manure, it is possible to estimate emissions into agricultural soil, taking into account the maximum nitrogen quantity permitted per surface and year, established in the Liquid Manure Regulations (GülleVO) to be 170 kg nitrogen ha^{-1} $year^{-1}$.

Liquid manure obtained per pig: $E_{manure} = 2829$ kg $year^{-1}$

Nitrogen portion: $N_{manure} = 17{,}12$ kg $year^{-1}$

Max permitted nitrogen application: $N_{max} = 170$ kg $year^{-1}$ ha^{-1}

Maximum application of liquid manure:

$$(E_{manure}/N_{manure})\, N_{max} = 28{,}029 \text{ kg } year^{-1}\, ha^{-1}$$

Before being spread onto agricultural soil, liquid manure is kept in containers. Depending on substance-specific properties, such as tendency to biodegradation and volatility, the concentration of contaminating substances may be reduced during storage. In Germany, liquid manure is applied once every 72 days, on average (Liquid Manure Ordinance). As an example, taking these points into account, the following estimation may be performed for phenol, a natural metabolite on the 1st EU priority list, assuming the maximum calculated annual quantity of liquid pig manure (28,092 kg ha^{-1}):

Maximum quantity use: 28,092 kg ha^{-1}

Phenol concentration in the manure (after 36 days): $C_{manure} = 4$ mg kg^{-1}

Soil depth (arable land): $RHO_{soil} = 1700$ kg m^{-3}

Soil quantity investigated:

$$M_{soil} = 100 \times 100 \text{ m } ha^{-1} \times 2 \text{ m} \times 1700 \text{ kg } m^{-3} = 3.4 \times 10^6 \text{ kg } ha^{-1}$$

Phenol concentration in the soil (PEC):

$$PEC = (28{,}092 \text{ kg } ha^{-1} \times 4 \text{ mg } kg^{-1})/3.4 \times 10^6 \text{ kg } ha^{-1} = 0.033 \text{ mg } kg^{-1}$$

This estimate allowed for an average dwell time of 36 days for the liquid manure in the tanks, so that, under a 'worst case' assumption, phenol (a readily biodegradable substance) had already been lowered in concentration by 90% in the liquid manure (40 mg kg^{-1} phenol in fresh liquid manure).

A comparison of the calculated soil concentration of 0.033 mg kg^{-1} with the predicted no-effect concentration (PNEC) of 0.136 mg kg^{-1} shows that no risk to the soil is expected from phenol via the liquid manure emission route.

19.3.5   Medicinal Residues in Surface Waters

The occurrence of medicinal residues in surface water is well documented in the scientific literature. Various studies have investigated the concentrations of active pharmaceutical ingredients and metabolites in wastewater, surface water and groundwater. These relate to residues of both veterinary and human medicines.

In investigations of 26 different active ingredients and four metabolites from various indication groups, such as antibiotics, anti-epileptics, lipid-lowering agents or antiphlogistics (pain and rheumatism drugs), the following maximum total concentrations were found in an international study (Ternes, 1997):

Wastewater	$54 \, \mu g \, L^{-1}$
Sewage plant outlet	$6 \, \mu g \, L^{-1}$
River	$3 \, \mu g \, L^{-1}$
Groundwater	$1 \, \mu g \, L^{-1}$

In the aqueous phases of the Glatt Valley Watershed, a densely populated region in Switzerland, the mass flow of antibacterial agents (fluoroquinolones) was analysed (Golet *et al.*, 1992). The fluoroquinolones, ciprofloxacin (CIP) and norfloxacin (NOR), were determined in municipal wastewater effluents and in the receiving surface water of the Glatt river. The results can be summarised as follows:

Raw sewage and final wastewater	$255-568 \, ng \, L^{-1}$ (CIP) and $36-106 \, ng \, L^{-1}$ (NOR)
Glatt river	$< 19 \, g \, L^{-1}$ (CIP, NOR)
Removal during wastewater treatment	79% (CIP) and 87% (NOR)

The risk quotient, measured as environmental concentration/PNEC, is less than 1. This indicates that there is little probability for adverse effects on the microbial activity in waste water treatment plants and on fish, daphnids and algae in the surface water.

Concentrations of iodinated X-ray contrast media used for radiological investigations were determined in a number of German municipal sewage treatment plants, effluents, and rivers and groundwater (Ternes, 1998). Four compounds, namely diatrizoate, iopamidol, iopromide and iomeprol were found in every case. Analyses showed that X-ray contrast media were not reduced significantly during sewage treatment. The following concentrations were measured:

Maximum concentration in sewage treatment plant effluent	$15 \, \mu g \, L^{-1}$ (iopamidol)
Median conc. in receiving waters	$0.49 \, \mu g/l$ (iopamidol), $0.23 \, \mu g \, L^{-1}$ (diatrizoate)
Maximum concentration in groundwater	$2.4 \, \mu g \, L^{-1}$ (iopamidol)

Since X-ray contrast media are used mainly in human medicine, polluted municipal sewage treatment plant effluents are the main source or maybe the only source for their contamination of the aquatic environment.

A number of polar pharmaceuticals have been detected in sewage influents and in effluents from various urban areas in Greece (Koutsouba *et al.*, 2003). The active ingredients found were the anti-inflammatory drugs, diclofenac and ibuprofen, the metabolite of the drugs clofibrate and clofibric acid and the analgesics phenazone and propyphenazone. The monitoring programme was performed in 1998/1999. Diclofenac was found in the influents and effluents of all sewage plants in concentrations between 12 and 560 $ng \, L^{-1}$ (influents) and 10 and 365 $ng \, L^{-1}$ (effluents). A few positive findings (3 of 11) occurred for propyphenazone in the influents where concentrations were measured in the range of 10–200 $ng \, L^{-1}$. Other parent compounds and metabolites were not detected.

BIBLIOGRAPHY

M. Bahadir, *Lehrbuch der Ökologischen Chemie*, F. Korte (ed), Thieme Verlag, Stuttgart, New York, 1987.

S. Baker, M. Herrchen, K. Hund-Rinke, W. Klein, W. Kordel, W. Peijnenburg and C. Rensing, Underlying issues including approaches and information needs in risk assessment, *Ecotoxicol. Environ. Safety*, 2003, **56**, 6–19.

M.J. Chapman and P.W. Scott, 'The role of experts in the EC's medicines licensing procedures', in *Chemotherapy in Aquaculture: From Theory to Reality*, C. Michel and D.J. Alderman (eds), Office International des Epizooties, Paris, 1992.

E.M. Golet, A.C. Alder and W. Giger, Environmental exposure and risk assessment of fluoroquinolone antibacterial agents in wastewater and river water of the Glatt Valley Watershed, Switzerland, *Environ. Sci. Technol.*, 2002, **36**, 3645–3651.

L. Haanstra and P. Doelman, An ecological dose–response model approach to short- and long-term effects of heavy metals on arylsulphatase activity in soil, *Biol. Fertil. Soils*, 1991, **11**, 18.

International Union of Pure and Applied Chemistry (IUPAC), 'Recommendation of the IUPAC-symposium on terminal residues of organochlorine pesticides and of workshop XII: chemistry and metabolism of terminal residues of organochlorine pesticides', in *Proceedings of the 2nd IUPAC International Congress of Pesticide Chemistry*, Tel Aviv, Gordon and Breach Science Publishers, New York, 1971.

V. Koutsouba, T. Heberer, B. Fuhrmann, K. Schmidt-Baumler, D. Tsipi and A. Hiskia, Determination of polar pharmaceuticals in sewage water of Greece by gas chromatography–mass spectrometry, *Chemosphere*, 2003, **51**, 69–75.

B.T. Lunestad, 'Fate and effects of antibacterial agents in aquatic environments', in *Chemotherapy in Aquaculture: From Theory to Reality*, C. Michel and D.J. Alderman (eds), Office International des Epizooties, Paris, 1992.

H. Parlar and D. Angerhöfer, in *Chemische Ökotoxikologie*, Springer, Berlin, Heidelberg, New York, 1991.

W.J.G.M. Peijnenburg, L. Posthuma, H.J.P. Eijsackers and H.E. Allen, A conceptual framework for implementation of bioavailability of metals for environmental management purposes, *Ecotoxicol. Environ. Safety*, 1997, **37**, 163–172.

H.H. Reber, Threshold levels of cadmium for soil respiration and growth of spring wheat and difficulties with their determination, *Biol. Fertil. Soils*, 1989, **7**, 152.

H.J. Schlotfeldt, 'Current practices of chemotherapy in fish culture', in: *Chemotherapy in Aquaculture: From Theory to Reality*, C. Michel and D.J. Alderman (eds), Office International des Epizooties, Paris, 1992.

G.S. Senesi, G. Baldassarre, N. Senesi and B. Radina, Trace elements inputs into soils by anthropogenic activities and implications for human health, *Chemosphere*, 1999, **39**, 343–377.

T. Steger-Hartmann, Analytik und Ökotoxikologie klinikspezifischer Abwasserinhaltsstoffe mit Schwerpunkt auf den Zytostatika Cyclophosphamid und Ifosfamid, PhD Thesis, University of Freiburg, 1995.

T.A. Ternes, 'Occurrence of drugs in German sewage treatment plants and rivers'. *Water Research*, **32**(11), 3245–3260.

T. A. Ternes and R. Hirsch, Occurrence and behaviour of X-ray contrast media in sewage facilities and the aquatic environment, *Environ. Sci. Technol.*, 2000, **34**, 2741–2748.

J. Tögel, Arzneimittel für Süßwasser-Nutzfische. Eine Analyse aus rechtlicher und pharmakologischer Sicht, PhD Thesis, University of Giessen, 1993.

R.R. Velagaleti, Behavior of pharmaceutical drugs (human and animal health) in the environment. *Drug Inform. J.* 1997, **31**, 715–722.

Chapter 20

Ecotoxicity – Effects of Toxicants on Ecosystems

MARTIN WILKINSON

20.1 INTRODUCTION

This brief overview introduces ecology to scientists with little or no previous knowledge of biology. Toxicology is most commonly concerned with effects of toxicants on humans. By contrast, ecotoxicology is concerned with effects on organisms other than the human species. This has three dimensions: toxicity to single species, toxic effects on interrelationships between species, and accumulation of toxicants by organisms and their movement between organisms and species. This requires basic knowledge of ecology before the toxic effects can be fully understood. Following an introduction to ecology and to the unifying concept of a balanced ecosystem, this chapter presents an overview of the general effects of human activities on ecosystems and the methods for monitoring such ecological effects.

To understand ecotoxicology requires knowledge of how organisms interact in nature with each other (the biotic environment) and with the physical and chemical aspects of the environment (the abiotic environment). This is the science of ecology which can be viewed at several levels of organisation at each of which there can be toxic effects. Examples of these levels in ascending order of complexity are given in Table 20.1.

The following account of ecology illustrates how these and other toxic effects can occur. A good starting point is an understanding of the sustainability of the ecosystem level. This requires considerable explanation, which is now given in a very simple way assuming no previous knowledge of biology.

20.2 UNDERSTANDING HOW ECOSYSTEMS WORK

To start with, we are going to make a very simple assumption that there are two requirements that organisms have from their environment to sustain life and that these take precedence over all other requirements: these are (i) a supply of carbon to form the organic molecules of which organisms are composed, and (ii) a supply of energy to power the chemical reactions that keep the organisms alive and the biosynthetic processes that provide the wide range of organic chemicals that organisms need for their structure and function. Carbon is freely available in the environment as carbon dioxide in the air and as various inorganic forms, including hydrogen

Table 20.1 *Levels of consideration in ecology*

Level of organisation	Description of level	Examples of toxicant effects
1. Individual organism or species	Concerned with how physical and chemical environmental factors control which species can occur in which place	Alteration of the physical and chemical factors can affect the growth or survival of particular species
2. Population	A group of individuals of a single species living together and having interrelationships through gene exchange by sexual reproduction	Effects on population size; adaptation of a species to toxicants by tolerant mutants spreading through its population
3. Community	A collection of populations of different species living together in one place (habitat) giving species assemblages characteristic of particular conditions, *e.g.* oak woodlands	Changes in species composition (presence or absence, or relative proportions) owing to selectively different effects of toxicants on different species
4. Ecosystem	Organisms in a particular habitat considered together with their physical and chemical environment, and with the processes linking the organisms and environment such as flow of energy and nutrients, and bio-geochemical cycles. Ecosystems are characterised by a degree of sustainability	Interference with nutrient recycling; concentration and accumulation of toxic substances in food chains; alteration of productivity; sustainability can be impaired by these alterations

carbonate (bicarbonate), dissolved in water. However, organisms require organic carbon and they can be divided into two major groups, heterotrophs and autotrophs, according to how they ensure their supplies of this, as shown in Table 20.2.

20.2.1 Energy and Carbon

In terms of numbers of species, autotrophs are very much in the minority, but they are of absolutely crucial importance because they make the organic matter that all organisms need. By far the biggest group of autotrophs, responsible for most of the

Table 20.2 *Nutritional types of organism*

Type of organism	Means of getting carbon	Means of getting energy
Heterotrophic e.g. animals, fungi, some bacteria	*Ready-made organic carbon.* By ingesting ready-made organic matter in the form of other living organisms or their dead remains or their waste products. Digestion to smaller molecules provides the building blocks for synthesis of other larger organic molecules using energy from respiration	*Chemical energy.* By breaking down (catabolism) some of the larger organic molecules ingested in the diet, in the process of respiration, and applying the chemical energy released to synthesis (anabolism) of other chemicals needed by the organism
Autotrophic mainly plants but also some bacteria	*Inorganic carbon.* Carbon dioxide (on land) or bicarbonate and other dissolved forms (in water) are reduced to organic carbon, primarily by photosynthesis in plants. Sugars resulting from photosynthesis can then provide an energy source in respiration or be used in biosynthesis of other organic molecules	*Light energy.* A physical form of energy, freely available in the environment, light, powers the anabolic reactions of photosynthesis in plants and some bacteria (but in a limited number of chemosynthetic bacteria chemical energy from inorganic reactions is used to reduce inorganic to organic carbon)

fixation of inorganic carbon into organic form on the earth, are the plants using the process of photosynthesis, which can be summarised as follows:

$$6CO_2 + 6H_2O + \text{light energy} \rightarrow C_6H_{12}O_6 + 6O_2$$

This equation is an oversimplification of many reaction steps but illustrates the basic principle. The other fundamental process, respiration, is a series of breakdown reactions which, unlike photosynthesis, is undertaken by all organisms:

$$C_6H_{12}O_6 + 6O_2 \rightarrow 6CO_2 + 6H_2O +$$
chemical energy available for use in the cell

The living cell couples anabolic and catabolic reactions to transfer energy from one type to the other.

Only autotrophs make new organic matter while all organisms use or consume it. Hence, growth of new body matter of autotrophs is called primary production. Production of heterotrophs, which simply recycle already existing organic matter, is called secondary production. Therefore, the production by the autotrophs must be sufficient to meet the needs of both autotrophs and heterotrophs for respiration.

Hence, in a balanced system there is a balance between production and respiration. The photosynthesis by plants will be approximately balanced by total community respiration.

20.2.2 Inorganic Nutrients

Our initial simplistic assumption has served to introduce the above concepts but energy and carbon alone are not enough for life. About 20 different inorganic nutrientions are needed because of their roles in biochemical reactions in living cells or because they are components of particular organic compounds *e.g.* nitrogen in proteins. Plants absorb these from water and soil and they are passed to heterotrophs in their diet.

Some nutrients, *e.g.* nitrogen and phosphorus (principally as nitrates and phosphates), may often be in low concentrations in the environment compared with the amounts needed and so may limit plant growth and primary production. Other nutrients such as various metal ions may be even less abundant but are needed in such smaller amounts. Some such trace elements, *e.g.* copper, may be toxic when available in more than trace quantities but their availablilty in soil or water may be regulated by natural chelators so reducing toxicity.

20.2.3 Food Chain

Organisms can be placed in a chain of dependence, known as a food chain, with several different trophic levels (levels at which organisms feed) with plants or primary producers absorbing light, inorganic carbon, and nutrients, and passing nutrients and organic molecules with their chemical energy to the higher trophic levels of herbivores and carnivores (Figure 20.1). Each trophic level produces waste material (as excretory products and dead matter) and carbon dioxide from respiration. The waste products are broken down by decomposer organisms (bacteria and fungi), which release nutrients back to the environment where they are available for re-use. Therefore, we can recognise that nutrients cycle between organisms and the environment. This is part of a more complex cyclic system – the biogeochemical cycle. For each element utilised by organisms there is such a cycle. The precise details differ between elements depending on the amount of the element available, the uses to which organisms put it, where they store it in their bodies, and the sinks for it in the environment. But all biogeochemical cycles incorporate the idea that, for any essential element at any one time, part of the total naturally occurring amount of the element is in the organisms and part is in different components of the physical environment. Individual atoms or molecules move between these compartments but the proportions in the different compartments remain roughly constant. These cycles must continue to function to ensure a supply of nutrients for organisms and to ensure continuing biological productivity.

Some organisms accumulate certain elements and compounds from the environment (bioaccumulation) causing them to have very high body loads relative to the outside concentrations (bioconcentration), *e.g.* metals in plant tissues. If the accumulated substance is conservative (not broken down by cellular processes) and stored, then

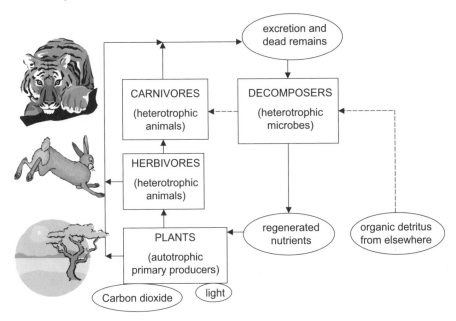

Figure 20.1 *Diagrammatic representation of a food chain. Organisms are contained in boxes and environmental requirements are encircled*

a high dose will be given to the organisms that eat the bioaccumulator. Because of losses of organic matter owing to respiration, each successive trophic level usually has a lower biomass (mass of living material in a given area at one time) or productivity than the levels below it. The body concentration of conservative substances passed up the food chain can therefore increase up the chain (biomagnification), sometimes resulting in toxic doses to organisms higher up the chain.

Nutrients and carbon are recycled. The only requirement not recycled directly is light energy. Energy is lost to the environment by organisms as heat. Consequently, primary productivity is dependent on the continuous input of energy from the sun. It is also controlled by the availability of all the other requirements for plant growth, carbon dioxide, water, and nutrients. Since the availability of all these substances differs between habitats, different levels of primary production are characteristic of different places (Table 20.3).

The rate of secondary production depends on the availability of energy, carbon, and nutrients from the primary producers and so factors affecting plant growth usually affect total production of the whole system. An exception is found in some detritus-based systems such as estuaries. In estuaries, the hydrographic conditions cause suspended particles from land drainage, the sea, and freshwater to accumulate, giving turbid water that restricts light penetration for photosynthesis. The accumulated suspended matter includes much organic detritus, which is instead used as a carbon, energy, and nutrient source by estuarine heterotrophs. There is so much detritus that there is high secondary production despite restricted photosynthesis in this system. The primary production has been done in other habitats from which the detritus

Table 20.3   *Generalised productivity of different habitat types (after Odum)*

Habitat type	Gross productivity (grams of dry matter per square metre per day) indicative of primary productivity
Deserts	Less than 0.5
Grasslands, deep lakes, mountain forests, some agriculture	0.5–3.0
Moist forests and secondary communities, shallow lakes, moist grasslands, most agriculture	3–10
Some estuaries, springs, coral reefs, terrestrial communities on alluvial plains, intensive year-round agriculture	10–25
Continental shelf waters	0.5–3.0
Deep oceans	Less than 0.5

has been transferred. This is illustrated in Figure 20.1. All ecosystems have two types of food chain – the grazing food chain based directly on plant photosynthesis within the system, and the detritus food chain based on consumption of organic detritus by detritus feeding organisms, which in turn are eaten by carnivores. Normally, the detritus chain is based mainly on the waste products of the system's own resident organisms. The estuarine system described above differs in its great reliance on imported waste from other systems.

20.2.4   Food Web

A food web is a more realistic concept than a food chain. Figure 20.2 presents a very simple food web based on imaginary species (most natural ones would contain many more species). Even with such a simple web, there can be a complex pattern of flow of energy, carbon, and nutrients, based on the feeding preferences of different species, as indicated by the lines on the diagram. For any particular habitat, there is a degree of stability by which the same assemblages of species are present in a food web in successive years, with the same dominant and rare species, with the same flow pathways important, and others less so. But what is it that determines which species shall be present? Here we must consider environmental factors.

20.2.5   Environmental Gradients

Organisms do not occur together wholly by chance. A particular habitat has its own set of environmental conditions to which an organism must be tolerant if it is to occur there. Different species have different tolerances to physical and chemical environmental factors (abiotic factors), *e.g.* temperature, rainfall, or soil nutrient status. The range of abiotic factors tolerated along a gradient of such factors (Figure 20.3) can be considered as the theoretical niche of the species. In practice, species usually occupy a narrower range of conditions than this – the realised niche. They do not occur at

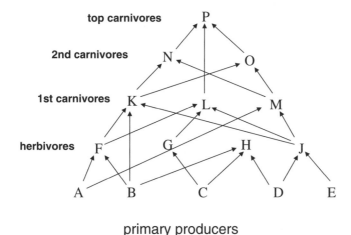

Figure 20.2 *Theoretical food web for a group of imaginary species, indicated by letters. Lines joining the imaginary species indicate feeding relationships and hence pathways for the flow of energy, carbon, and nutrients. Note the complexity of the diagram since species exhibit feeding preferences rather than feeding on all the species on the trophic level below, and also because some species feed at more than one trophic level*

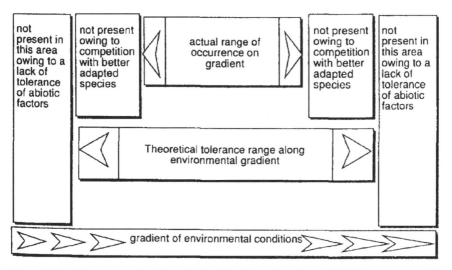

Figure 20.3 *Diagram to show the relationship of actual to theoretical tolerance ranges for a hypothetical species along a gradient of environmental conditions*

the extremities of the theoretical range because interactions there with other organisms (biotic interactions) inhibit them. For example, a species will be best adapted to the environment near to the middle of its tolerance range. Towards the extremities it might be under some stress. It will not compete there with other better adapted species, which are towards the middle of their tolerance ranges.

20.2.6   The Ecosystem

We are now in a position to arrive at a concept of an ecosystem. It consists of all the organisms in a particular place or habitat, their interrelationships with each other in terms of nutrient, carbon, and energy flows, and in terms of biotic determinants of community composition such as competition between species, the physical habitat, and the abiotic factors associated with it, which also play a role in determining community composition and in determining primary, and hence secondary production. This may seem a wooly concept to a physical scientist who is used to a rigid quantitative approach. This is only a preliminary descriptive account. Ecosystems can be quantified, for example, in terms of the fluxes of carbon, energy, and nutrients and the productivities of each trophic level. They can be quantitatively modelled using computers to enable predictions to be made about ecosystem performance.

For our purposes, the most important property of ecosystems is their dynamic stability – their capacity to remain broadly the same over time in species composition and abundance and in the magnitudes of processes despite environmental variations. Although the climate fluctuates from year to year, the structure of the ecosystem tends to be stable within limits, and therefore it is sustainable. One example of dynamic stability is in population sizes. Man's population does not fluctuate wildly from year to year because the generation time is about 20 years and several generations are overlapping. A contrast is in many insects where reproduction occurs every year and the lifespan is only one year or less. There can be fluctuations of several orders of magnitude in population size over several years but they fluctuate around a mean value. This may result from density-dependent factors – environmental factors whose intensity or effect depends on the population density. For example, at high density food may be short giving a population crash while at low density the abundance of food may allow population size to increase, thus fluctuating over several years about a mean.

Ecosystem stability is not rigid. Some systems change naturally; hence the dynamic nature of the stability. On a short timescale, this happens with winter and summer aspects of a community in a temperate climate. On a longer timescale, there is ecological succession where one community naturally replaces another on an area of land or water, usually as a result of the modification of the habitat conditions by the organisms that are replaced so that it is no longer suitable for their own survival. This particularly happens where an open area of land or water is available for colonisation; an example is the formation and growth of maritime sand dune systems. Near the high-tide mark on a beach is an inhospitable environment for plants, windswept with high water loss by evaporation and with sand abrasion, high sand surface temperatures in summer, and a low nutrient and higly saline soil, subject to erosion by waves and wind. Only a few species, the dune-building grasses, can tolerate this environment, forming an open community where, unusually, most ground area is not colonised. These grasses grow best through depositing sand which they stabilise, so building up dunes. The dune soil becomes less saline due to leaching by rainwater, nutrients accumulate from the grass litter aided by nitrogen-fixing bacteria associated with their roots, and the growing dunes provide shelter. Going inland, to dunes formed earlier, the habitat becomes progressively more normal, less inhospitable, and there is a progressive replacement of the dune-building grasses by a wider range of more normal, less tolerant plants. Ultimately, a closed (complete ground coverage) climax community is achieved in equilibrium with the climate and with any local conditions such as soil type.

Mature stable ecosystems are characterised by a preponderance of organisms referred to as K-strategists or climax species – species which succeed by having very precise adaptation to their environment. Earlier stages in a succession may have a greater proportion of r-strategists or opportunists – organisms with wide environmental tolerance, which do not survive so well in stable habitats in competition with more precisely adapted K-species. By contrast, r-strategists are higly reproductive, flooding the environment with their propagules, ready to opportunistically colonise any habitat space which may become available. In stressful environments, either man-made stress or naturally harsh conditions, tolerance to abiotic factors becomes a greater determinant of community composition than biotic interactions, and r-strategists predominate.

The above description of the ecosystem concept has stressed the ability of such systems to remain stable within limits in various ways. Maintenance of this stability is the key to understanding effects on ecosystems due to pollutants.

20.3   EFFECTS OF HUMAN ACTIVITIES ON ECOSYSTEMS

Human activities affect the dynamic balance of ecosystems in two ways, by pollution and by disturbance. This book is concerned with toxic effects and so here we shall consider only pollutants. Pollutants are hard to define but may be considered as substances which can potentially have an impact on ecosystems either because they are novel chemicals synthesised by man, which normal decomposer organisms are not accustomed to dealing with or because they are discharged in unusually high amounts and/or to a system from which they did not come, *e.g.* human waste (from food grown on land) discharged in concentrated form through sewer outlets to rivers or the sea.

Ecosystems become unbalanced through pollutant effects. Their stability is disturbed and the productivity and recycling impaired so that they are no longer sustainable systems. This results from the selective action of toxicants, affecting different species in different ways, or to different extents, or at different concentrations. There may be lethal effects where species are killed but more commonly there are sublethal effects where species remain alive but with reduced growth or reduced reproductive ability or modified development, all leading to ecosystem alteration. A summary of ways in which toxic pollutants may affect organisms at the different levels of consideration in ecology is given as a flow diagram in Figure 20.4.

At an ecosystem level, the above effects can give rise to various symptoms of stress in the system. However, stress can be due not only to toxicants but to non-toxic pollutants, to physical disturbance, and to natural stress in extreme habitats. Part of the art of measuring biological effects of pollution (summarised later) is in distinguishing man-made from natural stress effects. The symptoms of stress in ecosystems are given in Table 20.4.

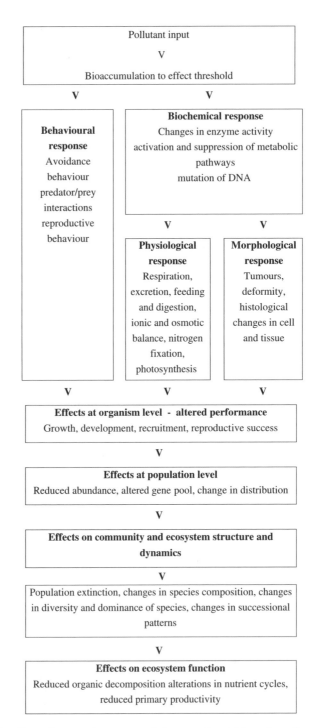

Figure 20.4   *Flow chart to show some of the ways in which toxic pollutants can affect natural ecosystems at different levels (after Sheehan et al., 1984)*

Table 20.4 *Trends expected in stressed ecosystems (after Odum, 1985)*

Energetics

1. Community respiration increases
2. Production to respiration ratio becomes unbalanced
3. Primary production exported to other systems or remaining unused increases

Nutrient cycling

4. Nutrient turnover increases
5. Horizontal transport of nutrients (*i.e.* to other systems) increases
6. Vertical cycling (*i.e.* internal recycling) of nutrients decreases
7. Nutrient loss increases

Community structure

8. Size of organisms decreases
9. Lifespans decrease
10. Species diversity decreases and dominance increases
11. Food chains become shorter

Ecosystem-level trends

12. Ecosystem becomes more open (*i.e.* more space available for colonisation)
13. Successional trends reverse
14. Efficiency of resource use decreases

As mentioned above, not all pollutants are directly toxic. Nonetheless some of the non-toxic ones are relevant to this account because they can have a secondarily toxic effect. An example is enrichment of a water body with plant nutrients such as nitrogen and phosphorus (eutrophication), which can enter as pollutants from sewage, fertiliser run-off, or some industry. Assuming adequate supplies of carbon and light, plant growth will be limited by nutrients. Nutrient pollution then can have a fertilising rather than a toxic effect. Considerable enrichment can give massive unchecked growth of plants, which outstrips the ability of herbivores to graze on it. The decay of the excess plant biomass by bacterial activity then creates a demand for oxygen for bacterial respiration, which may exceed its rate of supply from the overlying atmosphere. The resulting deoxygenation of water can have a lethal effect on animals because they require respiratory oxygen more critically than plants, which can produce their own oxygen through photosynthesis. Some of these effects on ecosystems can be used in biological measurement of pollution. The next section gives an overview of the techniques that are used.

20.4 MEASUREMENT OF TOXIC EFFECTS ON ORGANISMS AND ECOSYSTEMS

Measurement can be made by direct toxicity assessment (DTA) of effluents or by assessment of changes in the ecosystem (ecological monitoring).

Direct measurement of toxicity or toxicity testing is a laboratory procedure carried out with a single species using toxicants as single chemicals or as effluents before or after mixing with the receiving environment. The organism is incubated under standard conditions for a fixed time in various dilutions or with various doses of the toxicant and with controls with no added toxicant. The concentration in aquatic media that brings about the death of 50% of the individuals in the test population is the median lethal concentration (LC_{50}). Alternatively, the dose that brings about 50% mortality in the test population is the median lethal dose (LD_{50}). Such lethal toxicity tests have been popular because they are straightforward to carry out but they are an extreme measure of toxicity and may have little relevance to more normal situations in which immediate lethality does not occur.

Most toxic effects in polluted ecosystems are sublethal and so sublethal tests should be used in order to assess the likely effects of pollutants. This can be done by determining the median effective concentration (EC_{50}). This is the concentration of added toxicant that in a given time under given conditions brings about a specified sublethal effect in 50% of the test population. Such an effect might be a 20% reduction in growth rate relative to a control with no added toxicant. It could also be a 20% change in any physiological process, such as a 20% reduction in photosynthetic or respiratory rate relative to a control, or a 20% change in a developmental process such as the formation of reproductive bodies in an alga.

A more stringent measure is the no-observed-effect concentration (NOEC). This is the highest concentration of added toxicant, which has no measurable inhibitory sublethal effect on the test organism under the specified conditions in the prescribed time. This is the most useful measure for risk assessments but it is very difficult to determine.

Results of toxicity tests are used by regulators because they can give a single easily determined and more or less repeatable numerical measure. But they should not be extrapolated out of context. Problems exist in the selection of suitable test organisms and in the extrapolation of toxicity test results to field conditions when they have been determined under highly artificial laboratory conditions, which are unlike those pertaining in the natural environment.

Ideally, test organisms should be chosen to represent all of the main trophic levels – a plant (autotroph), a herbivore, and a carnivore. Fulfilling these criteria alone is not enough. The particular species, chosen should be appropriate to the environment where the toxicant is to be discharged. There is a tendency to use a restricted range of species, which can be found in culture collections, which might have evolved over long periods of repeated subculturing so as not to have the same environmental response as the original isolate. An example of an inappropriate choice that has occurred is the use of the marine oyster embryo bioassay to test a susbstance to be discharged to a freshwater river, presumably because the oyster embryo test was a newly developed test, which was popular and fashionable at the time. There is a need for the development of a wider range of tests.

An apparently inappropriate test might be suitable, used as a standard reference method, to rank the general toxicity of many different chemicals. This may enable the choice of the least toxic substance for any particular process. What cannot be done easily from laboratory tests is the prediction of effects on the structure or functioning of ecosystems. It is inherent in the nature of a toxicity test that it is done under constant laboratory conditions, which cannot mimic the complex and fluctuating field environment and the biotic interactions between species in the field, and therefore cannot elicit the same response from the organisms and their assemblages. One approach being taken to remedy this is the development of *in situ* tests, which are carried out in the field. The organism is grown captive in a polluted location and some measure of its growth, physiology, biochemistry, or survival is compared to similarly treated captive organisms in a similar but less polluted control environment. These methods are in their infancy and do not always find favour because of the undefined nature of the conditions and uncertainty that the control environment is similar in all features except the pollution. They are not, therefore, toxicity tests in the strict sense, but have a role to play in the elucidation of toxic effects in the field. Such techniques have sometimes been termed 'environmental bioassay'.

The above is not to decry the use of toxicity tests but simply to counsel their wise use. For example, in the last decade, the British water industry has started to use toxicity criteria in determining consents given to discharge liquid effluents into watercourses or coastal waters. Consents have traditionally contained only physical and chemical limits on effluent composition. DTA of effluents has been introduced against the background of the concept of integrated pollution prevention and control (IPPC). The use of toxicity criteria based on DTA makes discharge consents more effective for complex effluents where there may be synergistic effects between components or where there may be so many components that they cannot all be regulated in detail in the consent. Using DTA of the effluent, it is the total toxicity of the effluent that is being assessed rather than the possible additive effect of the specific chemicals present based on their separate toxicity data.

Applying toxicity criteria in defining consents to discharge can be a problem. Routine application of tests on a wide range of organisms with a large number of effluents is desirable but can be very costly, especially if vertebrates are used. For example, fish require complex facilities and (in the United Kingdom) Home Office approval. Alternative quick screening techniques have been devised such as those based on bacterial luminescence, of which Microtox is one proprietary example. The Microtox test is based on light emission by a culture of luminescent bacteria following their activation from dormancy. The light emission is reduced, relative to a control in clean media, when the bacteria are in toxic solutions. Hence, an EC_{50} can be calculated in terms of a 50% reduction in luminescence relative to the control. This might be thought to be an example of an inappropriate test organism but it is only used as a first screening. If serious toxicity is shown in the relatively quick and cheap Microtox test, more relevant though time consuming and expensive tests with the full range of organisms can be carried out.

Ecological monitoring gives a broader assessment of the ecological effects of toxicants that can be obtained from conventional toxicity testing. It is defined as the assessment of effects of toxicants and pollutants in an ecological context, either by means of their accumulation in organisms other than man, or by looking for abnormal ecological effects at the level of species, community, or ecosystem. It performs a different role from that of chemical analysis of toxicants in the environment. Chemical analysis usually relies on occasional instantaneous sampling. It does not necessarily indicate average, maximum, or minimum environmental values of the toxicant. Ecological monitoring avoids the extremely frequent chemical sampling necessary to get over this problem. Indigenous organisms integrate concentrations of toxicant over time. Furthermore, they show what chemical sampling cannot do – the effects of the toxicants on the natural communities – a very good reason for carrying out such monitoring. Ecological methods do not give numerical estimates of toxicant concentrations and so both chemical and ecological approaches should be taken.

Ecological monitoring can use naturally occurring organisms in the field or organisms transplanted to the field for the purpose, and may be supported by laboratory tests. Table 20.5 presents an illustrative selection of approaches to ecological monitoring, with a bias towards aquatic assessment.

Table 20.5   *An overview of selected measures used in ecological monitoring*

1. Assessments carried out in the field		
Using organisms occurring naturally in the environment	Pollutant accumulation by organisms (bioaccumulation monitoring)	Some organisms accumulate metals, radionuclides, and some hydrocarbons to high levels in their tissues in proportion to the external concentration. Gives higher, more detectable concentrations. Integrates concentration over time. May indicate biologically active fractions of the substance. Algae may indicate dissolved fraction while animals feeding on suspended matter (*e.g.* mussels) may indicate particulate fraction
	Assessments using single species	Presence or absence of indicator organisms. There are few genuine indicator species; must be used with care
	Biochemical measurements (sometimes called 'biomarkers')	Biochemical measurements on single species – measurement of activity or amount of substances induced by presence of pollutants, *e.g.* enzymes or metal-binding proteins
		Pathology – presence of tumours induced by pollutants
	Assessments using communities and populations	Age structure – in a species that can be aged and which recruits annually, abnormal age structure may indicate a failure to recruit in one year due to pollution or to natural climatic factors
		Life-forms and successions – successions regressed to earlier stages with abnormal abundance of opportunists may indicate stress
		Numerical structure: (i) Species richness – fewer species may occur under stress (although there may be a temporary increase under slight stress)

Table 20.5 *An overview of selected measures used in ecological monitoring (continued)*

---

1. Assessments carried out in the field

	(ii) Diversity – there are many numerical indices that are mathematical formulations of species number, numbers of individuals, and the distribution of individual numbers between species. Used as general assessments of community structure in ecology but variations from expected values can indicate toxicant induced stress. Also specially developed indices such as the Trent Biotic Index or BMWP score which indicates degree of sewage stress on animal communities in rivers based on numbers of taxa and presence of key species or groups
Using organisms planted out at test site	*In situ* toxicity assessment using measurements of the growth of organisms at a test site compared with a control site
	Colonisation of artificial substrata – provides a uniform substratum that can be compared between different sites using numerical indices (see above) of the communities of small organisms that develop
	Colonisation of cleared natural substrata – again using numerical indices of community structure – may also show whether an alternative community can develop under pollutant influence when the established one is dislodged
	Bioaccumulation monitoring using monitors artificially placed at a variety of test sites to enable comparison

2. Tests carried out in the laboratory

Toxicity testing	LC_{50}, LD_{50}, EC_{50} and NOEC. Limitations as described in text
Growth potential	Testing of survival or growth rate of organisms in laboratory culture under standard conditions in waters from test sites in comparison with water from clean control sites to see the extent to which test sites might support growth of certain species
Biostimulation	Measurement of the growth of algae in natural water samples spiked with various concentrations of various added nutrients to determine its potential for eutrophication

---

20.5 CONCLUSION

Toxic pollutants can disturb the sustainability of natural ecosystems by a variety of effects on species, populations, communities, and ecosystem processes. However, such systems are characterised by dynamic stability and have some capacity to absorb pollutants. Toxicity testing has limitations in predicting such effects, and chemical measurement of environmental toxicants should be accompanied by ecological monitoring. A difficulty requiring specialist knowledge is the distinction of ecological effects due to pollution or other human disturbance from those due to naturally occurring difficult environmental conditions.

BIBLIOGRAPHY

J.M. Anderson, *Ecology for the Environmental Sciences: Biosphere, Ecosystems and Man*, Edward Arnold, London, 1981.

A. Beeby, *Applying Ecology*, Chapman and Hall, London, 1993.

F. Moriarty, *Ecotoxicology*, 2nd edn, Academic Press, London, 1988.

E.P. Odum, Trends in stressed ecosystems, *Bioscience*, 1985, **35**, 419–422.

P.J. Sheehan, D.R. Miller, G.C. Butler and P. Bourdeau (eds), *Effects of Pollutants at the Ecosystem Level*, SCOPE, Wiley, Chichester, 1984.

M. Wilkinson, T.C. Telfer, R. Cruz, S. Conroy-Dalton, E. Cunningham and S. Grundy, 'The utility of field transplants of seaweeds in the study of polluted estuaries', in *'Changes in Fluxes in Estuaries', from 'Science to Management'*, K.R. Dyer and R.J. Orth (eds), Olsen & Olsen, Fredensborg, 1995, 257–260.

World Health Organisation, *Environmental Toxicology and Ecotoxicology: Proceedings of the Third International Course*, Environmental Health Series No 10, WHO Regional Office for Europe, Copenhagen, 1986.

Chapter 21

Radionuclides

MILTON V. PARK

21.1 INTRODUCTION

The phenomenon of the natural radioactivity of some chemical elements was first appreciated by Becquerel in 1896. He observed that photographic emulsions wrapped in black paper and placed near a uranium compound, potassium uranyl sulfate, were blackened. This effect was subsequently attributed to the emission of a radiation by the uranium with properties not dissimilar to those of the already known X-rays, in that it was capable of ionising air, and the activity of a uranium compound could be measured by the rate at which a known quantity could bring about the discharge of an electroscope. The emission of these rays was a fundamental property of the uranium atom, the activity being independent of the nature of the compound, of its valence state, of the temperature or of the previous history of the material. The spontaneous emission of radiation of this type is now known as *radioactivity*.

This *ionising radiation* differs from non-ionising radiation, such as light or radio waves, in its possession of sufficient energy to remove electrons from the atoms of matter through which it passes, and therein lies its particular hazard. This is in contrast to non-ionising radiation, which does not normally possess this property.

Following Becquerel's discovery, the work of Rutherford and Soddy, and of P. and M. Curie, established that the nuclei of some natural elements were not completely stable. These unstable elements were found to emit radiation of three main types, two having the properties of charged particles (α and β radiation), and the third having the characteristics of high-energy electromagnetic radiation.

21.2 TYPES OF IONISING RADIATION

The following types of ionising radiation are associated with radioactivity.

21.2.1 Alpha (α) Radiation

Alpha (α) radiation has been shown to be composed of helium nuclei, consisting of two protons and two neutrons bound together very tightly to give a very stable unit. Consequently each particle possesses a positive charge of 2 units, and a mass of 4 mass units.

These particles possess single kinetic energies of the order of several MeV, characteristic of the radionuclide emitting them. One electronvolt (eV) is defined as the energy gained by an electron passing through an electric potential of 1V. Because of their comparatively high charge, α particles interact intensely with matter, and in consequence impart energy to the medium along their path to a much greater extent than β particles.

21.2.2 Beta (β) Radiation

This comprises high-speed electrons of kinetic energy up to more than 3 MeV originating in the nucleus. They are identical in properties to atomic electrons in mass (1/1840 unit), and in charge (1 unit of negative charge). Although in normal use the term β particles or radiation refers to these high-speed negative electrons, a further type of β particle is also known. This has the same mass as an electron, but is positively charged, and is known as *positron* radiation. The particle is indicated by the symbol β^+ to distinguish it from the more common β^-.

An emitted positron ultimately combines with an electron, its *anti-particle*, resulting in the annihilation of both, and the conversion of their masses into *annihilation radiation*, which appears as two γ-ray photons of energy 0.51 MeV. Positron emitters can be detected not only through their positron emission, but also through this characteristic emission of these 0.51 MeV γ photons. Sodium-22 and fluorine-18 are examples of positron emitters and have found wide use in tracer studies.

The energies of particles from a single radionuclide exhibit a continuous distribution of energies from zero to a maximum value (E_{max}). Only a very small fraction of the emitted β particles have energies close to the maximum, most having much lower energies. A representation of the distribution of the energies of β particles from tritium ($E_{max} = 0.018$ MeV) is shown in Figure 21.1.

Although the shapes of all β spectra are broadly similar, their precise shapes and E_{max} values are characteristic of each radionuclide. The E_{max} value is the characteristic usually given to describe β energies. However, in dosimetry a more useful quantity

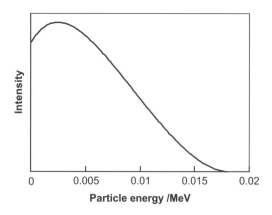

Figure 21.1   *The energy spectrum of tritium*

is the average energy. This quantity is a function of E_{max} and of the atomic number of the element, but it approximates to $E_{max}/3$.

21.2.3 Gamma (γ) Radiation

γ radiation is emitted only in conjunction with other types of decay, and belongs to the class known as electromagnetic radiation, like radio waves and visible light, but of very much shorter wavelength and higher energy. It is emitted when the nucleus produced following radioactive decay is in an excited state, and then returns to the ground state by emitting this radiation to carry away excess energy. γ energies range from a few keV to several MeV, and although one radionuclide may emit γ rays of several different energies, their energies and relative intensities are specific to that nuclide and can be used to identify it.

21.2.4 X-rays

X-rays are also a form of electromagnetic radiation, but differ from γ radiation in that they result from *extra-nuclear* loss of energy of charged particles, for example, electrons, but have shorter wavelengths than ultraviolet radiation. They may be emitted when an orbital electron of an atom jumps to another orbit of lower energy. The difference in energy is radiated as electromagnetic radiation. If the energy is high enough for the radiation to cause ionisation, the emission is called an X-ray. Since the energy levels of X-rays are determined by differences between the energy levels of orbits, they have fixed values for any particular transition.

These vacancies in electron orbits can arise from the phenomenon known as *electron capture*, in which the nucleus captures one of the innermost K shell electrons, resulting in the conversion of a nuclear proton into a neutron. In filling the resulting vacancy in the K shell, a rearrangement takes place with other electrons dropping to lower energy orbitals and emitting their excess energy as X-rays. A complex series of X-rays of different energies results, but one or two energies usually predominate.

A further source of X-rays is the phenomenon known as *bremsstrahlung* ("braking radiation"). This occurs when high-energy electrons are slowed down in an absorber. Part of the energy lost by the particles is radiated as X-rays with a broad spectrum of energies. This is always present when β particles are absorbed and can present problems in shielding against β radiation. Note that an X-ray tube is mainly an emitter of bremsstrahlung, its spectrum comprising a combination of a continuous spectrum of the bremsstrahlung produced by the braking effect of the electrons in the target material, for example, copper, along with X-rays produced by the orbital electrons of the target material dropping into vacant orbitals at lower energies, as mentioned above. These latter transitions occur at specific energies characteristic of the target element, and are known as the *characteristic X-rays* of that element.

21.2.5 Neutron Radiation

The final particle that will be mentioned here, although briefly, is the *neutron*. This is a very common particle, being a basic constituent of the nucleus and having a mass almost identical to the proton but carrying no charge. There are no significant naturally occurring neutron emitters. Radionuclides that emit neutrons can be produced artificially, however, and the neutron is of great importance both in nuclear fission reactors and in the production of radionuclides not available naturally.

21.3   RADIONUCLIDES

Each element can exist in the form of several *nuclides*, the same nuclide containing a given number of protons and of neutrons in the nucleus, *i.e.* all atoms possess the same atomic number and mass number. Atoms possessing the same number of protons in the nucleus, and therefore chemically identical and of the same element, but differing numbers of neutrons are called *isotopes*. Thus, ^{12}C is a nuclide containing a nucleus composed of six protons and six neutrons. Hydrogen has three isotopes, from ^{1}H to ^{3}H.

Most naturally occurring nuclides are stable, but some possess unstable nuclei and transform spontaneously into the nuclide of another element, emitting radiation in this process. These unstable nuclides subject to *radioactive decay* are described as *radionuclides*. Unstable nuclides are more common in the heavier elements found in nature. All nuclides with atomic number greater than 83 (bismuth) are radioactive. Of these, only uranium (^{235}U and ^{238}U) and thorium (^{232}Th) exhibit half-lives long enough to enable them to exist naturally. They decay by different sequential reactions, with emission of α and β radiation, until the nuclei have broken down to stable nuclides of lead. Several lighter elements also have naturally occurring radionuclides, in particular ^{14}C and ^{40}K. This latter isotope, ^{40}K, is the main source of radioactivity within the body, constituting 0.0117% of the total potassium in the body. It is not being replenished, and presumably is the residue of material formed at the time the earth was created, which still remains because of its very slow rate of decay. A different situation is found with ^{14}C. This decays at a much faster rate and is being continuously produced in the upper atmosphere by the interaction of cosmic ray neutrons with nitrogen in the air. The nitrogen nucleus absorbs a neutron and releases a proton, resulting in its conversion into ^{14}C. This constant level of ^{14}C in the environment has been made use of in the technique known as radio carbon dating.

Over the last 50 years or so, a number of radionuclides of stable elements have been produced by artificial means, and there is today a considerable industry for their production for a variety of scientific, medical and industrial purposes. Thus, radioactive cobalt-60 can be made by irradiating non-radioactive cobalt-59 with neutrons in a nuclear reactor. The nucleus may capture a neutron with the emission of a γ photon, referred to as a neutron/gamma (n,γ) reaction, resulting in its conversion into cobalt-60. Cobalt-60 is unstable and decays to the stable nickel-60 with the emission of a β particle and γ radiation:

$$^{59}_{27}Co + n \rightarrow {}^{60}_{27}Co \rightarrow {}^{60}_{28}Ni + \beta^{-}$$

21.4   THE UNIT OF RADIOACTIVITY

Radioactive decay is a random process, so it is impossible to predict when a particular nucleus will decay. On the other hand, since large numbers are involved, it is possible to forecast when a proportion of the nuclei will have decayed. As a consequence of this random behaviour, radioactive decay is found to follow first-order kinetics, *i.e.* the rate of decay of a particular radionuclide is proportional to the number of nuclei of that nuclide present, or

$$-\frac{dN}{dt} = \lambda N$$

where N is the number of nuclei of the radionuclide present, and λ the rate constant for this process, the *radioactive decay constant*. This equation also indicates that measurement of the rate of disintegration can be used to determine the amount of radioactive material present.

Integration of this equation between an initial time 0 and a time t, with N_0 the number of nuclei at time 0, gives:

$$N = N_0 \exp(-\lambda t)$$

The decay constant, λ, is characteristic of a particular radionuclide, but a quantity more commonly used is the *half-life*, $t_{1/2}$ or $t_{0.5}$, of a radioactive species. This is defined as the time it takes for one half of the nuclei in a sample to decay. As the decay process follows first-order kinetics, the half-life, like the decay constant, is also a constant characteristic of a particular radionuclide. Its relationship to the rate constant can be shown by substituting the appropriate values of N ($N_0/2$) and t ($t_{1/2}$) in the above equation, when the following equation is obtained:

$$t_{1/2} \times \lambda = \ln 2$$

Some examples of radioactive half-lives with values ranging from less than a microsecond to millions of years are given in Table 21.1.

Since the disintegration rate or *activity*, A, is proportional to the number of unstable nuclei, we have the corresponding relationship between activity and time:

$$A = A_0 \exp(-\lambda t)$$

where A_0 is the initial activity at time zero.

Or, in terms of $\ln 2$ and $t_{1/2}$

$$A = A_0 \exp(-0.693t/t_{1/2})$$

Hence, if the half-life of the radionuclide is known, the activity of a sample at anytime can be calculated, as shown by example in Table 21.2.

The current SI unit of radioactive activity, or quantity of radioactive material, is the *becquerel* (Bq), 1 Bq corresponding to an activity of one nuclear disintegration per second (1 dps), which usually involves the emission of one or more charged particles (α or β) and possibly X- or γ radiation. An older unit, which is also used is the *curie* (Ci), originally related to the activity of 1 g of radium. It is equivalent to 3.7×10^{10} Bq, or 37 GBq.

Table 21.1 *Some examples of half-lives of radionuclides*

Radionuclide	^{14}C	^{60}Co	^{125}I	^{40}K	^{3}H	^{212}Po	^{90}Sr
Half-life	5730 years	5.27 years	60.1 days	1.28×10^9 years	12.3 years	2.98×10^{-7} s	29.1 years

Table 21.2   *The decay of a sample of ^{32}P ($t_{1/2}$ 14.29 days), initial activity 100 disintegrations s^{-1}*

Time(days)	Number of half-lives	Activity (disintegrations s^{-1})
0.0	0	100.0
14.3	1	50.0
28.6	2	25.0
42.9	3	12.5
57.2	4	6.3
71.5	5	3.1

21.5   INTERACTION OF RADIATION WITH MATTER

Radiation can be divided into two classes: that comprising charged particles such as α and β particles, *i.e. directly ionising radiation*; and that comprising electromagnetic radiation and uncharged particles such as neutrons, *i.e. indirectly ionising radiation*. They each interact differently with the medium through which they pass.

21.5.1   Directly Ionising Radiation

Charged particles (*e.g.* α and β) lose energy when passing through a medium mainly by interacting with the electrons of the medium. These are either excited to higher energy levels or are ionised from the parent atom. A further possibility is the emission of bremsstrahlung mentioned earlier. The charged particles emitted from radionuclides have only a very limited penetration range in the body; at the most a few millimetres.

α particles move at significantly slower speeds than β particles; they are also much heavier (over 7300 times the mass of an electron) and possess double the electrical charge. Hence they impart energy to the medium at a markedly greater rate than β particles, producing a much denser track of ionisation along their path than an electron of equivalent energy. Consequently, they have a much shorter range (in water the range of a 1 MeV β particle is about 4.3 mm, whereas that of an α particle of the same energy is only about 7 μm). The rate at which charged particles impart energy to a medium is known as the *linear energy transfer* (LET). The LET values of particles, typically expressed in keV μm^{-1}, are functions of their velocities and energies, increasing at lower energies and velocities. The LET of typical α particles is much greater than that of β particles.

A consequence of this is that, because of their very limited range, the hazard from α emitters outside the body, *external exposure*, is very small; on the other hand, the hazard is particularly great if they are inhaled or ingested, thereby placing them in close proximity to cell surfaces, *internal exposure*.

21.5.2   Indirectly Ionising Radiation

γ and X-rays, being uncharged, do not lose their energy in the same way. They can be considered as continuing until they collide with a nucleus or electron of the medium through which they are passing. All or part of the energy of the photon is then transferred to the particle. Collision with an electron can result in its being ejected from the parent atom with either complete absorption of the photon, or with absorption of only part of the photon energy and scattering of the beam with reduced energy. A further possibility is that, particularly close to a nucleus, an energetic γ photon may be converted into a positron–electron pair.

Effectively, γ and X-rays transfer all or part of their radiation to charged particles, which then interact as discussed for α and β particles above. However, they are much less attenuated as they pass through than are α or β particles, and in consequence a proportion of their radiation may not be absorbed and pass right through a medium through which it is passing. A further consequence of this is that they can bring about ionising events much deeper within a medium than can α or β particles.

The deposition of energy from X- and γ rays occurs along the tracks of the secondary electrons resulting from the absorption process. However, the degree of ionisation for X- and γ rays is much less when compared with α particles.

21.6 BIOLOGICAL EFFECTS OF IONISING RADIATION

The particular hazards associated with radionuclides arise from their property of emitting radiation of sufficient energy to bring about the ionisation of atoms in the medium through which it is passing. The first stage involves the generation of free *radicals*, highly reactive species bearing an odd electron, which can react with, and covalently modify, other molecules as illustrated in Figure 21.2.

In the next or *chemical stage*, lasting only for a few seconds, these free radicals can attack components of the cell, attaching themselves to molecules or causing breaks in long-chain molecules. Of particular importance are their effects on the chromosomes and DNA of the nucleus, and on the permeability of membranes in general. At the low-dose levels associated with radiological protection, the most radiosensitive structure in the cell is the DNA of the nucleus.

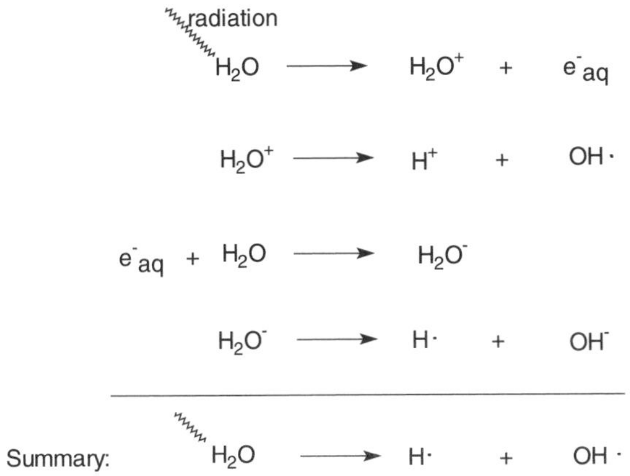

$$\text{radiation}$$
$$H_2O \longrightarrow H_2O^+ + e^-_{aq}$$

$$H_2O^+ \longrightarrow H^+ + OH\cdot$$

$$e^-_{aq} + H_2O \longrightarrow H_2O^-$$

$$H_2O^- \longrightarrow H\cdot + OH^-$$

Summary: $$H_2O \longrightarrow H\cdot + OH\cdot$$

Figure 21.2 *The reaction that takes place in water (the main component of animal tissue) as a result of the passage of ionising radiation. Here, H· and OH· represent the hydrogen and hydroxyl free radicals, respectively. Some combination of like free radicals with each other can occur giving rise to hydrogen and hydrogen peroxide*

In the final, or *biological, stage*, these chemical reactions may result in three outcomes, as follows.

21.6.1   Death of the Cell

If sufficient cells are killed in the general irradiation of a person or animal, particularly the precursor (stem) cells providing a supply of mature functional cells in the blood, gastrointestinal tract lining, skin and gonads, the condition known as radiation sickness appears after a few hours or days. If the dose is sufficiently high, it will result in death.

21.6.2   Cell Survival but with Permanent Molecular Modification

The most likely result in this case is modification of the cell DNA. This modification may be harmless, or it may give rise at a later stage in daughter cells to a malignant transformation, resulting in the *development of a cancer*. If the damage occurs in a cell whose function is to transmit genetic information, such as cells in the reproductive organs, this may result in a *hereditary effect*, or *defect*, being passed on to future generations of the organism.

21.6.3   Repair

Living organisms have always been exposed to ionising radiation from the natural environment, and there are repair mechanisms present in the cell to counter this type of damage. Particularly at low-radiation doses, it is most probably repaired with no deleterious effects on the organism. However, the repair may be imperfect, resulting in a damaged but viable cell, *i.e.* a cell survival but with permanent molecular modification.

These repair mechanisms appear to be highly efficient. It has been estimated that many million ion pairs are created in the total mass of the DNA of an individual in any year by the exposure of the body to natural background radiation alone. Despite this, the incidence of deaths from cancer is no more than one in four of all deaths, and only a small proportion of these is attributable to radiation. Thus, the probability of one of these ionisations giving rise to a cancer is very small indeed.

21.7   UNITS OF RADIATION DOSE

Ionising radiation cannot be detected by the human senses, but a number of instrumental methods are available for this purpose. These include photographic film, which is blackened by ionising radiation, thermoluminescent material, Geiger tubes and scintillation counters. Measurements made by such methods can be interpreted in terms of the radiation dose absorbed by the body, or a particular part of the body.

21.7.1   Radiation-Absorbed Dose

The original unit of exposure was the *röntgen* (R), which was defined in terms of the radiation producing a certain amount of electrical charge in air. This has been replaced by the concept of the *radiation-absorbed dose*, a measure of the energy deposition in any medium by ionising radiation. The SI unit of this is the *gray* (Gy), with 1 Gy corresponding to an energy deposition of $1 \, J \, kg^{-1}$. An older unit of absorbed dose, the *rad*, is also found. This is approximately the absorbed dose in tissue exposed to 1 R, and is equivalent to $0.01 \, J \, kg^{-1}$. The centigray (cGy), which is equal to 1 rad, is sometimes used.

21.7.2 Equivalent Dose

It was found that the same amount of absorbed dose of different types and energies of radiation could produce very different amounts of biological damage, *i.e.* this was a function of both the absorbed dose and the type of radiation. Thus 1 Gy of α radiation causes much more tissue damage than 1 Gy of β radiation, as a result of the much higher LET of α radiation. Consequently when using different types of radiation, the *equivalent dose* is obtained by multiplying the absorbed dose by an empirical factor, the *radiation weighting factor*, w_R, the value of which is dependent on the type and energy of the radiation incident on the body, or in the case of sources within the body, emitted by the source. The unit of equivalent dose is the *sievert* (Sv), and this is related to the gray as follows:

$$sievert = gray \times w_R$$

The values of w_R are broadly related to the LET of the different radiations. All radiations of low LET have been given a value of w_R of unity. For other radiations it is based on the observed values of *relative biological effectiveness*, defined as the inverse ratio of the absorbed doses producing the same degree of a defined biological effect. The values of the w_R for some types of radiation are given in Table 21.3.

Where a tissue T, is irradiated with different types of radiation with different w_R, the equivalent dose in that tissue (H_T) is obtained by summing the products of the absorbed dose in that tissue ($D_{T,R}$) and the w_R for each type of radiation:

$$H_T = \sum_R w_R \times D_{T,R}$$

The equivalent dose thus provides an index of harm to a particular tissue from various radiations, *e.g.* 1 Sv of α radiation to the lung is deemed to create the same risk of inducing a fatal lung cancer as 1 Sv of γ radiation, although the absorbed dose (Gy) is much greater in the latter case.

Table 21.3 *Radiation weighting factors for some types of ionising radiation (ICRP 60)*

Type and energy range	Radiation weighting factor (w_R)
γ-radiation	1
X-radiation	1
β-particles	1
α-particles	20
Neutrons, energy < 10 keV	5
Neutrons, energy 10–100 keV	10
Neutrons, energy >100 keV–2 MeV	20

Note: In a recent publication of the ICRP, ICRP 92, proposals have been made for modifications to the w_R values for neutrons and some other types of radiation. Those for α, β, X and γ radiations remain unchanged.

21.7.3 Effective Dose

A further complication is that different tissues exhibit different risks of the development of a fatal cancer. For a given dose of a particular type of radiation to the lungs and to the skin, the risk to the lungs is much greater than that to the skin. This becomes of particular significance when irradiation is non-uniform, such as might be produced by a source of radiation within the body. Only certain tissues near the source might be subjected to significant radiation dose. This is taken into account by summing the equivalent doses to each of the tissues of the body multiplied by a weighting factor related to the risk associated with that organ:

$$E = \sum_T w_T \times H_T$$

where E is the *effective dose*, H_T the equivalent dose in a tissue or organ T and w_T the *tissue weighting factor* for tissue T. The w_T values are shown in Table 21.4.

The sum of the w_T values has been normalised to unity. This allows a variety of non-uniform distributions of dose in the body to be expressed as a single number, broadly representing the risk to health from any of the different distributions of equivalent dose or from a similar dose received uniformly throughout the whole body. Table 21.5 summarises the different dose quantities discussed above.

Table 21.4 *Tissue weighting factors for the calculation of the effective dose (ICRP 60)*

Tissue or organ	Tissue weighting factor (w_T)
Gonads	0.20
Bone marrow (red)	0.12
Colon	0.12
Lung	0.12
Stomach	0.12
Bladder	0.05
Breast	0.05
Liver	0.05
Oesophagus	0.05
Thyroid	0.05
Skin	0.01
Bone surface	0.01
Remainder	0.05
Whole body total	1.00

Table 21.5 *Summary of dose quantities*

Dose quantity	Explanation and units
Absorbed dose	Energy imparted by radiation to unit mass of tissue (Gy or $J\,kg^{-1}$)
Equivalent dose	Absorbed dose weighted for harmfulness of different radiations (radiation weighting factors) (Sv)
Effective dose	Equivalent dose weighted for susceptibility to harm of different tissues (tissue weighting factors) (Sv)

21.8 EFFECTS OF RADIATION IN MAN

For some considerable time now, it has been recognised that short-term harmful effects could be produced by over-exposure to ionising radiation. It is only in the last 40 or 50 years that there has been a realisation of the long-term adverse effects of much lower doses of radiation, effects which are not related directly to dose.

The sources of information about the effects of irradiation on man are very meagre and unsatisfactory, the information not being obtained under carefully controlled conditions in most cases, resulting in considerable uncertainty in much of the data. These sources include data from the atomic bomb victims of Hiroshima and Nagasaki, from victims of fall-out from nuclear tests and from radiation accident and therapy cases.

For radiological protection, two types of health effects are often defined, *deterministic* and *stochastic effects*.

21.8.1 Deterministic Effects

With deterministic effects, the *severity* of the effect is dependent upon the dose, *e.g.* the acute radiation syndromes associated with substantial whole-body irradiation. In these cases, some *threshold dose* has to be exceeded before the effect becomes apparent. Examples of these effects include radiation sickness, cataracts and damage to the skin. Acute radiation syndromes following whole-body exposure occur after exposure to doses of radiation well above those of interest in the setting of dose limits. Irradiation affects particularly the precursor (stem) cell pools of those tissues, such as bone marrow, the gut lining, skin and germinal epithelium, with a high turnover rate of cells. Exposure to a few grays will result in a sudden loss of cell-replacement capacity. From the meagre data available, no individuals would be expected to die at doses below 1 Gy, and the estimated median dose lethal to 50% of the population (LD_{50}) for a 60-day acute exposure ($LD_{50/60}$) has been estimated to be between 3 and 5 Gy (Table 21.6).

21.8.2 Stochastic Effects

With stochastic effects, in contrast, the *probability* of the effect is dependent on the dose, and there is assumed to be *no threshold*. Examples of these are fatal cancers and serious hereditary diseases. With the notable exception of the embryo or fetus, which is particularly sensitive to ionising radiation, it would appear that no stochastic effects other than cancer (and benign tumours in some organs) are induced by radiation in an exposed individual.

Table 21.6 *Range of doses associated with specific radiation-induced syndromes and death in human beings exposed to acute low LET uniform whole-body radiation (ICRP 60)*

Whole body absorbed dose (Gy)	Principal effect contributing to death	Time of death after exposure (days)
3–5	Damage to bone marrow ($LD_{50/60}$)	30–60
5–15	Damage to the gastrointestinal tract and lungs	10–20
>15	Damage to nervous system	1–5

In contrast to the relatively high doses required before deterministic effects are noted, stochastic effects occur at much lower values of equivalent dose. Based on an analysis of the type of data already mentioned, the International Commission on Radiological Protection (ICRP) has estimated that the lifetime risk for fatal cancer for a population of all ages and each sex and assuming uniform irradiation, is $0.1\,\mathrm{Sv}^{-1}$, *i.e.* 1000 deaths per 10^4 persons exposed to 1 Sv. For lower doses or dose rates, less than 0.2 Gy or $0.1\,\mathrm{Gy\,h}^{-1}$, typical of normal industrial exposures such as in the nuclear industry, the ICRP recommends for individuals in a whole population of all ages a reduction in this value by a factor of 2, to $0.05\,\mathrm{Sv}^{-1}$. For the working population of age 20–64 years, a lower value of $0.04\,\mathrm{Sv}^{-1}$ is proposed. Thus, using the latter risk factor for a person whose working life extended over 40 years and was subjected to an annual dose of 10 mSv in the course of his or her work, *i.e.* a cumulative dose of 0.4 Sv, the probability of death from cancer attributable to radiation exposure is 0.4×0.04, *i.e.* 0.016, or 1.6%. This corresponds to an annual risk of 1/40th of that, *i.e.* 0.04%, or 1 in 2500. At this dose rate no deterministic effects would be apparent, although the risk of a fatal cancer is significant. Obviously, if the annual occupational dose were only 1 mSv per year, the risk would be correspondingly reduced to a tenth of that. To place these risk factors in perspective, the annual risk of death from smoking is 1 in 200; from all cancers, 1 in 400; from all natural causes in a 40-year old, 1 in 700; from accidents in the home, 1 in 15,000; and from accidents on the road, 1 in 17,000.

Typical exposures of the general public in the UK, to different sources of radiation are given in Table 21.7. Of the total annual effective dose of 2.6 mSv, about 85% is from natural sources and 15% from man-made sources; of that 15%, the major proportion is from exposures for medical purposes (a chest X-ray, for instance, would involve an effective dose equivalent of 0.02 mSv). There can be large variations in individual doses, particularly from natural radon exposure indoors

Table 21.7 *Average annual effective radiation dose received by individuals in the UK from all sources (see Hughes, 1999)*

Source	Annual effective dose (mSv)	Total exposure (%)
Natural sources		
Cosmic radiation	0.32	12
Terrestrial γ rays	0.35	13.5
Radon decay products	1.30	50
Other internal radiation	0.27	10
Sub-total		85.5
Artificial origin		
Medical procedures	0.370	14
Weapons fall-out	0.004	0.2
Discharges to the environment	0.0003	<0.1
Occupational exposure	0.006	0.2
Products	0.0001	<0.1
Sub-total		14.5
Total (rounded)	2.6	100

(see also 14.7.3 Radon), which gives the largest single contribution to the overall dose and is dependent on the geology of the area of domicile.

The ICRP has published recommended dose limits for exposures arising from human activities, other than medical exposures, both for the occupational workers and for the general public. For occupational workers, this is 20 mSv year^{-1}, and for the general public, 1 mSv year^{-1}. These limits are based on the risk estimates for stochastic effects, and represent a level of dose above which the consequences for the individual would be regarded as unacceptable.

21.9   ROUTES OF EXPOSURE

Radiation exposure of an individual can be of two types, an *external radiation exposure* and/or an *internal exposure*.

21.9.1   External Exposures

These arise from sources of radiation outside the body, which would normally be safely held within a suitable container. The dose received from such sources can be minimised by a combination of *time, distance* and *shielding* as follows:

– Minimising the *time* in which one is exposed to the source *i.e.* dose = dose rate $\times$ time
– Maximising the *distance* from the source *i.e.* the inverse square law applies
– Placing *shielding* material between the source and the individual attenuates the radiation.

By a combination of these and with measurement of the dose received, the exposure can be kept to an acceptable level.

21.9.2   Internal Exposure

The situation can be more complex where there is an internal exposure. While this can arise from internalisation of a radionuclide for medical purposes, when the amount taken will be known, it can also arise from inadvertent intake of material by ingestion, by inhalation, through a wound in the skin, or contamination of the skin itself. Since this situation usually arises from inadvertent release of radioactive material, referred to as *contamination*, it can be difficult to determine the amount of radionuclide absorbed.

Once internalised, the radionuclide will be in intimate contact with the cells of the body, and will continue to irradiate tissues until either the radioactivity has decayed or it is excreted. The rate of excretion will depend on the chemical characteristics of the compound, its *biological half-life*, as distinct from its radioactive half-life. There is a particular hazard associated with radionuclides that possess long half-lives and which become fixed in particular tissues. An example of this is strontium-90, which is accumulated in the bone, being in the same period of the Periodic Table as calcium, and in consequence is only excreted slowly. Additionally, it has a radioactive half-life of 29 years.

Because of this internal hazard, there is a need for *secondary limits* in terms of the amounts of individual radionuclides internalised, limiting the dose obtained from an internal radionuclide to the recommended dose limits discussed earlier. The *annual limit of intake (ALI)* of a radionuclide is the amount (in Bq) of that radionuclide that would give a harm commitment to the organs it irradiates equal to that resulting from whole-body irradiation of the annual effective dose limit. In determining the radiation burden, the *committed effective dose, E(50)*, is used. This is the dose from the radionuclide integrated over a period of 50 years, the likely maximum life span of the individual following the intake, and is of particular importance where the radionuclide is retained within the body for a long time.

Examples of ALIs of some radionuclides are given in Table 21.8. Apart from depending on the individual radionuclide, the ALI also can vary according to the mode of entry into the body, whether by inhalation or ingestion, and on the fraction of inhaled or ingested activity that is absorbed (f_1). Clearance of inhaled material from the body may depend on its chemical form. Compounds of radionuclides are therefore classified into three groups depending on how rapidly they are cleared from the lungs:

- Class D – Retained for days, *e.g.* K, Na, Ca
- Class W – Retained for weeks, *e.g.* Am, Cm
- Class Y – Retained for years, *e.g.* PuO_2.

The ALI will depend on the category to which the compound of the radionuclide belongs. The values of ALI can also be used to set *derived limits of concentration* for radionuclides in air or water, given particular volumes inhaled or ingested in a year.

21.10 METABOLISM OF RADIONUCLIDES

The ability of a radionuclide to be absorbed and distributed around the body depends on its solubility and ability to pass through different membrane barriers. This in turn is related to both the nature of the element concerned and its chemical speciation.

Table 21.8 *Some annual limits of intake (ICRP 61, Appendix A)*

Radio-nuclide	$t_{1/2}$	Inhalation			Ingestion	
		Class	f_1	ALI (kBq)	f_1	ALI (kBq)
^{42}K	12.36 h	D	1.0	50,000	1.0	50,000
^{137}Cs	30.0 year	D	1.0	2000	1.0	1000
^{45}Ca	163 days	W	0.3	10,000	0.3	20,000
^{90}Sr	29.12 years	D	0.3	400	0.3	600
		Y	0.01	60	0.01	5000
^{238}Pu	87.74 years	W	0.001	0.3	0.001	40
		Y	0.00001	0.3	0.0001	300
					0.00001	2000
^{239}Pu	24065 years	W	0.001	0.3	0.001	40
		Y	0.00001	0.3	0.0001	300
					0.00001	2000

If a radionuclide absorbed into the body is an isotope of an element normally present, *e.g.* Na, K or Cl, it will behave like the stable element; this is the basis of tracer studies. If it has similar chemical properties to an element normally present, it will tend to follow the metabolic pathways of the natural metabolite, *e.g.* ^{137}Cs and K, or ^{90}Sr and Ca. For other radionuclides, their metabolism will depend on their affinity for biological ligands and for membrane transport systems. "Labelled" organic compounds will, of course, follow the metabolic path of the corresponding "cold" compound.

The final stage in the metabolism of radionuclides in the body is their discharge to the exterior. The two main routes of this elimination are by the exhalation of gas-phase compounds and by excretion in the urine, faeces, sweat, saliva and, potentially, in milk. Exhalation is the main pathway for volatile compounds such as ^{3}H$_2$O vapour or gaseous derivatives of ^{14}C, and also elemental radon (^{220}Rn or ^{222}Rn) produced by the radioactive decay of internally deposited thorium or radium. Radionuclides in the faeces originate either from species not absorbed during passage through the gut or which have been previously absorbed and subsequently excreted back into the gastrointestinal tract, usually in the bile.

21.11 SOME EXAMPLES OF RADIONUCLIDE METABOLISM

After absorption into the body, the different inorganic elements can be classified into three groups according to their distribution in the body:

- Elements that distribute throughout the body tissues
- Elements that concentrate in a particular organ or tissue
- Elements that concentrate in a number of tissues.

21.11.1 Elements that Distribute throughout the Body Tissues

Examples of these are ^{42}K and ^{137}Cs, which predominantly exist in solution in the form of the corresponding simple ions. Both are readily absorbed from the gut ($f_1 = 1.0$), and because of their similar chemistry both will follow the behaviour of body potassium, although in the case of caesium with some expected differences in quantitative rates of transfer across cell membranes. Both are excreted in the urine and the faeces.

21.11.2 Elements that Concentrate in a Particular Organ or Tissue

Calcium has an important function as a major component of bone, although bone also acts as a reservoir of calcium in the body. In man about 17% of calcium in the skeleton is recycled each year. About 40% of ^{45}Ca is absorbed from the gut, and around 65% of that is deposited in the skeleton. A similar route is followed by ^{90}Sr, although a smaller proportion is absorbed from the gut and the urinary excretion is greater.

The thyroid gland normally concentrates iodine to form an iodinated thyroid hormone, and consequently ^{131}I follows this pathway. It is readily absorbed from the gut ($f_1 = 1.0$) and from the lungs if inhaled, and although the major part is rapidly excreted in the urine, about 30% is accumulated by the thyroid, later being lost from the body with a $t_{1/2}$ of about 100 days. The proportion taken up by the thyroid decreases if the intake of stable iodide is increased.

21.11.3   Elements that Concentrate in a Number of Tissues

Plutonium is an example in this group. Being an α emitter, it has a particularly high potential for damage to tissues. Its compounds can range from being soluble in water, *e.g.* plutonium nitrate or chloride, to being chemically inert and insoluble, *e.g.* plutonium(IV) oxide. The soluble compounds readily hydrolyse in water at near neutral pH forming an insoluble hydrated oxide. They can also complex with other molecules *in vivo*, thus remaining in solution.

Plutonium mainly enters the body by inhalation. The soluble component is rapidly absorbed from the lungs, transported in the blood either to be excreted through the kidneys or deposited in tissues, mainly in the bone and liver. The material remaining is in the form of the hydrated oxide polymer or of insoluble compounds, such as PuO_2. This particulate material may be ingested by macrophages, which may then migrate to lymph nodes, or be removed by ciliary action, to be swallowed and excreted in the faeces. Some solution in lung fluids with transfer to the blood circulation may also occur.

Of plutonium activity entering the blood, about 45% is deposited in the liver, 45% in the skeleton and the remainder either excreted or deposited in other tissues. Biological half-lives in the bone and liver are about 100 and 40 years, respectively. From animal studies it would appear that the lungs, the cells of the inner surface of bone, the bone marrow and the liver are most at risk from accidental intakes of plutonium.

BIBLIOGRAPHY

A. Martin and S.A. Harbison, *An Introduction to Radiation Protection*, 4th edn, Chapman & Hall, London, 1996.

International Agency for Research on Cancer, IARC monographs on the evaluation of carcinogenic risk to humans, *Ionizing Radiation, Part 1: X- and Gamma (γ)-Radiation, and Neutrons*, Vol. 75, IARC Press, Lyon, 2000.

International Agency for Research on Cancer, IARC monographs on the evaluation of carcinogenic risk to humans, *Ionizing Radiation, Part 2: Some Internally Deposited Radionuclides*, Vol. 78, IARC Press, Lyon, 2001.

International Commission on Radiological Protection, Limits for intakes of radionuclides by workers (ICRP publication 30), *Annals of the ICRP*, 1979, **2(3–4)**, 1–116.

International Commission on Radiological Protection, 1990 recommendations of the International Commission on Radiological Protection (ICRP publication 60), *Annals of the ICRP*, 1991, **21(1–3)**, 1–197.

International Commission on Radiological Protection, Annual limits on intake of radionuclides by workers based on the 1990 recommendations (ICRP publication 61), *Annals of the ICRP*, 1991, **21(4)**, 1–41.

International Commission on Radiological Protection, Relative biological effectiveness (RBE), quality factor (Q), and radiation weighting factor ($w(R)$). A report of the International Commission on Radiological Protection (ICRP publication 92), *Annals of the ICRP*, 2003, **33(4)**, 1–117.

J. Shapiro, *Radiation Protection. A Guide for Scientists, Regulators and Physicians*, 4th edn, Harvard University Press, Cambridge, Mass, 2002.

J.S. Hughes, *Ionising Radiation Exposure of the UK Population: 1999 Review (Report No NRPB-R311)*, National Radiological Protection Board*, Didcot, England, 1999.

National Radiological Protection Board, *Living with Radiation*, 6th edn, National Radiological Protection Board*, Didcot, England, 2000.

*Now the Radiation Protection Division of the Health Protection Agency.

Chapter 22

Biocides and Pesticides

BIRGER HEINZOW AND HELLE RAUN ANDERSEN

22.1 INTRODUCTION

About 75% of animal species in the world are insects. Some are beneficial predators and pollinators, but many are pests, acting as competitors for food. Others are vectors of infectious and parasitic disease. Not surprisingly, man has always tried to control such pests. In the past, preparations containing sulfur, arsenic compounds, extracts of tobacco and chrysanthemum, and strychnine were used, but only the synthetic pesticides produced by application of modern chemistry has been really successful. The agricultural yield has increased dramatically over the last 50 years and biocides have played a major role in this. Unfortunately, many of the compounds used may be harmful to the environment when used carelessly.

A biocide is any substance used with the intention of killing living organisms. A pesticide is defined as any substance or mixture intended for preventing, repelling, and destroying or mitigating any pest. The ideal pesticide is one that is toxic primarily to the target pests and is rapidly inactivated in the environment but few such pesticides exist. The development of more selective, less persistent, and safer pesticides is one of the great demands on modern chemistry. Based on presumed selectivity, biocides and pesticides may be classified into the groups shown in Table 22.1.

Although these groupings suggest selectivity, it must be emphasized that biocides may be harmful to any living organism, to ecosystems, and to man, depending upon the route of exposure and the dose. Table 22.2 lists toxic effects on humans that have been identified following acute, relatively high-level exposure.

There is at present great concern about pesticide residues in food and in the environment, and the possibility of harm to humans following long term low-level exposure with the development of chronic poisoning. These problems will not be considered here, but it should be remembered that a risk may be tolerable if sufficient benefit is obtained from the use of a pesticide. For example, a small risk from chronic exposure may be accepted for a pesticide that helps to eradicate diseases like malaria and river blindness.

Recent years have seen the introduction of biological pest control agents. These agents may be chemicals or living microorganisms. The chemical agents may be put into the following four classes:

    (i) Semiochemicals (chemicals emitted by living organisms that modify the behaviour of other organisms of the same or other species)
    (ii) Hormones

Table 22.1 *Detailed classification of biocides and pesticides*

Biocide/Pesticide	Pest
Rodenticides	Rodents (rats, mice, *etc.*)
Avicides	Birds
Nemat(od)icides	Nematodes (pin-worms)
Molluscicides	Molluscs (slugs, snails, *etc.*)
Insecticides	Insects
Acaricides	Acarians (mites)
Arachnicides	Spiders
Larvacides	Larvae
Miticides	Mites
Scabicides	Scabies
Pediculicides	Lice
Herbicides	Plants (used against weeds and unwanted vegetation)
Fungicides	Fungi (mildews, moulds)

Table 22.2 *Toxic effects of biocides on humans*

Compound	Symptoms of intoxication
Aniline	Methaemoglobinaemia, irritation of skin and mucous epithelium, sensitization
Chlorinated hydrocarbons	Central nervous system: neurotoxicity, seizures, tremor, hypotension
Dinitrocresol	Thirst, sweating, hyper-thermia, vertigo, vomiting, diarrhoea, cyanosis, arrhythmias, pulmonary oedema, liver and kidney failure
Dithiocarbamate	Irritation of skin an mucous epithelium, vertigo, vomiting, diarrhoea
Phenoxycarbonic acid	Irritation of skin and mucous epithelium (ulceration), muscle fibrillation, peripheral neuropathy (pain), hyperglycaemia
Carbonic acid	Irritation of skin and mucous epithelium, ulceration, necrosis
Dipyridinium	Irritation of skin and mucous epithelium, vomiting, diarrhoea, hyper-thermia. After latency, renal failure, liver damage
Paraquat	Lung fibrosis after latency period
Organophosphate esters	Headache, blurred vision, sweating, dyspnoea, vomiting, cyanosis, muscle fibrillation, seizures, miosis, bradycardia
Carbamate	Similar symptoms to organophosphate esters, faster onset and decline
Pyrethrum, pyrethrin	Central and peripheral neurotoxicity, irritation, paraesthesia, seizures, sensitization
Pyrethroids	Tremor, salivation, chorea, paraesthesia
Organotin	Central nervous system: neurotoxicity, depression, anorexia, diarrhoea, headache, vertigo, vomiting, blurred vision
Coumarin	Blood-clotting disturbance

  (iii) Natural plant regulators and insect growth regulators
  (iv) Enzymes.

Semiochemicals may be subclassified into pheromones, allomones, and kairomones. Pheromones are substances emitted by members of one species that modify the behaviour of others within the same species. Allomones are chemicals emitted by one species that modify the behaviour of a different species, to the benefit of the emitting species. Kairomones are chemicals emitted by one species that modify the behaviour of a different species to the benefit of the receptor species.

Examples of pesticides of natural origin include the insecticide azadirachtin, derived from the neem tree (*Azadirachta indica*), an insect growth regulator that interferes with the moulting hormone ecdysone and the insecticidal toxins produced by the vegetative forms of *Bacillus thuringiensis*. Insect growth regulators of the juvenile hormone type such as azadirachtin appear to act specifically on insects and thus have low mammalian toxicity.

Microbial pest control agents include naturally occurring organisms such as bacteria, fungi, viruses, and protozoa, as well as genetically modified microorganisms.

22.2   ORGANOCHLORINE INSECTICIDES

Dichlorodiphenyltrichloroethane (DDT) was the first of a variety of contact organochlorine (chlorinated hydrocarbon) insecticides, which include aldrin, dieldrin, endrin, chlordane, and hexachlorobenzene. These compounds were used extensively in the mid-1940s to 1960s. Their properties of low volatility, chemical stability, and environmental persistence led to their bioaccumulation (bioconcentration, biomagnification) in the food chain of fish, birds, and mammals owing to their lipophilicity and slow metabolic degradation. It was then demonstrated that these compounds (notably DDT, DDE (Dichlorodiphenyldichloroethylene), and cyclodienes) possess hormone disrupting and cytochrome P450 inducing properties that interfere with the reproductive system, although the presenting symptoms are varied and non-specific (Table 22.3). In avian species of a high trophic level like pelicans, seagulls, and eagles, the adverse effects of the dichlorodiphenyl derivatives is related to induction of the steroid-metabolizing enzymes and the inability of the reproductive organs to mobilize enough calcium in the production of the eggshell. This eggshell-thinning leads to cracks allowing bacteria to infiltrate with resultant death of the foetus even if there is not complete

Table 22.3   *Symptoms of organochlorine poisoning*

Compound/group	Acute symptoms	Chronic symptoms
Dichlorodiphenyls (*e.g.* DDT, DDD, dicofol, perthane, methoxychlor)	Paraesthesia, ataxia, dizziness, headache, nausea, vomiting, fatigue, lethargy, tremor	Anorexia, weight loss, anaemia, tremor, weakness, hyper-excitability, anxiety
Benzene and cyclohexane-types (*e.g.* HCH, lindane)	Headache, tremor, convulsion	
Cylodienes (*e.g.* aldrin, dieldrin, heptachlor, chlordane, endosulfan, mirex, toxaphene)	Dizziness, headache, nausea, vomiting, hyper-excitability, hyper-reflexia, myoclonic jerking, convulsion	Dizziness, headache, hyper-excitability, muscle twitching, myoclonic jerking, anxiety, insomnia, irritability, epileptiform convulsions
Chlordecone, mirex		Arthralgia, rash, ataxia, slurred speech, blurred vision, irritability, loss of memory, weakness, tremor, decrease of sperm count

breakage of the egg in the nest. Behavioural changes occurring in some avian species can interfere with courting behaviour and cause hyperactivity on the nest accentuating egg breakage.

Recently, the hormone-like activity of dichlorodiphenyl dichloroethylene has raised concern in connection with the steady decline of sperm counts in man over the past 30 years. This arises partly because DDT contamination of the aquatic environment has been associated with feminization of male alligators. Further, the main metabolite of *p,p'*-DDT, *p,p'*-DDE has a high affinity for the androgen receptor (AR) with anti-androgenic activity. In contrast, *o,p'*-DDT and *o,p'*-DDE bind to the oestrogen receptor (ER) and act like estrogens. Other organochlorine pesticides with identified hormone-like activity include endosulfan, toxaphene, dieldrin, DBCP (Dibromochloropropane), and methoxychlor.

An interesting characteristic of these compounds is their partitioning into the atmosphere from land and water. Depending on their vapour pressure, compounds like DDT and toxaphenes volatize in hot climate zones. As the organochlorine vapour moves into low-temperature zones, condensation occurs depositing the organochlorine in regions where they have never been used as pesticides. This global transfer explains why organochlorine residues are found in the tissues of wildlife species in colder areas.

22.2.1 Mechanism of Toxic Action of Organochlorines

DDT-type insecticides interact with the neuronal membrane by altering the membrane permeability (transport) for potassium and sodium and the calcium-mediated processes. By inhibiting these functions, the repolarization of the nerves is disturbed resulting in hyper-excitability.

Cyclodienes and cyclohexane compounds have a central nervous system-stimulating mode of action. These compounds antagonize the neurotransmitter, gamma-aminobutyric acid (GABA), permitting only partial repolarization of the neurone and, thus, uncoordinated nervous excitation. The inhibition of repolarization reflects the inhibition of the nerve Na-, K-ATPase (adenosine triphosphatase) and also of the Ca-Mg-ATPase. This inhibition results in the accumulation of intracellular free calcium.

Owing to their lipophilicity, organochlorine compounds are partitioned and stored largely in the adipose tissue, where they are biologically inactive. There is an equilibrium between body fat and free-circulating compounds. Redistribution and mobilization of fat, for example due to disease, ageing, or fasting, may result in mobilization of stored organochlorines in quantities that may lead to manifest toxicity. Owing to the different patterns of use of organochlorine pesticides in industrialized and developing countries, the pattern of distribution of the residues in human fat (including breast milk) differs among countries.

Today most organochlorine pesticides are considered obsolete and have been banned. Exceptions are DDT, which is permitted in some regions for vector-borne malaria control, and endosulfan, which is non-persistent.

22.3 ORGANOPHOSPHATES AND CARBAMATES

A large number (>200) of insecticides are – derived from esters of phosphoric acid, phosphorothioic acid, and carbamic acid. These potent insecticides produce their biological action by blocking the nervous tissue enzyme acetylcholinesterase

(AChE) in ganglia and in the parasympathetic nervous system (see Chapter 17). This enzyme hydrolyses the neurotransmitter acetylcholine (ACh) and, if it is inhibited, ACh accumulates at the nerve endings (see Chapter 17) causing continual and unco-ordinated ACh stimulation of the muscarinic receptors of the parasympathetic nervous system, and the stimulation and blocking of nicotinic receptors in the ganglia of the autonomic nervous system, the motor muscles, and the central nervous system. The clinical symptoms are described in Table 22.4.

Two further manifestations of toxicity may occur with some compounds:

  (i) Intermediate syndrome, a paralytic condition mainly of cranial nerves (caused by fenthion, dimethoate, monocrotophos, methamidophos)
  (ii) Organophosphate-induced delayed neurotoxicity (caused by leptophos), with symptoms of pyramidal tract damage such as spasticity.

Intermediate syndrome is a paralytic condition mainly affecting the cranial nerves. Organophosphate-induced delayed neurotoxicity reflects damage to the afferent fibres of the peripheral and central nerves and is related to inhibition of 'neuropathy target esterase' (NTE). The delayed syndrome has been termed organophosphate-induced delayed neuropathy (OPIDN), and is manifested chiefly by weakness or paralysis and paraesthesia of the extremities, predominantly the legs. This syndrome may persist for years.

Epidemiologic studies also suggest that a proportion of patients acutely poisoned by any organophosphate may experience some long-term neuropsychiatric symptoms.

22.3.1 Mechanism and Site of Action

Organosphosphorus esters react with the active site of AChE (a serine hydroxyl group) resulting in a phosphorylated and inhibited enzyme. With some insecticides (and particularly with the chemically related war nerve gases), this process is irreversible and the duration and severity of toxicity are prolonged. New generation organophosphates have been developed with reversible phosphorylation allowing spontaneous dissociation from the nervous tissue.

Table 22.4 *Symptoms of organophosphate and carbamate inhibition of acetylcholine, esterase*

Acute symptoms	Chronic symptoms
Miosis, hypersecretion, bronchoconstriction, diarrhoea, cramps, urination, bradycardia, cardiac arrest, muscle fasciculation, tremor, weakness, paralysis, restlessness, ataxia, lethargy, confusion, loss of memory, convulsion, respiratory depression, coma	Loss of ambition and libido, autonomic dysfunction, cephalgia, gastrointestinal symptoms, delayed and lasting neuropathy and neuropsychological dysfunction, emotional lability, confusion, memory loss, anxiety, depression, insomnia, ataxia, speech difficulty, muscle weakness, neurological deficits

Carbamate esters bind to the active site of AChE forming a carbamylated and inhibited protein. Subsequently, the enzyme becomes decarbamylated and active again. In principle, the only difference between organophosphate and carbamate pesticides lies in the rate constants for the dephosphorylation and decarbamoylation processes. The rate constant for dephosphorylation is several orders of magnitude slower than that for the natural substrate ACh and also much slower than that for decarbamylation.

22.3.2 Specific Treatment of Acetylcholine Esterase Poisoning

Choline esterase inhibitor poisoning by organophosphates or carbamates is a serious, and potentially fatal, medical emergency and requires immediate treatment with the antagonist atropine and, in the case of organophosphate poisoning only, reactivation of AChE with oxime therapy. The degree and time course of the intoxication can be monitored by analysis of the activity of non-specific serum choline esterase and of specific erythrocyte AChE. The life-threatening symptoms of respiratory depression and hypoxia require immediate medical and supportive intervention.

22.3.3 Dithiocarbamates

Dithiocarbamates such as the fungicides maneb and mancozeb are non-cholinesterase-inhibiting substances and are generally of low toxicity. However, exposure to these pesticides combined with alcohol ingestion may produce headaches, palpitations, nausea, vomiting, and a flushed face.

22.4 NICOTINOID INSECTICIDES

Imidacloprid is a systemic, chloro-nicotinyl insecticide interfering with the transmission of stimuli in the insect nervous system. Specifically, it causes a blockage of the nicotinergic neuronal pathway (again with accumulation of ACh). This pathway is more abundant in insects than in warm-blooded animals, giving a degree of selective toxicity for insects associated with only moderate toxicity in mammals. Although fatal human poisoning has not been described, signs and symptoms of poisoning would be expected to be similar to those of nicotine poisoning. Signs and symptoms would include fatigue, twitching, cramps, and muscle weakness.

22.5 PYRETHROID INSECTICIDES

Synthetic pyrethroids are currently among the most widely used pesticides. Historically, natural pyrethroids were extracted from chrysanthemum flowers, which contain the insecticidal esters called pyrethrines. Pyrethrum is the name given to the oleoresin extract of dried chrysanthemum flowers. The keto-alcoholic esters of chrysanthemic and pyrethroic acids are known as pyrethrins, cinerins, and jasmolins. These lipophilic esters rapidly penetrate the outer integument of many insects and paralyze their nervous systems. They are often used in combination with the synergists, piperonyl butoxide and *n*-octyl bicycloheptene dicarboximide. The synergists enhance the toxicity of pyrethrins by inhibiting the enzymes that can destroy them.

Table 22.5 *The two classes of pyrethroids*

Chemical characteristics	Toxicity class	Symptoms
Class 1		
Pyrethrin, allethrin, tetramethrin, resmethrin, phenothrin, permethrin	T-syndrome Action on the central and peripheral nervous systems	Hyperexcitation, tremor, prostration
Class 2		
a-cyano substituted: Cypermethrin, deltamethrin, fenvalerate, fluvalinate	CS-syndrome Action on the mammalian central nervous system	Cutaneous paraesthesia, salivation, chronic seizures, dermal tingling

The synthetic insecticides are known as pyrethroids (chemically similar to pyrethrins) and are more stable. Being very potent insecticides with low mammalian toxicity, they have gained a strong market share. Based on the characteristics of poisoning, two classes of pyrethroids are distinguished, as shown in Table 22.5.

Allergic sensitization occurs with natural pyrethrum, but there are rarely any other toxic effects. Some pyrethroids may cause local facial sensations, which are not associated with systemic poisoning. The effects are reversible and no treatment is necessary.

22.5.1 Mechanism of Action

The type I esters affect the sodium channels in nerve membranes with prolongation of sodium influx causing repetitive neuronal discharge and prolonged after-potential but no severe membrane depolarization. The type II α-cyanoesters lead to greater and more prolonged sodium influx with persistent membrane depolarization and eventually nerve blockade. Whereas the first group exerts its main effects on synaptic transmission causing hyper-excitability and tremor, the second group shows its first effects on the sensory nervous system. Other actions include inhibition of Na^{2+}/Mg^{2+} ATPase and alteration of calcium and chloride ion homeostasis.

22.6 OTHER PESTICIDES

22.6.1 Fungicides

Fungicides consist of a diverse range of chemical substances ranging from inorganic compounds such as sulfur and copper to complex organic compounds. The main groups are dithiocarbamates, azoles and benzimidazoles, and organophosphates. Most currently used fungicides have low acute toxicity, but some fungicides possess biological characteristics that could cause chronic effects. Fungicides are cytotoxic and most show positive results in *in vitro* mutagenicity testing, a characteristic inherent to this group of biocides.

Several classes of fungicides have been developed to inhibit fungal membrane synthesis and growth by inhibiting specific CYP450 enzymes in the sterol pathways. The process of steroidogenesis seems to be conserved throughout living organisms and so these fungicides may also inhibit mammalian steroidogenesis. One of these enzymes, aromatase CYP450 converts C19 androgens to aromatic C18 oestrogens. Several fungicides *e.g.* fenarimol and prochloraz inhibit aromatase activity in mammals and affect mating behaviour and other reproduction parameters.

One of the first chemicals reported to be an anti-androgen was the fungicide vinclozolin. Since then several other pesticides have been demonstrated to possess anti-androgenic activity *e.g. p,p'*-DDE, procymidone, linuron, methoxychlor, fenitrothion, and prochloraz.

Signs and symptoms of poisoning with dinitrophenolic compounds (*e.g.* DNOC (dinitro-ortho-cresol), binapacryl, dinoterb) include tremors, increased respiratory rate, sweating, lethargy and insomnia, nausea, restlessness, thirst, raised body temperature, tachycardia, fatigue. The dinitrophenols affect oxidative phosphorylation and poisoning will thus lead to sudden increase in metabolic rate, as will be the poisoning with pentachlorophenol (PCP). PCP is readily absorbed through the skin and poisoning results in uncoupling of oxidative phosphorylation. Severe poisoning will result in profuse sweating, dehydration, accelerated heart rate, elevated temperature (42 °C), nausea and vomiting, coma and death.

22.6.2 Herbicides

Chlorophenoxy-compounds (2,4-D, 2,4,5-T, MCPA) have, according to the WHO classification (see below), moderate toxicity to humans. Symptoms like dizziness, nausea, diarrhoea, weakness, fatigue, kidney failure are unspecific. A specific effect related to the production of 2,4,5-T was a type of contact dermatitis: chemical worker chloracne. The underlying cause was exposure to 2,3,7,8-TCDD (dioxin), a by-product formed during the synthesis of the compound. This substance was released in quantity during the Seveso explosion in 1976. Chloracne was the most obvious effect on those exposed as a result of this explosion and an ongoing study of other possible effects including cancer continues. There is also continuing concern about the possible effects of the dioxin present in the herbicide 'Agent Orange', a defoliant sprayed during the Vietnam conflict in large quantities (>10 million gallons).

The bipyridyl herbicides (*e.g.* paraquat, diquat) can cause severe cases of poisoning with delayed development after exposure. When ingested, paraquat has life-threatening effects on the gastrointestinal tract, kidney, liver, heart, and other organs. The lung is the primary target organ of paraquat, and pulmonary effects represent the most lethal and least treatable manifestation of toxicity. The underlying mechanism is through the generation of free radicals with oxidative damage to lung tissue. While acute pulmonary oedema and early lung damage may occur within a few hours of severe acute exposures, death, most commonly occurs 7–14 days after the ingestion. Initially (within hours), irritation of mouth and throat, with nausea, vomiting, abdominal pain, and diarrhoea (often bloody) may occur and later (1–3 days) signs of kidney and liver damage. For paraquat, 5–14 days after poisoning, the delayed toxic damage of pulmonary fibrosis occurs with progressive dyspnoea, resulting in death from respiratory failure. Severe poisoning from both chemicals can result in shock and death within a few hours of intake.

22.6.3 Rodenticides

22.6.3.1 Anticoagulants (warfarin, bromadiolone, difenacoum, chlorophacinone).
Warfarin and related compounds such as coumarins and indandiones are the most
commonly used rodenticides. Anticoagulants act through inhibition of blood
clotting. Coumarins and indandiones depress the hepatic synthesis of vitamin K
dependent blood-clotting factors (II (prothrombin), VII, IX, and X). Signs and symptoms of poisoning are nausea, vomiting, diarrhoea upon ingestion. Bleeding from
nose and gums, blood in excretions; internal bleeding leading to shock and coma.
Treatment with vitamin K is necessary even in the absence of symptoms because of
increased tendency to bleed.

22.7 PESTICIDE RESIDUES IN FOOD AND DRINKING WATER

Pesticide residues in food and drinking water must be considered in any adequate
risk assessment of the effects of the use of pesticides. Pesticides are currently
perceived by the public as posing a major health threat to the well-being of
the population. In consumer surveys, usually about 80% of those questioned
consider pesticide residues as a major concern.

Processes of risk assessment for pesticide residues are based on three components:
estimation of residues in food, estimation of food-consumption patterns from
national surveys, and lexicological characterization by comparison of exposure
estimates with data from animal toxicology studies. Commonly the theoretical
maximum legal residue level is used in the calculation. For extrapolation from animal experiments to man and the calculation of acceptable daily intake (ADI) of the
United Nations Food and Agriculture Organisation/World Health Organisation
(WHO) or the reference dose (RfD) of the US Environment Protection Agency, two
main assumptions are made: (i) effects in animals can be used to predict effects in
man and (ii) effects of high doses can be mathematically related to effects (or lack
of effects) of low doses. To accommodate the uncertainty, safety (or uncertainty)
factors are often applied. These range from 10 to 10,000, 100 being the most commonly
used. If exposure is below the ADI or RfD (the amount of a substance a human can
consume daily without harm), the risk associated is considered insignificant.

In most cases, the maximum residue concentrations found in food are far below
the ADI and RfDs. However, owing to misuse or inadvertent addition of pesticides
to food, cases of intoxication may occur. For example, the illegal application of
aldicarb to watermelons resulted in several hundred cases of acute cholinergic intoxication in western USA in 1985. A similar illegal application of aldicarb to cucum-
bers caused acute gastrointestinal illness in 400 school children in London in 1994.

The most commonly involved pesticides in non-occupational human poisoning
are organophosphates, carbamates, chlorinated hydrocarbons, and organic mercurials.
Food is the most common vehicle of exposure, followed by skin contact, and respiratory exposure. Pesticides have made their contribution to the quantity and quality
of food available and their use will continue for many years, but there is a need for
monitoring environmental impact and long-term health effects in man of low-
level exposure to ensure that the benefits of pesticides can be enjoyed without any
accompanying harm.

22.8   PESTICIDE EXPOSURE IN THE OCCUPATIONAL SETTING

Workers involved in the production or use of pesticides or post application re-entry activities may be exposed to pesticides. The most important occupational exposure routes are through the skin and by inhalation of fumes or aerosols. The exposure is usually highest by the dermal route, but the inhalation route can be important for pesticides with high vapour pressure, especially if they are applied in confined spaces such as greenhouses. In addition, some application techniques generate a high proportion of respirable particles or aerosols, especially during mixing and application of pesticides. Proper use of appropriate protective equipment is essential to reduce personal exposure and to prevent acute poisoning. People working near by areas during pesticide application may be exposed to drift and vapour. Such exposure is called 'bystander exposure'.

At re-entry activities, the exposure will depend on the crop types and amounts of pesticides applied, the degradation time combined with the time elapsed between application and handling of the treated plants as well as the specific work functions, the skin area exposed and duration of the work process. External dermal exposure can be determined by several methods such as the patch method (use of absorbent cloth or paper patches attached to different body regions inside and outside clothing), the whole body method (clothing, usually two layers of cotton, that act as the pesticide collection media) or use of fluorescent tracers or visible dyes added to the pesticide formulation. However, none of the methods is practical for routine use, and they will give no information about the internal exposure of the pesticides as this depends on the ability of the pesticide to penetrate the skin.

Healthy skin is a remarkably good barrier with complex properties that distinguish it from a simple membrane. Very lipophilic chemicals may pass slowly through the skin or even remain in the stratum corneum and form a reservoir.

The ability of a pesticide to penetrate the skin depends not only on the physico-chemical properties of the active substance but also on the total chemical composition of the formulated product. Several co-formulants, as for example, solvents and surfactants, can enhance the uptake of the active pesticide ingredient. A hydrophilic vehicle may promote penetration of lipophilic substances by compensating for the lipid solubility in the stratum corneum, while a lipophilic vehicle may promote the penetration of a water-soluble substance that cannot itself penetrate the fatty layer.

Organophosphates constitute the pesticide group, which has been most studied. Many organophosphates are readily absorbed through the skin and fatal poisonings have occurred during their application owing to lack of protective equipment and unsafe handling.

22.9   EXPOSURE ASSESSMENT

22.9.1   Biological Monitoring (Biomarkers)

Biomarkers are biologically relevant measurements that can relate to internal exposure (such as urinary excretion or plasma concentration data) or effect (such as erythrocyte acetyl cholinesterase activity). The major advantages of biomarkers are that they evaluate the actual absorption and integrate absorption from all routes of exposure.

The absorbed dose of a pesticide can only be quantified if the metabolism and toxicokinetics of the parent compound are understood, ideally from human studies. Pesticides that are extensively metabolized to a large number of metabolites may not be good candidates for biological monitoring, although a minor specific metabolite might be used. Metabolites common to groups of pesticides can be used, such as the urinary alkyl phosphate metabolites as biomarkers of exposure to organophosphates. Urine samples are usually analyzed for the presence of five dialkylphosphate (DAP) compounds produced by the metabolism of most organophosphate (OP) pesticides: dimethylphosphate (DMP), dimethylthiophosphate (DMTP), dimethyldithiophosphate (DMDTP), diethylphosphate (DEP), and diethylthiophosphate (DETP).

Measurement of urinary alkyl phosphates and specific urinary metabolites of pyrethroids have been used to estimate the exposure of the general population to organophosphates and pyrethroids. These measurements are biomarkers of exposure and they do not reflect the toxic effect of the exposure. The toxicity of the parent organophosphates may vary markedly, while producing the same pattern of alkyl phosphate excretion.

Inhibition of blood cholinesterase activity (plasma cholinesterase (ChE) and erythrocyte AChE) may be measured as a biomarker of total exposure to organophosphate and carbamate pesticides. Inhibition of erythrocyte AChE could also be considered as a biomarker of effect, reflecting the inhibition of AChE in neural tissue and neuro-muscular junctions. However, reduction of ChE activity in blood is a rather insensitive indicator of the absorbed dose due to the large variability within and between individuals and therefore, these measurements cannot be used to investigate small differences in exposure in the general population. At the individual level, ChE activity is only useful in cases of severe exposure to organophosphates or carbamates and a pre-exposure baseline ChE activity from the individual is necessary. The ChE activity must show at least 15–20% depression from an individual's normal level to be considered indicative of pesticide over-exposure.

Persistent organochlorine insecticides, such as DDT, and their metabolites may be measured in human serum, tissue, and breast milk to determine exposure. Samples from all over the world illustrate the global distribution of these pesticides and their accumulation in the food chain. The levels of these substances are now declining because most have been banned for more than 20 years. An exception is DDT, which is still in use for malaria control in some countries until a more effective method to control the mosquito vectors is found.

22.10   RISK ASSESSMENT AND RISK MANAGEMENT

As a guide for the classification of the toxicological hazards of pesticides, WHO has developed the following classes based on the acute toxicity (LD50 for the rat) of the compound in $mg\,kg^{-1}$ body weight by the oral or dermal route (Table 22.6).

Although the ADI or RfD for individual pesticides are seldom exceeded, the presence of multiple pesticide residues on food crops has raised public concern and has triggered work both in the USA and EU to look at the potential effects of pesticide mixtures. Mixtures of similarly acting pesticides are likely to show additivity of effects, and this has been demonstrated experimentally in several studies. In the past, the effects of combinations of pesticides have not been considered as a normal part

Table 22.6 *WHO classification of pesticide toxicity*

Class of toxicity hazard	Solids		Liquids	
	Oral route LD_{50} $mg\ kg^{-1}\ bw^{-1}$	Dermal route LD_{50} $mg\ kg^{-1}\ bw^{-1}$	Oral route LD_{50} $mg\ kg^{-1}\ bw^{-1}$	Dermal route LD_{50} $mg\ kg^{-1}\ bw^{-1}$
Ia Extreme	≤5	≤10	≤20	≤40
Ib High	5–50	10–100	20–200	40–400
II Moderate	50–500	100–1000	200–2000	400–4000
III Slight	>500	>1000	>2000	>4000
III+ Unlikely	>2000	–	>3000	–

of risk assessment, despite the fact that some, such as organophosphates and carbamates, and some pesticides possessing estrogenic or anti-androgenic effects, share a common mode of action and additivity is to be anticipated. Recently, pesticides exhibiting anti-ChE activity have been subject to cumulative risk assessment by the US Environmental Protection Agency.

22.11 SUSCEPTIBLE AND VULNERABLE GROUPS

Age, sex, genetic make-up, health status and previous or concurrent exposures may influence individual sensitivity towards pesticide exposures. Certain groups in the population, notably pregnant women and young children, may be at higher risk from some pesticides than adults. Neurotoxic and endocrine disrupting pesticides may affect the developing brain and endocrine system of foetuses and of young children. Exposure during pregnancy or early childhood to endocrine disruptors is associated with reproductive, neurological, and behavioural problems. These chemicals can disrupt normal cellular communication, limit the production of chemical messengers, and interfere with development of organs in the immune and nervous systems, reproductive function, and growth processes as well as increasing the incidence of specific diseases (*e.g.* childhood diabetes, childhood cancer, and thyroid diseases).

Young children have a higher food-intake per unit body weight than adults and so they are often the critical group in the population to be considered for risk assessment in relation to poisons in food. Measurements of urinary alkyl phosphates have demonstrated a higher internal exposure level of organophosphates in children than in adults. In addition, metabolizing enzymes are not fully developed in foetuses and newborns making them more susceptible to some pesticides.

In relation to genetic factors, defects or polymorphism in metabolizing enzymes may increase the sensitivity to certain pesticides. For example, low activity of para-oxonase may increase the sensitivity to organophosphates.

22.12 CONCLUSION

Pesticides are generally designed to kill higher organisms and thus pose special toxicological problems. However, despite the development of some non-chemical methods of pest control, there is at present no generally satisfactory alternative to chemical control. Between total ban and careless use, there is a pathway of safe, careful application following rational risk–benefit assessment and this is the aim of appropriate pesticide management, *i.e.* – to optimize the benefits obtained while minimizing the risks.

In general, the production, use, and disposal of biocides is regulated to safeguard human health and to limit environmental impact. This involves detailed (and costly) pre-registration procedures to provide information on compound and residue chemistry, environmental fate and kinetics, toxicity, environmental impact, product efficacy, and application performance and safety. The information collected is generally available in databases to facilitate relevant decision-making. Further, in most countries residues in crops and food are routinely monitored to ensure concentrations are below tolerance levels established by government agencies.

Many of the present concerns about pesticides relate to pesticides that were introduced some time ago without the extensive toxicity and environmental impact studies required today. An ongoing concern is the occurrence of pesticide residues in ground- and drinking water in the EU where the drinking water limit value of $0.1 \mu g L^{-1}$ has been exceeded, most often by residues of the herbicides atrazine and diuron. However, this limit value for pesticides is not derived as toxicological tolerable concentration but rather as hygienic and preventive position that no pesticides whatever should contaminate ground- and drinking water.

National governments have introduced defined residue limits and guideline levels for pesticide residues in water, most commonly developed for drinking water, but values have also been proposed for environmental waters, effluent waters, irrigation waters, and livestock drinking waters. It is important to check on the definitions of terms used in legislation since the same terminology may have different meanings in different systems. For example, the term 'guideline value' (GV) as used by WHO means a recommended maximum value calculated from health considerations and has no legal force, whereas in some countries, a GV is at or about the analytical limit of determination or a maximum level that might occur if good practices are followed and may be legally enforceable.

BIBLIOGRAPHY

J.C. Aldous, G.A. Ellam, V. Murray and G. Pike, An outbreak of illness among school children in London: toxic poisoning not mass hysteria, *J. Epidemiol, Commu. Health*, 1994, **48**, 41–45.

D.B. Barr, R. Bravo, G. Weerasekera, L.M. Caltabiano, R.D. Whitehead Jr., A.O. Olsson, S.P. Caudill, S.E. Schober, J.R. Pirkle, E.J. Sampson, R.J. Jackson and L.L. Needham, Concentrations of dialkyl phosphate metabolites of organophosphorus pesticides in the U.S. population, *Environ. Health Perspect.*, 2004, **112**, 186–200.

Compendium of Pesticide Common Names website (This Compendium is believed to be the only place where all of the ISO-approved standard names of chemical pesticides are listed. It also includes approved names from national and international bodies for pesticides that do not have ISO names) http://www.hclrss. demon.co.uk/index.html.

M. Eddleston, L. Karalliedde, N. Buckley, R. Fernando, G. Hutchinson, G. Isbister, F. Konradsen, D. Murray, J.C. Piola, N. Senanayake, R. Sheriff, S. Singh, S.B. Siwach and L. Smit., Pesticide poisoning in the developing world – a minimum pesticides list, *Lancet*, 2002, **360**, 1163–1167.

P. Grandjean, *Skin Penetration. Hazardous Chemicals at Work*, Taylor & Francis, London, 1990.

D.J. Hamilton, A. Ambrus, R.M. Dieterle, A.S. Felsot, C.A. Harris, P.T. Holland, A. Katayama, N. Kurihara, J. Linders, J. Unsworth and S.-S. Wong, Regulatory limits for pesticide residues in water (IUPAC Technical Report), *Pure Appl. Chem.*, 2003, **75**, 1123–1155.

J. Hardt and J. Angerer, Determination of dialkyl phosphates in human urine using gas chromatography-mass spectrometry, *J. Anal. Toxicol.*, 2002, **4**, 678–684.

U. Heudorf, J. Angerer and H. Drexler, Current internal exposure to pesticides in children and adolescents in Germany: urinary levels of metabolites of pyrethroid and organophosphorus insecticides, *Int. Arch. Occup. Environ. Health*, 2004, **77**, 67–72.

J. Jeyaratnam, Acute pesticide poisoning: a major global health problem, *World Health Stat. Q.*, 1990, **43**, 139–144.

C.D. Klaasen (ed), *Casarett and Doulls Toxicology – The Basic Science of Poisons*, 6th edn, Pergamon Press, New York, 2001.

D. Koh and J. Jeyaratnam, Pesticides hazards in developing countries, *Sci. Total Environ.*, 1996, **188** (Suppl 1), S78–S85.

R.I. Krieger, *Handbook of Pesticide Toxicology*, Academic Press, New York, 2001.

B.W. Lee, L. London, J. Paulauskis, J. Myers and D.C. Christiani, Association between human paraoxonase gene polymorphism and chronic symptoms in pesticide-exposed workers, *J. Occup. Environ. Med.*, 2003, **45**, 118–122.

B. Mackness, P. Durrington, A. Povey, S. Thomson, M. Dippnall, M. Mackness, T. Smith and N. Cherry, Paraoxonase and susceptibility to organophosphorus poisoning in farmers dipping sheep, *Pharmacogenetics*, 2003, **13**, 81–88.

C. Nellemann, M. Dalgaard, H.R. Lam and A.M. Vinggaard, The combined effects of vinclozolin and procymidone do not deviate from expected additivity *in vitro* and *in vivo*, *Toxicol. Sci.*, 2003, **71**, 251–262.

R. Reigart and J.R. Roberts, Recognition and management of pesticide poisonings, 5th edn, USEPA, 1999, www.epa.gov/oppfead1/safety/healthcare/handbook/handbook.htm.

J.R. Richardson, H.W. Chambers and J.E. Chambers, Analysis of the additivity of *in vitro* inhibition of cholinesterase by mixtures of chlorpyrifos-oxon and azinphos-methyl-oxon, *Toxicol. Appl. Pharmacol.*, 2001, **172**, 128–139.

C.D.S. Tomlin, *The Pesticide Manual*, 13th edn, British Crop Protection Council, Farnham, 2003.

G.W. Ware and H.N. Nigg (eds), Minimizing human exposure to pesticides, Symposium Proceedings, ACS, San Francisco 1992, *Rev. Environ. Contam. Toxicol.*, 1992, **128–129** (two complete volumes).

C.K. Winter, Dietary pesticide risk assessment, *Rev. Environ. Contam. Toxicol.*, 1992, **127**, 23–67.

Chapter 23

Toxicology in the Clinical Laboratory

ROBIN A. BRAITHWAITE

23.1 INTRODUCTION

The nature of poisoning including its laboratory diagnosis and medical management has changed dramatically over the last 30 years. There is now a wider range of drugs and other substances taken in "overdose" or involved in cases of suspected abuse or "accidental" poisoning (see Table 23.1). Many of the cases in the UK admitted to hospital Accidental & Emergency Departments involve ingestion of alcohol, non-opioid analgesics, *e.g.* paracetamol; antidepressants, *e.g.* selective serotonin re-uptake inhibitors (SSRI) and tricyclic antidepressants; benzodiazepines, *e.g.* diazepam and illicit drugs. However, the length of hospital stay is relatively short in most cases, <24 h, and mortality is very low, <1%. The majority of deaths from poisoning occur outside of hospital care and in recent years there has been a significant increase in deaths due to illicit drug use, particularly heroin. This situation is in sharp contrast to the problem in many developing countries where a far larger proportion of poisoned patients may have ingested pesticides or corrosive chemicals, or occupationally or environmentally exposed to many other hazardous substances including natural toxins.

Over recent years, there has been an increased emphasis on supportive care in the clinical management of acutely poisoned patients; also the acute management of most, but not all patients, requires access to routine biochemical and haematological investigations, which are shown in Table 23.2.

The use of diuresis, haemodialysis and haemoperfusion to speed up the elimination of drugs and poisons from the body has greatly decreased, whereas the use of both single and repeated doses of activated charcoal has become a common, safe and effective alternative. Antidotes can be valuable, both to help establish a diagnosis and in the treatment of specific poisons, *e.g.* naloxone in the treatment of opioid poisoning and acetylcysteine (Parvolex) in paracetamol poisoning.

Although diagnostic problems have generally become more complex, knowledge of the identity of the poison(s) does not usually influence the immediate management of most patients, except for a few specific agents such as paracetamol, methanol or ethylene glycol. The decision to carry out toxicological investigations is generally influenced by specific diagnostic and clinical management problems, as well as availability and pressure on clinical resources (see Table 23.3).

Table 23.1 *Types of compounds that may be involved in a case of poisoning*

Over the counter medicines (OTC)
Prescription only medicines (POM)
Illicit drugs and other abused substances
Household products
Beverages and food supplements
Agrochemicals
Industrial chemicals
Herbal and "traditional" medicines
Poisonous plants
Cosmetics

Table 23.2 *Quantitative biochemical investigation of blood and plasma*

	Blood
Glucose concentration	Hypoglycaemia
	Hyperglycaemia
Acid–base status	pH
	pCO_2
	pO_2
Carboxyhaemoglobin/methaemoglobin	
Acetylcholinesterase (red cell)	
Prothrombin time (INR)	

	Plasma
Electrolytes (Na, K, Ca, Mg)	
Urea, creatinine	
Osmolarity	
Anion gap	
Cholinesterase	
Creatinine kinase (CK)	

Table 23.3 *Clinical situations where laboratory investigations may be useful in the diagnosis and management of poisoning*

Where the diagnosis of the patient is uncertain
Assessment of the severity of poisoning
When administration of an antidote depends on the identification of a poison and/or its concentration in blood
Estimation of risk of complications or fatal outcome to poisoning
Use of an active elimination technique, *e.g.* haemofiltration, and monitoring its efficacy
Investigation of non-accidental poisoning in children, *e.g.* Munchausen Syndrome by Proxy
Confirmation of "recreational poisoning" due to substance abuse
Investigation of new drugs or an unusual clinical presentation
Investigation of adverse drug reactions and iatrogenic poisoning
Confirmation of brain death
Occupational or environmental exposure to chemicals and the investigation of chemical incidents
Investigation of deaths due to suspected drug abuse or poisoning

If toxicological investigations are to influence acute patient management they are required relatively urgently, within 1–2 h. For most non-urgent cases, laboratory investigations for diagnostic purposes are required within 1–2 days. In the United Kingdom in recent years, there has been an increased interest in clinical audit and the epidemiology of poisoning, including drug abuse, associated with patients admitted to hospital, as well as trends in fatalities reported to Her Majesty's Coroners. However, there are no simple tests or single techniques that are able to detect all the drug and chemical poisons that might be involved in a case of suspected poisoning. Many patients these days have coincidentally taken a cocktail of drugs including alcohol and/or illicit drugs, which can complicate the diagnostic picture. Patients from some countries and some ethnic communities may have access to both unusual pharmaceutical products and "traditional" medicines such as herbal remedies. Because of the diverse physical and chemical nature of such "poisons" and their disparate pharmacological and toxicological effects, schemes of analysis, for an "unknown" drug or poison, are difficult to undertake and are generally only undertaken by specialist toxicology laboratories.

23.2   SPECIMEN COLLECTION FOR TOXICOLOGICAL ANALYSIS

It is vitally important to collect appropriate specimens from the patient for any laboratory investigation of poisoning. In the investigation of poisoning the following specimens are recommended:

Urine: 20 mL in a plain plastic sterile container, avoiding the use of preservatives, *e.g.* boric acid.

Blood: 10 mL in an anticoagulated container, heparin or EDTA, but avoid the use of tubes containing gel. Plasma should be separated and stored at 4 °C prior to analysis. In some cases, however, whole blood may be required for analysis, *e.g.* cyanide, lead. Retention of the original specimen container can be helpful in medico-legal cases.

Stomach aspirate: Stomach aspiration or washout is no longer a recommended practice. If available, they are not normally of value unless urine is not available or tablets/capsules are visible or identifiable. Analysis of stomach contents can, however, be useful in post-mortem investigation of suspicious deaths.

Scene residues: Any drug(s) or syringes found at the scene, or contents of any container associated with the poisoning should be retained.

Separate recommendations are given in the case of specimen collection as part of a post-mortem investigation of suspected poisoning or drug or alcohol ingestion in road traffic fatalities. In such cases it is important to specify the site(s) of blood collection and quantitative measurements should only be undertaken on peripheral blood, *e.g.* femoral vein. The collection of vitreous fluid from the eye, stomach contents and certain tissue specimens, *e.g.* liver, brain, muscle, may also be important in certain cases.

Specimens should be stored in a refrigerator (4 °C) if there is delay in transport to the laboratory. Most drugs and poisons are fairly stable when stored at 4 °C for several days. However, some poisons may be unstable or volatile and require special storage or stabilisation, *e.g.* volatile solvents, gases, cyanide and alcohol. Best practice is to

Table 23.4 *Important techniques for the laboratory investigation of poisoning*

Simple "spot" tests
Immunoassay (IA)
Thin-layer chromatography (TLC)
Gas chromatography (GC)
GC-mass spectrometry (GC-MS)
Liquid chromatography (LC)
LC-mass spectrometry (LC-MS)
Atomic absorption spectrometry (AAS)
Electrothermal atomic absorption spectrometry (ETAAS)
Inductively coupled plasma-mass spectrometry (ICP-MS)

fill blood specimen containers to the top and secure the top, to reduce the volume of "dead space" or chance of leakage. Additional care in sampling and storage, and also use of chain of custody procedures, *e.g.* a detailed log of receipt of specimen(s) and its subsequent handling by the laboratory, are required in all medico-legal cases, particular those that might result in criminal charges.

23.3 CHOICE OF LABORATORY TECHNIQUES

A wide range of common techniques may be employed in the qualitative and quantitative analysis of drug and chemical poisons in biological fluids (see Table 23.4). Immunoassay, gas and liquid chromatography and gas chromatography–mass spectrometry and more recently liquid chromatography–mass spectrometry, are the recommended modern techniques for reliable qualitative screening and quantitative analysis of most drugs and chemical poisons and their metabolites. Less common chemical poisons, *e.g.* metals, are best analysed by atomic absorption spectrometry or inductively coupled plasma–mass spectrometry (ICP-MS).

23.4 BIOCHEMICAL AND HAEMATOLOGICAL INVESTIGATION IN THE INVESTIGATION OF POISONING

Metabolic or blood coagulation disturbances occur frequently in critically ill patients, and those who have been severely poisoned are no exception. Simple, readily available biochemical and haematological tests are carried out initially and are usually of more value than toxicological tests, since the immediate care of poisoned patients is largely supportive. These biochemical and haematological tests may be helpful in diagnosis, and will sometimes be required to monitor the patient's clinical progress (see Table 23.2).

23.4.1 Blood Gases

Clinical and biochemical assessment of the adequacy of ventilation, lung function, is absolutely essential in any patient with an impaired level of consciousness or breathing difficulty. This is determined by measuring the oxygen (pO_2) and carbon dioxide (pCO_2) partial pressure in arterial blood. A low arterial pO_2 is termed hypoxia and indicates respiratory impairment.

23.4.2 Acid Base

Abnormalities, particularly metabolic acidosis, are common in critically ill patients, including those who are poisoned. Such abnormalities usually require treatment in their own right, and may be of diagnostic help in overdosage involving so-called "metabolic poisons" such as methanol or ethylene glycol.

Metabolic acidosis occurs either because the production of acids in the body exceeds the patient's capacity to buffer or eliminate hydrogen ions or because of excessive loss of HCO_3^- from the body. These two mechanisms may be differentiated by measuring what is known as the "plasma anion gap", which is elevated when the metabolic acidosis is due to accumulation of hydrogen ion, and normal when the acidosis is due to renal loss of bicarbonate. In overdosage, high anion gap acidosis, usually lactic acidosis, occurs more commonly than a "normal" anion gap acidosis.

$$\Delta AG = [\text{plasma sodium}] - \sum [\text{plasma chloride} + \text{plasma bicarbonate}]$$

where ΔAG is the anion gap, with reference values 8–16 mmol L^{-1}.

23.4.3 Fluid and Electrolyte Disorders

23.4.3.1 Sodium and Water. Abnormalities of the plasma sodium ion concentration are relatively uncommon in cases of poisoning. The major exception to this rule is with ingestion of "ecstasy" methylenedioxymethamphetamine (MDMA) or related drugs in which hyponatraemia, low plasma sodium concentration, has been described in a number of cases, and may in itself have contributed to a fatal outcome by causing cerebral oedema. The mechanism of hyponatraemia in MDMA poisoning is probably due to the syndrome of inappropriate antidiuretic hormone secretion (SIADH), exacerbated by the ingestion of excessive quantities of water or hypotonic fluids.

Significant hypernatraemia, *i.e.* plasma sodium concentration greater than 200 mmol L^{-1}, may sometimes be seen in cases of salt poisoning, usually those involving infants and young children including suspected Munchausen Syndrome by Proxy. Determining the concentration of electrolytes, sodium and potassium are sometimes useful in the investigation of fatalities, but can only be carried out on blood plasma collected before death. However, after death, specimens taken from the eye, vitreous fluid, can be useful. Measurement of vitreous potassium concentrations is sometimes used to determine the likely time of death; potassium concentrations in vitreous fluid slowly rise after death.

23.4.3.2 Plasma Osmolality and Osmolal Gaps. Since the plasma sodium concentration is rarely abnormal in overdose it follows that plasma osmolality is usually normal. However, the ingestion of any alcohol or glycol will increase the plasma osmolality, which when measured will be significantly greater, >10 mOsmol kg^{-1} than that calculated from measurement of the plasma electrolyte, glucose and urea concentrations. Such elevated osmolal gaps can be useful, although not infallible, indicators of alcohol or glycol ingestion, including ethylene, propylene and diethylene glycol.

23.4.3.3 Calcium. Disorders of calcium homeostasis are relatively unusual in poisoning. However, hypocalcaemia is characteristically found in ethylene glycol poisoning, which causes insoluble complexes to be formed between calcium and oxalate due to the metabolism of ethylene glycol to oxalic and other acidic products.

Calcium oxalate crystals are deposited in the tissues, and crystalluria can sometimes be seen on microscopic examination of urine. Hypocalcaemia and a high-anion-gap metabolic acidosis with an elevated osmolal gap is diagnostically strongly indicative of ethylene glycol poisoning.

23.5 SUBSTANCES OF CLINICAL OR MEDICO-LEGAL INTEREST

23.5.1 Alcohols and Glycols

Measurement of blood or urine alcohol in the clinical laboratory is not required in most cases in which excess alcohol intake is suspected. However, blood alcohol measurements can be useful in the assessment of unconscious or very intoxicated patients, particularly when binge drinking is suspected. For reference purposes, the legal limit for alcohol in blood in the UK is $80 \, \text{mg} \, \text{dL}^{-1}$ $(0.8 \, \text{g} \, \text{L}^{-1})$. The equivalent concentration in urine, assuming complete absorption of the alcohol, is $107 \, \text{mg} \, \text{dL}^{-1}$ $(1.1 \, \text{g} \, \text{L}^{-1})$. Breath alcohol measurement devices may sometimes be used in accident and emergency departments or other clinics to detect drinkers. The legal limit for breath alcohol in drivers is $35 \, \mu\text{g} \, \text{dL}^{-1}$ in the UK. The clinical effect of any given blood, urine or breath concentration of alcohol depends on an individual's previous experience of alcohol use; chronic alcoholics tolerate far higher alcohol levels than the social or novice drinker. Individuals also vary in the rate at which they eliminate alcohol, which generally ranges between 12 and $25 \, \text{mg} \, \text{dL}^{-1} \, \text{h}^{-1}$ with an average population elimination rate of $18 \, \text{mg} \, \text{dL}^{-1} \, \text{h}^{-1}$. Regular heavy drinkers and young children eliminate alcohol at a faster rate, *i.e.* $25–35 \, \text{mg} \, \text{dL}^{-1} \, \text{h}^{-1}$. Measurement of blood or urine alcohol can be of value in cases involving children or under-age adolescents, where medico-legal issues may be involved or where ingestion is strongly denied, and retrospective measurement might be useful for future follow-up of suspected alcohol abuse.

Measurement of alcohol is commonly carried out in hospital laboratories by enzymatic procedures that will measure ethyl alcohol, but not methyl or isopropyl alcohol. In screening for poisons particularly when there is evidence of metabolic acidosis, alcohol is best measured by gas liquid chromatography, which is able to resolve the presence of different alcohols. Simultaneous measurement of methanol and ethanol is important in the active management of methanol poisoning where ethyl alcohol can be used as an antidote to prevent the metabolism of methanol by liver enzymes to produce formic acid; also when haemodialysis or haemofiltration may be used to enhance elimination of methanol or ethylene glycol.

In cases of fatalities involving suspected alcohol ingestion or road traffic accidents it is important to ensure proper specimen collection, documentation and storage. In the case of post-mortem investigations, it is important to take blood specimens from a peripheral site, *e.g.* femoral vein; urine and vitreous humour samples may also be useful in certain cases, particularly if there has been extensive injuries to the body in a traffic accident.

In cases of suspected (ethylene) glycol ingestion, or of severe metabolic acidosis, it is useful to measure ethylene glycol in blood plasma as an emergency investigation by using gas liquid chromatography, using a method capable of differentiating between different glycols, particularly ethylene and propylene glycol. It is also important to be able to differentiate between ethylene glycol and other substances in cases of suspected non-accidental poisoning.

23.5.2 Amphetamines and related Drugs of Abuse

The abuse of stimulants such as amphetamine and methamphetamine has been well known for many years (see Figure 23.1). However, methamphetamine abuse is largely confined to North America, Japan and Far East. Amphetamine, but not methamphetamine, is a common inexpensive drug of abuse in the UK. However, the abuse of some of their newer analogues such as methylenedioxyamphetamine (MDA), MDMA ("ecstasy") and methylenedioxyethyl-amphetamine (MDEA) is increasingly common, although users rarely require hospital care. Admission may be required following "overdosage" with such drugs or as a result of adverse reactions, which are most commonly seen as delirium, with cardiovascular manifestations such as heart rhythm disturbances, hypertension and hyperpyrexia, increased body temperature and heat stroke. Other adverse effects that have been described include liver toxicity and neuropsychiatric illness such as depression and suicide linked to destruction of serotonergic pathways in the brain.

Reliable immunoassay techniques are used by clinical laboratories for the qualitative detection of the amphetamine class of drugs, but are unable to identify individual drugs, which may include closely related analogues, also some OTC such as pseudo ephedrine. Identification of individual amphetamines and their metabolites is best carried out using gas-liquid chromatography with nitrogen-specific or mass spectrometric detection on blood or urine specimens. Quantitative measurements are only of use in the retrospective investigation of unusual or severe presentations with "ecstasy" and related amphetamine analogues, particularly in the case of fatalities. However, the metabolism of these drugs is very complex due to the presence of different enantiomers. There is also an increase in the abuse of some older, less commonly used anesthetic agents such as ketamine and γ-hydroxybutyrate (GHB), which have been linked to the rave or dance music scene. These are not amphetamines, but are associated with the abuse of amphetamine derivatives such as "ecstasy". These drugs have also been linked to cases of "date rape", drug facilitated sexual assault (DFSA), where the victim's drink may have been deliberately spiked. These drugs are structurally unrelated to the amphetamines and difficult to detect by routine screening procedures, particularly those based on immunoassay. In addition,

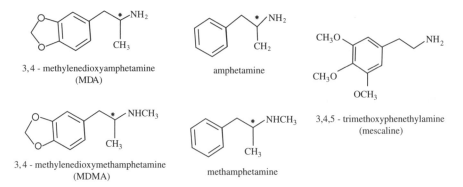

Figure 23.1 *Structure of ecstasy (MDMA) and MDA in comparison with amphetamine and methamphetamine and mescaline. (* chiral centre)*

such drugs are rapidly eliminated from the body and if ingestion is suspected it is advisable to obtain specimens, particularly urine, within 12 h of ingestion.

23.5.3 Cannabis

Cannabinoids represent a wide range of 60 or more pharmacologically active and inactive compounds derived from the plant *Cannabis satvia*, also known as marijuana in North America. The main active component is tetrahydrocannabinol (THC), which exerts its effects by interacting with specific receptors, CB1 and CB2, in the brain (Figure 23.2). The plant has been used for its medicinal and intoxicant properties for several thousands of years. It is widely recognised as a drug of abuse, but may have medicinal properties that could be of benefit to some patients suffering from particular chronic illnesses such as multiple sclerosis and other neurological disorders. For this reason, a number of synthetic cannabinoid-like drugs have been synthesised, *e.g.* nabilone, for clinical trial.

Cannabis is mainly abused by smoking on its own or mixed with tobacco. Generally, the effects last for about 2–4 h, but may be prolonged for up to 6 h. There is also evidence to suggest that the drug impairs driving performance and that chronic use may increase the risk of mental illness, particularly psychoses.

Laboratory investigations may be carried out in order to detect the use of cannabis in patients and some employees also in the case of traffic accidents or acute psychosis. The presence of cannabinoids can be detected in body fluids, blood, urine, saliva, using widely available immunoassay techniques; however, these are unable to differentiate between the active drug and its several metabolites. Confirmation of the presence of various cannabinoids, particularly that of THC and its main carboxylic acid metabolite (see Figure 23.2) is generally carried out using GC–MS or LC–MS with limits of measurement of different cannabinoids of $<1\,\mu g\,L^{-1}$.

Following use or abuse of cannabis products there is a long residence time in the body. Whereas the main active component, THC, may be detected in blood for a relatively short period of time after dosing, generally up to 12 h or less, its inactive carboxylic acid metabolite (see Figure 23.2) may be detected in urine for several weeks following last use in heavy or regular users of the drug.

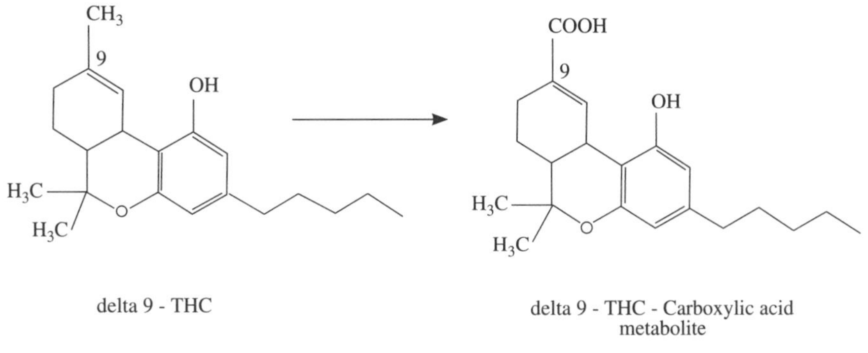

delta 9 - THC

delta 9 - THC - Carboxylic acid metabolite

Figure 23.2 *Structure of important cannabinoid compounds: pharmacologically active drug Δ^9-tetrahydrocannabinal (THC) and its inactive carboxylic acid metabolite*

23.5.4 Carbon Monoxide

Carbon monoxide (CO) is the product of incomplete combustion of organic material and can cause serious injury or death in fires, as well as a wide variety of other sources of exposure (see Table 23.5). In the case of fires, many deaths may be caused by the inhalation of CO and other fumes, particularly in those people trapped in burning buildings. Accidental carbon monoxide poisoning, often unsuspected, may occur at home with open fires, or portable heaters without proper ventilation. Historically, carbon monoxide poisoning was associated with suicide in homes with coal gas ovens. However, modern domestic gas supplies, "natural" gas, contain no carbon monoxide. But, inhalation of fumes from car engines or other combustion power tools can be the cause of accidents, also fatal self-poisoning. But, most modern cars, particularly if fitted with a catalytic converter, do not produce much carbon monoxide, making suicide much more difficult. Carbon monoxide can also be produced following exposure to the common solvent dichloromethane, particularly following inhalation of fumes over a prolonged period with poor ventilation.

Carbon monoxide poisoning may sometimes be difficult to diagnose as it can cause a wide variety of signs and symptoms that can sometimes mimic other common disease states (see Table 23.6). If exposure to carbon monoxide is not suspected, it is possible that a diagnosis of carbon monoxide poisoning may be missed or delayed, particularly in "low-level" chronic exposure.

Table 23.5 *Common causes of carbon monoxide poisoning*

Fires and explosions
Faulty gas water heaters
Blocked fireplaces or chimneys
Gas cookers
Poorly ventilated portable paraffin and gas heaters
Coal or wood burning kitchen stoves due to faulty operation or poor ventilation
Car exhausts; running engine in confined area or pipe attached to exhaust (suicide attempt)
Use of combustion power tools, *e.g.* lawn mowers, chain saws in confined areas without ventilation
Cooking or heating from charcoal indoors without ventilation
Methylene chloride used in confined areas

Table 23.6 *Signs and symptoms of carbon monoxide poisoning*

Headache
Pink skin and mucosae
Hearing loss
Nausea and vomiting
Weakness
Confusion
Loss of consciousness
Ataxia
Neurological signs; late signs or following chronic exposure
Psychiatric symptoms; late signs or following chronic exposure
Exacerbation of pre-existing disease, *e.g.* angina

Table 23.7 *Relationship between carbon monoxide saturation (COHb) and toxicity*

Less than 3% reference value; non-smokers
Less than 10% reference values; smokers
10–20% headache, nausea, and unstable angina in coronary heart disease
20–30% headache, nausea, dizziness, and digestive problems
30–50% severe headache, vomiting, and impaired consciousness
Greater than 50% coma, convulsions, respiratory depression, and death[a]

[a] Wide variation in COHb in fatalities.

Carbon monoxide has an affinity for haemoglobin more than 200 times that of oxygen, causing a reduction in the O_2-carrying capacity of the blood. At the cellular level, CO binds to cytochrome A_3, blocking the electron transport chain, which leads to a reduction in oxygen to vital tissues and organs such as the brain, causing a reduction in aerobic respiration. Measurement of blood carboxy haemoglobin can be essential in the confirmation of a diagnosis of carbon monoxide poisoning, in cases where there has been recent exposure to the gas or source of combustion. However, in cases of "low-level" exposure it is important to take into account any history of smoking which can also lead to increased carboxyhaemoglobin values. The reference value in non-smokers is $<3\%$, and in smokers, 10% or less. The relationship between carbon monoxide in blood and toxicity is shown in Table 23.7.

A variety of techniques can be used by the clinical laboratory in the measurement of carboxyhaemoglobin, and include spectrophotometry and gas chromatography. In the case of emergency investigations some types of blood gas analyser are able to give a reading of oxygen, carbon dioxide and CO saturation in blood within minutes. If suitably stoppered, blood specimens may not lose CO if stored in a refrigerator, even for prolonged periods.

23.5.5 Chlorphenoxy Pesticides

The phenoxy herbicides are chemical analogues of plant growth hormones and include compounds such as 2,4-dichlorophenoxyacetic acid (2,4-D) and 2,4,5-trichlorophenoxyacetic acid (2,4,5-T). These compounds were historically used as war agents, defoliants. In Vietnam, they were also associated with an increased incidence of birth defects and certain cancers, possibly due to "contamination" with dioxins.

This group of pesticides, although relatively uncommon in overdose, may be important to detect in blood and urine, particularly in cases where there is deep coma and metabolic acidosis. Such compounds would be missed by most protocols to screen for the presence of "unknown" poisons. Liquid chromatographic procedures are generally used in order to confirm the ingestion of this group of compounds, particularly if active elimination treatment using forced alkaline diuresis is being undertaken.

23.5.6 Cocaine

Cocaine is a very potent naturally occurring stimulant, and local anaesthetic, derived from *Erythroxylon coca* a shrub that is a native of South America. The plant's stimulant properties have been known to local indigenous populations for many thousands of years. Cocaine itself was not isolated until the mid-19th century and it was subsequently developed as a local anaesthetic, which is still used today. Abuse of cocaine developed in the early 20th century with nasal insuflation (snorting) of the powder, cocaine hydrochloride. In the last 30 years abuse of cocaine in the form of its free base (crack) has caused a much greater problem of addiction. The free base form of cocaine is absorbed more rapidly than nasal absorption of the salt; as a consequence it is far more addictive.

Cocaine is rapidly hydrolysed in the body by various esterases in the blood, or by direct chemical hydrolysis. Its main, inactive, metabolite in the body is benzoylecogonine, although a large number of other cocaine metabolites are known (see Figure 23.3). When cocaine is used with concurrent alcohol ingestion a unique cocaine metabolite, cocaethylene, may be formed which is thought to have greater toxicity than cocaine itself. In addition, a specific pyrolysis product of cocaine may be detected when "crack" cocaine is smoked.

Chronic abuse of cocaine is associated with a wide range of morbidity and mortality affecting both the central nervous and cardiovascular systems. This may result in sudden death in relatively young users of the drug, or exacerbation of pre-existing disease in older patients. Laboratory investigation of suspected cocaine abuse or intoxication is helpful in detecting such abuse or the investigation of suspected cocaine related fatalities. However, concentrations in blood relate poorly to signs and symptoms of toxicity. Cocaine and its metabolites can be detected in body fluids, blood, urine, saliva, using commonly available immunoassay techniques. However, this technique is unable to differentiate between cocaine and its various metabolites. Specific confirmation of cocaine usage is generally carried out using GC–MS, usually for the presence of the main, inactive, metabolite benzoylecognine in blood, urine or saliva.

23.5.7 Cyanide

Cyanide poisoning is relatively rare and mostly involves suicidal ingestion of cyanide salts. In such cases there is often access to cyanide salts in the workplace. Hydrogen cyanide (HCN), or prussic acid is a highly toxic volatile liquid and fumes of HCN are given off when cyanide salts are mixed with acids. Although HCN has a characteristic almond-like odour, but due to genetic differences, only about 50% of the population is unable to detect its smell. Soluble salts of cyanide, include potassium and sodium cyanide, are widely used industrially in electro-plating and metal processing, also as laboratory reagents. Hydrogen cyanide can also be formed as a combustion product in fires from nitrogen containing materials such as wool and silk or synthetic polymers such as polyurethanes, polyamides and polymides. Less common sources of cyanide include the accidental or intentional ingestion of cyanogenic plants or their seeds.

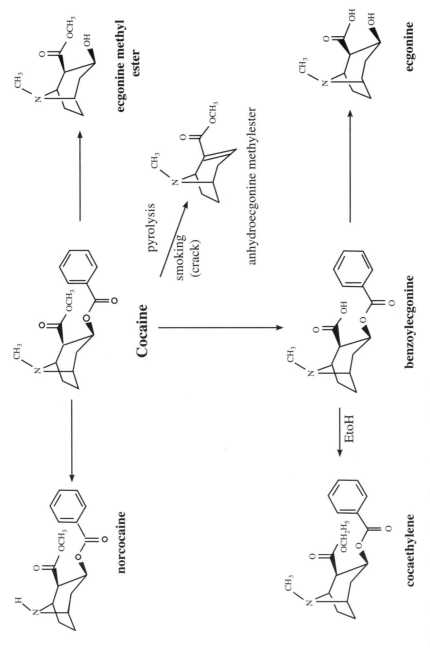

Figure 23.3 *Structure of cocaine and its major metabolites*

Table 23.8   *Signs and symptoms of acute cyanide poisoning*

Mild–moderate	Severe
Dizziness	Apnoea
Headache	Cardiovascular failure
Palpitations	Respiratory arrest
Anxiety	Metabolic acidosis
Confusion	Smell of bitter almonds[a]
Dyspnoea	Brick-red colour skin[a]

[a] Not every case.

The signs and symptoms of cyanide acute poisoning are shown in Table 23.8 and appear rapidly after inhalation of HCN or significant ingestion of cyanide salts; the estimated fatal doses are approximately 100 mg HCN or 300 mg potassium cyanide. The mechanism of cyanide toxicity is blockage of electron transport in cytochrome a–a^3 complex, leading to a dramatic decrease in oxidative metabolism and cellular hypoxia, which most directly affects the brain and heart. Cyanide is rapidly metabolised in the liver by the enzyme rhodanase to thiocyanate (SCN), which is largely non-toxic. As a consequence, blood cyanide concentrations decline rapidly following exposure, or ingestion, with an estimated elimination half-life of 1–2 h. A number of antidotes are useful in the treatment of cyanide poisoning, *e.g.* cobalt EDTA and hydroxocobalamin.

The measurement of blood cyanide concentrations may be useful where the diagnosis is uncertain, or when trying to judge the efficacy or effective dose of antidotes. Very small quantities of cyanide are found in blood due to normal metabolic processes, but reference values are $<0.05\,\text{mg L}^{-1}$. Minor signs and symptoms of cyanide toxicity are associated with blood cyanide concentrations up to $1\,\text{mg L}^{-1}$. Severe symptoms are generally associated with blood cyanide concentrations in excess of $1\,\text{mg L}^{-1}$. Concentrations in excess of $10\,\text{mg L}^{-1}$, following suicidal ingestion or industrial exposure, are associated with a fatal outcome. In post-mortem cases following suicidal ingestion of cyanide salts very high blood cyanide concentrations, $>20\,\text{mg L}^{-1}$, may be observed, which may be due to diffusion of unabsorbed cyanide from the stomach into blood after death, particularly for specimens taken from "central" areas of the body such as the heart.

23.5.8   Metals, Metalloids and their Salts

Acute poisoning with metals and their inorganic compounds is relatively rare. However, chronic exposure to such compounds, occupationally or environmentally, may present a much bigger problem. More often the laboratory may be asked to carry out "screening tests" on patients who present with neurological or unusual gastrointestinal symptoms that might be associated with chronic exposure or ingestion of toxic metals such as lead or mercury. Unusual sources of metal poisoning have included ceramic kitchen ware, "traditional" or herbal medicines, cosmetics, metal objects and children's toys. The elements for which screening is often carried out are lead, cadmium, arsenic, mercury and thallium. However, with the introduction of improved technology such as ICP–MS it is possible to screen for the presence of "elevated" concentrations of a much wider range of elements in blood, urine, hair and nails, *e.g.* antimony, tin, nickel. Requests for investigation of particular metals may arise from abnormal biochemistry or haematology investigations. These investigations include: lead in anaemia; basophilic stippling of red cells, raised red-cell zinc protoprophyrin (ZPP); mercury in proteinuria; arsenic in abnormal liver function tests and cadmium in increased urinary excretion of low-molecular-weight proteins (β_2 microglobulin, retinol-binding protein).

23.5.8.1 Aluminium. Measurement of plasma or whole blood aluminium is commonly undertaken in the management of patients with end-stage renal failure receiving treatment with haemodialysis or haemofiltration. Common sources of aluminium include the water used in dialysis treatment and the failure of water treatment devices such as reverse osmosis designed to prevent transfer of aluminium into blood. In addition, aluminium containing drugs may be given to such patients in order to control absorption of phosphate. Aluminium may also be measured in water used for dialysis, and also in dialysate fluids. Aluminium overload leads to a variety of severe life-threatening problems such as osteomalacia, anaemia and encephalopathy, known as dialysis dementia.

23.5.8.2 Arsenic. Arsenic has historically always been associated with poisoning, particularly in murder cases, up until the late 19th century. This was the first time that reliable chemical tests were established for the detection of specific poisons and is still associated with the names of their inventors, *e.g.* James Marsh in England and Hugo Reinch in Germany. Toxicologically, arsenic is still an important element and it may be found in many different organic and inorganic forms. It is also widely distributed in the environment, particularly rocks and sediments, leading to widespread contamination of drinking water supplies in different parts of the world, particularly in West Bengal and Bangladesh. Arsenic has three main oxidation states, As(O), As(III) and As(V). The oxidation III state is particularly toxic, *e.g.* arsenic(III) oxide (As_2O_3). The hydride of arsenic(III), arsine (AsH_3), is extremely toxic causing haemolysis of red blood cells, liver and kidney failure; a number of serious industrial accidents have been associated with its release. Organic forms of arsenic, *e.g.* arsenocholine and arsenobetaine, are almost non-toxic and are naturally occurring in seafood; complex arseno sugars are found in edible seaweed. An added complexity in the investigation of suspected arsenic poisoning is that inorganic forms of arsenic are metabolised by the liver, forming mono- and dimethyl species which are less toxic.

A wide range of signs and symptoms are associated with arsenic poisoning and depend on the form of arsenic ingested, the dose and duration of exposure. In cases of acute poisoning there may be severe gastroenteritis with bloody diarrhoea, vomiting and coma. Chronic poisoning is also associated with characteristic skin changes, and an increased incidence of cancer of the liver, skin and lungs. Laboratory investigation of suspected arsenic poisoning, particularly in symptomatic cases, or recent exposure, involves the analysis of arsenic and its metabolites in blood or urine. In cases of

Table 23.9 *Measurement of arsenic species in biological fluids*

Arsenic(III), Arsenic(V)	Inorganic arsenic species
Monomethyl arsonic acid (MMAA) ⎫	
Methyl arsonate ⎬	Inorganic methylated arsenic
Dimethyl arsinic Acid (DMAA) ⎭	metabolite species
(cacodylic acid)	
Complex organo-As compounds,	Complex dietary organic arsenic
e.g. arsenobetaine, arsenocholine and	species from seafood, *e.g.* fish,
arsenosugars	shellfish and seaweed

historical arsenic poisoning the determination of arsenic in samples of hair or nails can be useful. However, the choice of analytical technique is very important, particularly the need to be able to differentiate between different forms of arsenic (see Table 23.9). Reference values for the presence of arsenic in non-exposed populations are $<10\,\mu g\,L^{-1}$ in blood and urine. Modern methods of arsenic analysis are based on ETAAS or ICP–MS. Determination of different arsenic species in biological specimens and food sources may be carried out using Liquid chromatography linked to ICP–MS.

23.5.8.3 Cadmium. Cadmium and its salts and alloys are widely used in many industries. Major uses include nickel–cadmium batteries, colour pigments and special alloys. Many of the risks associated with occupational and environmental exposure to cadmium have been known for many years. Important health risks include emphysema from the inhalation of cadmium fumes and renal impairment, which is the major target organ for toxicity. For those not occupationally exposed, the major source of cadmium exposure remains the diet, although cadmium is poorly absorbed from the gut. Inhalation of tobacco smoke represents a continuing and important source of exposure to cadmium, as the bioavailability of cadmium by the lung is very high. Following absorption, cadmium has an extremely long residence time in the body of decades and is mainly stored in liver and kidney, where it is bound to a protein metallothionein. The most important effect of excessive cadmium exposure is its effects on the kidney with a characteristic increase in excretion of low-molecular-weight proteins such as β_2-microglobulin (B$_2$M) and retinol-binding protein (RBP), which are indicative of renal tubular rather than glomerular damage. Assessment of cadmium exposure can be carried out by measurement of blood or urine cadmium concentration. The most common technique is that of ETAAS or ICP–MS.

Reference values for cadmium in blood and urine in non-exposed populations are extremely low. Blood cadmium concentrations are $<1\,\mu g\,L^{-1}$ in non-smokers and $4\,\mu g\,L^{-1}$ or less in smokers. Urine concentrations of cadmium are generally $<1\,\mu g\,L^{-1}$ in both smokers and non-smokers. Current occupational guidance values for the assessment of cadmium exposure are $5\,\mu g\,L^{-1}$ in both blood and urine. Cadmium-induced renal impairment is generally related to urine cadmium concentrations in excess of $15\,\mu g\,L^{-1}$. However, elevated blood and urine cadmium concentrations may be observed many years after the cessation of excessive exposure.

23.5.8.4 Lead. Acute and chronic lead poisoning still presents a significant problem that is caused by both occupational and environmental exposure. Some of the most common environmental sources of lead exposure include lead water pipes and lead-based paints in older houses, cosmetics, imported ceramics, herbal and traditional medicines. The absorption of lead by the gastrointestinal tract is influenced by diet and nutritional status, which is particularly important in pregnancy, nursing mothers and young children, particularly those that might be deficient in iron and calcium. Inhalation of lead fumes is a more important route of exposure in factory workers, or those using indoor shooting ranges or involved in removing lead paint by burning.

Signs and symptoms of lead poisoning are relatively non-specific and include metallic taste, tiredness, lethargy, abdominal pain, weight loss, learning difficulties in children, anaemia and renal impairment; of particular concern is the effect of lead exposure on brain development in young children. Extensive evidence indicates that lead may have a damaging effect on child behaviour and intelligence at relatively low levels of exposure. Laboratory investigations, particularly measurement of lead in blood, play an important part in the diagnosis of lead poisoning. Lead is known to have an influence on haemoglobin synthesis that leads to anaemia following chronic exposure. This may be suspected in patients presenting with unexplained low haemoglobin values, increased red-cell zinc protoporphyrin or basophilic stippling of red cells.

The toxicokinetics of lead is particularly complex, which can make interpretation of lead measurements in biological fluids quite difficult. In the blood, lead is almost entirely bound to the red cells, but is in equilibrium with lead in soft tissues and bone. Analysis of lead in biological fluids and tissues is generally undertaken using ETAAS and ICP–MS. This latter technique also has the ability to determine the different isotopes of lead present in a specimen, which can give some indication of the likely source of the lead in the environment.

Measurement of lead in blood is the single most useful index of recent lead exposure. Measurement of lead in urine is most useful in the assessment of recent lead exposure where chelation therapy may be carried out to reduce the body burden of lead. Reference value for blood lead in non-occupationally exposed populations including children is $<10\,\mu g\,dL^{-1}$ ($100\,\mu g\,L^{-1}$; $0.5\,\mu mol\,L^{-1}$). Those who are occupationally exposed to lead may undergo workplace surveillance under the Health and Safety at Work regulations specifically the lead at work regulation. This permits exposure to lead in the workplace, subject to medical guidance, up to permitted levels. Currently, in male workers in the UK, the occupational "action level" for blood lead is $50\,\mu g\,dL^{-1}$ ($500\,\mu g\,L^{-1}$) and the "suspension level" is $60\,\mu g\,dL^{-1}$ ($600\,\mu g\,L^{-1}$). Lower limits are in operation for female workers and for younger males. Laboratories undertaking monitoring of workplace blood lead concentrations have to comply with strict criteria for monitoring of their analytical performance in external quality assessment schemes.

23.5.8.5 Mercury. Mercury and its compounds have been known to man for more than 2000 years and are still widely used in the chemical industry, also in the manufacture of drugs and pesticides. Mercury containing dental amalgams are also in widespread use in many countries. Also some foods, *e.g.* fish, may contain high concentrations of organomercury, *e.g.* methyl mercury, due to pollution of the seas. The toxicity of mercury and its compounds is greatly influenced by its chemical form and oxidation state. The most acutely toxic form of mercury is as its Hg(II) salts, particularly Hg(II) chloride, used as a disinfectant, which was historically seen in cases of murder and suicide. However, there is significant toxicity associated with inhalation of (elemental) mercury vapour, which has a very high vapour pressure at normal room temperature. Chronic exposure to mercury vapours is a well-known problem in industries where mercury is used, and also in occupations such as scientific instrument repair. It is also associated with accidental poisoning in children, where mercury may be played with and spilt on carpets. Organo forms of mercury, such as methyl mercury, are strongly neurotoxic and being relatively lipid soluble may accumulate in fatty tissues of the body such as brain. Important environmental disasters, *e.g.* Minimata in Japan, are associated with a disease of epidemic proportions (Minimata disease), which was caused by mercury in industrial effluent contaminating river water and rice fields. The signs and symptoms of acute and chronic mercury poisoning are well established and mainly affect the central and peripheral nervous system, kidney or skin.

Collection of both blood and urine specimens can be useful in the investigation of any suspected poisoning or exposure to mercury. Hair analysis can also be helpful in assessment of historical exposure. Mercury is commonly determined in biological specimens, particularly urine, blood, hair, using "cold vapour" AAS. More recently, ICP–MS has become the technique of choice and is able to detect the presence of mercury at very low concentration in biological fluids, $<0.1\,\mu g\,L^{-1}$. Reference values for inorganic mercury in non-exposed populations are $<4\,\mu g\,L^{-1}$ in blood and $5\,\mu g\,L^{-1}$ in urine.

23.5.8.6 Thallium. Thallium and its salts were historically used as depilatory agents in the treatment of ringworm of the scalp. The most recent use of thallium compounds has been as rodenticides and manufacture of special alloys and optical equipment. The soluble salts of thallium such as the sulfate, acetate and carbonate are highly toxic and ingestion of such compounds has been associated with severe poisoning including murder. The signs and symptoms of thallium poisoning are very complex and may be quite confusing. Initial symptoms may include gastrointestinal disorder and fever; neurological and cardiovascular signs of poisoning may develop more slowly, particularly that of peripheral neuropathy. The most characteristic sign associated with thallium poisoning is hair loss, including total alopecia, which may take 2–3 weeks to develop. Thallium may be determined in specimens of blood or urine using either electrothermal atomic absorption spectrometry or ICP–MS. Reference values for thallium in blood or urine specimens are $<1\,\mu g\,L^{-1}$. Measurement of thallium excretion is useful in cases of thallium poisoning where antidotal therapy with Berlin (Prussian) Blue, potassium ferric hexacyanoferrate, has been used successfully.

23.5.8.7 Opioids. This includes a very wide and important range of drugs that are extensively used in the management of moderate to severe pain, but is also associated with abuse (see Table 23.10).

Opioids are defined as drugs that act as agonists at the opioid receptor sites present in brain and other tissues, particularly the gut. This includes two groups of compounds:

1. Compounds that are classically known as opiates, *i.e.* derived from opium, but generally meaning those structurally related to morphine or derivatives of morphine such a heroin (see Figure 23.4).
2. Synthetic drugs with opioid activity that are structurally unrelated to morphine, *e.g.* methadone, pethidine, dextropropoxyphene as shown in Figure 23.5.

These drugs have a wide range of pharmacological effects, influencing the central nervous system, particularly pain pathways and are associated with misuse and addiction (see Table 23.11).

Table 23.10 *Opioids*

Opiates (structurally related to morphine)
   Morphine
   Codeine
   Dihydrocodeine (DF 118)
   Heroin (Diamorphine)
   Hydromorphone (Palladone)
   Oxycodone (Oxynorm, Oxycontin)

Synthetic opioids (structurally unrelated to morphine)
   Buprenorphine (Temgesic)
   Dextromoramide (Palfium)
   Dextropropoxyphene (Coproxamol with paracetamol)
   Dipipanone (Diconal with cyclizine)
   Methadone (Physeptone)
   Nalbuphine (Nubain)
   Pentazocine
   Pethidine
   Tramadol (Tramake, Zamadol, Zydol)

Table 23.11 *Main pharmacological effects of opioids*

Anaesthesia[a]
Analgesia; reduced sensation to pain and its perception
Coma[a]
Constipation
Convulsions[a]
Development of tolerance and dependence
Drowsiness
Miosis (pin-point pupils)
Nausea and vomiting; less common on recumbent patients
Reduced cough reflex
Reduced gut motility; delayed stomach emptying and digestion of food
Respiratory depression; reduces rate of breathing and sensitivity to carbon dioxide at respiratory centre within brain stem
Respiratory failure and apnoea; suppression of breathing reflex within brain stem[a]

[a] Associated with high doses, including overdosage.

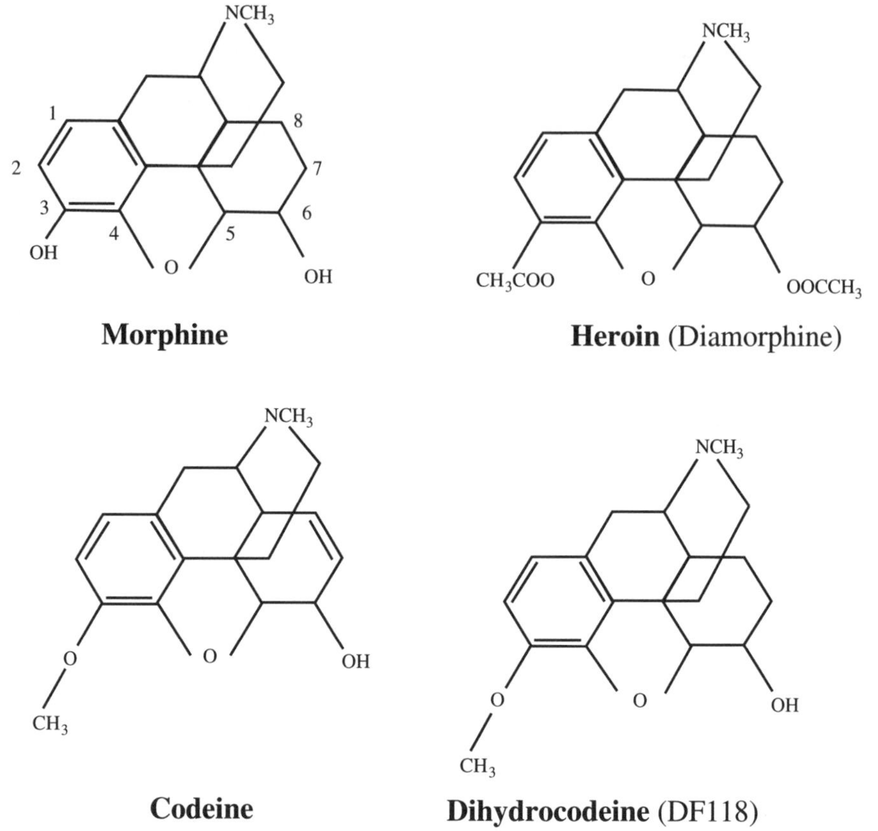

Morphine

Heroin (Diamorphine)

Codeine

Dihydrocodeine (DF118)

Figure 23.4 *Structure of morphine and related opiates: codeine, dihydrocodeine (DHC) and heroin (diamorphine)*

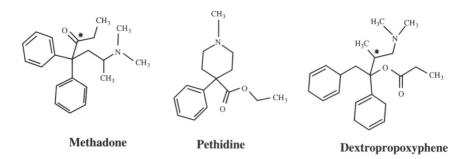

Methadone   **Pethidine**   **Dextropropoxyphene**

Figure 23.5 *Structure of important opioids: methadone, pethidine and dextropropoxyphene (* denotes chiral centre)*

Laboratory investigations can play an important role in the investigation of suspected abuse of these drugs, including cases of suspected poisoning. The mechanism of fatal opioid poisoning, particularly in heroin users is particularly complex. Techniques such as immunoassay are widely used in the detection of opiate type drugs in blood or urine but are generally unable to differentiate between individual drugs or their metabolites. Opiate immunoassays are not able to detect structurally unrelated opioids such as methadone or dextropropoxyphene, although separate immunoassays are available for these drugs. The metabolism of important opiates, such as heroin, is particularly complex, with the formation of both active and inactive metabolites (see Figure 23.6).

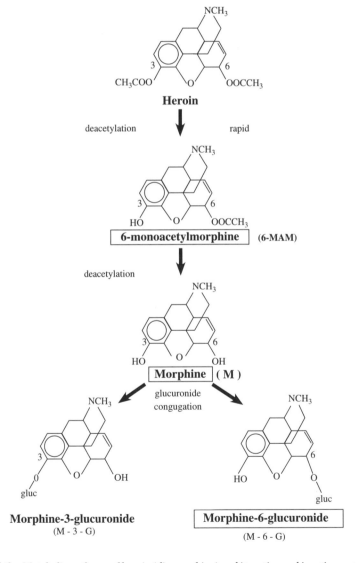

Figure 23.6 *Metabolic pathway of heroin (diamorphine) and its active and inactive metabolites*

Laboratory investigation of suspected poisoning or abuse is carried out frequently for important opioids, such as methadone and dextropropoxyphene. Methadone is widely used in the treatment of heroin abuse, and dextropropoxyphene is present in many preparations containing paracetamol, *e.g.*, coproxamol, and may cause severe symptoms of opioid poisoning in acute overdosage. It is associated with rapid death when combined with alcohol.

23.5.8.8 Organophosphate and Carbamate Pesticides. Organophosphate and carbamate pesticides are probably some of the most widely used pesticides throughout the world and have a broad spectrum of application in both agriculture and the home environment. These compounds often have similar structures and a common mechanism of action and toxicity, involving inhibition of cholinesterase enzymes in the body. However, different pesticides inhibit neural, red-cell and plasma cholinesterase enzymes to varying extents, and with differing time courses. Some organophosphate pesticides and some related compounds used as warfare agents cause a potent and very prolonged inhibition of cholinesterase actively, causing severe toxicity and rapid death.

Acute ingestion of organophosphate or carbamate pesticides is relatively rare in the UK, but they are commonly taken for self-harm in many developing countries where there is also much accidental poisoning in children. However, chronic exposure to such pesticides may be seen in certain agricultural occupations, *e.g.* sheep farmers. Initial investigation of suspected poisoning may involve the measurement of plasma (pseudo) cholinesterase and or red-cell (acetyl) cholinesterase enzyme activity that may be depressed following acute or chronic exposure (see Table 23.12). Measurement of red-cell acetylcholinesterase is a particularly useful procedure to assess the severity of poisoning and a guide to antidotal therapy.

Identification and measurement of individual pesticides or their metabolites is technically difficult and is of no clinical value in acute patient management, but can be of interest in the longer term management and prognosis or severe cases, particularly if the identity of the pesticide is unknown. Detection of organophosphate metabolites in urine may be useful in the assessment of occupational and environmental exposure to such pesticides and extensive biomonitoring data have been published, particularly from the USA.

Table 23.12 *Diagnosis of organophosphate and carbamate pesticides toxicity*

Assays	Finding	Implication
Plasma cholinesterase ⟶	"Normal" ⟶	Poisoning excluded
	"Low" ⟶	Possibility of poisoning
Red-cell acetylcholinesterase ⟶	"Normal" ⟶	Poisoning excluded
	"Low" ⟶	Likelihood of poisoning very strong

23.5.8.9 Paraquat and Diquat. Paraquat and diquat are the most well-known members of the bipyridyl group of herbicides (Figure 23.7).

Severe paraquat poisoning is now relatively uncommon in the UK and most cases involve the non-fatal ingestion of retail products having a low concentration of herbicide in a solid granular formulation, *e.g.* Weedol, Pathclear. Ingestion of liquid concentrates containing high concentrations of paraquat (20%) manufactured for agricultural purposes, *e.g.* Gramoxone, is unusual, but remains a common problem in many developing countries, and ingestion of such products often results in a fatal outcome. Paraquat, unlike diquat, undergoes an energy-dependent uptake into the lungs. This causes a destruction of type I and II lung cells, resulting in an acute alveolitis, followed by an irreversible fibrosis. Toxicity and outcome of paraquat poisoning is related to the formulation available and the doses ingested (see Table 23.13).

Paraquat and diquat are easily detected in urine using a simple dithionite spot test, which is recommended in all cases of suspected paraquat ingestion. This spot test is extremely sensitive, about $2 \, mg \, L^{-1}$ in urine, and is also a reliable simple way of

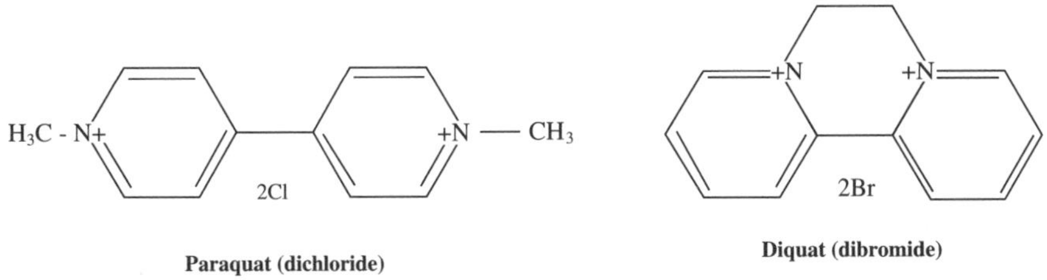

Paraquat (dichloride)

Diquat (dibromide)

Figure 23.7 *Structure of quaternary ammonium bipyridyl pesticides, paraquat and diquat*

Table 23.13 *Toxicity of paraquat in relation to ingested dose*

Dose	Outcome
Less than 1 g	Nausea, vomiting Renal damage possible Most patients survive
1–2 g	Nausea, vomiting, diarrhoea Mild renal impairment Delayed lung fibrosis and respiratory failure Death following 10–20 days
2–5 g	Hepatic and renal damage Lung fibrosis Respiratory failure Death within 1 week
Greater than 5 g	General organ failure Death within 24–36 h

assessing the likely severity of ingestion in those patients who present early, *i.e.* within 12 h. Measurement of plasma paraquat is technically much more difficult and requires use of a specific immunoassay or GC–MS or LC–MS. Measurement of plasma paraquat concentrations is useful where the history of ingestion is uncertain, particularly where there is a late presentation. It is of greatest value in the early prediction of severity and prognosis of a likely fatal outcome. Prognostic graphs based on the plasma paraquat concentration in the first 24 h of ingestion are available, against which the individual patient can be compared. Such quantitative measurements have only limited value in the active management of patients, since there is as yet no effective treatment for severe paraquat poisoning, and care is solely palliative.

23.5.8.10   Rodenticides.   A wide range of different chemical poisons may be used in rodent control. In general, there is relatively strict control on the use and availability of such compounds in most developed countries, including the UK. As a consequence, poisoning, either accidental or intentional, due to ingestion of rodenticides is now relatively uncommon. However, it remains very common in many developing countries, particularly in rural areas where products are freely available with little or no regulation.

Historical rodenticides based on phosphorus, arsenic or thallium are banned in many countries because of their very high acute toxicity. Older compounds based on alkaloids, such as strychnine or brucine, are not so commonly used, but are extremely toxic and can cause rapid death if ingested. Laboratory detection of such compounds in body fluids is relatively easy using chromatographic techniques. More commonly used compounds are aluminium and zinc phosphide, which produce highly toxic phosphine gas (PH_3) on contact with moisture. The very large number of fatalities related to the ingestion of phosphides in some developing countries are related to suicidal ingestion. Laboratory investigation of such cases is extremely difficult, but may be undertaken by measurement of aluminium or zinc in plasma or urine. Other commonly used compounds are those based on warfarin or longer acting coumarin derivatives such as brodifacoum, bromadiolone and difenacoum. Laboratory investigation of suspected poisoning involves measuring blood clotting time and prothrombin time, *i.e.* the International Normalised Ratio (INR). The effects of these drugs can be reversed by the administration of vitamin K. The presence of courmarin type anticoagulants and their metabolites can be confirmed by liquid chromatography using fluorimetric or mass spectrometic detection.

23.5.8.11   Volatile Substances.   There may be exposure to volatile substances particularly organic solvents in the workplace and the home that may cause toxicity, particularly if exposure is excessive or there is inadequate ventilation. In addition, intentional inhalation of gases or vapours is now a well-recognised problem of substance abuse. This was historically termed "glue sniffing" when it was associated with toluene-based adhesives; however, a much wider range of products and volatile substances are known to be abused, volatile substance abuse (VSA), and is well recognised (see Table 23.14).

Laboratory investigation of workplace exposure or abuse may be helpful in the assessment of exposure or detection of suspected abuse. It is also invaluable in the investigation of factory accidents or fatalities associated with exposure. In some

Table 23.14 *Some common products that may be associated with VSA*

Product	Volatile component
Adhesives	Toluene, hexane, xylenes
Aerosols	Butane, propane, fluorocarbons
Cigarette lighter fuel refils	Butane, propane
Dry cleaning or	Dichloromethane, tetrachloroethylene
degreasing agents	trichloroethylene
Laboratory solvents	Chloroform, dichloromethane,
	diethylether, hexane
Paint stripper	Dichloromethane, methanol, toluene
Typewriter corrector fluid	1,1,1-Trichloroethane
Vasodilators	Butylnitrite, isobutylnitrite, isoamylnitrite
Whipped cream	Nitrous oxide

cases, the parent compound may be detected in blood, *e.g.* toluene, relatively easily. However, in the case of very volatile gases, *e.g.* propane and butane, detection in blood or other fluids is difficult, but can be found in tissues such as brain or lungs in fatalities. In some cases, measurement of a convenient metabolite in urine, *e.g.* trichloroacetic acid from trichloroethylene or hippuric acid from toluene, may be useful in the assessment of chronic exposure to certain solvents. In all such cases, particularly when collecting blood or tissue specimens, special care is required in the use of suitable specimen containers and storage conditions. The analysis of biological fluids for volatile compounds requires the use of head-space analysis and GC, flame ionisation detection (FID) and electron capture detection (ECD), or mass spectrometric detection. In such cases it is helpful to have access to the commercial product, particularly when investigating cases of suspected VSA or factory accidents.

23.6 MISCELLANEOUS DRUGS

A wide range of over the counter and prescribed drugs may be associated with any case of suspected overdosage or differential diagnosis of poisoning. Those drugs most commonly associated with such cases are outlined below.

23.6.1 Analgesics

Over the counter preparations containing paracetamol or aspirin (salicylate), may also contain codeine or dihydrocodeine. Certain prescribed preparations containing paracetamol may contain potent opioids such as dextropropoxyphene, *e.g.* Coproxamol. A number of non-steroidal antiinflammatory drugs, *e.g.* Ibuprofen, may also be widely used in pain relief.

23.6.2 Antidepressants

A wide range of drugs may be used in the treatment of depressive disorders including tricyclic antidepressants such as imipramine, amitriptyline; specific serotonin re-uptake inhibitors (SSRI) such as fluoxetine, *e.g.* Prozac, and paroxetine, *e.g.* Seroxat; or newer novel agents such as mirtazapine, *e.g.* Zispin, or venlafaxine, *e.g.* Efexor. Generally, older generation of antidepressants are much more toxic in overdose than newer drugs used to treat depression.

23.6.3 Antiepileptics

A wide range of drugs used to control epileptic seizures may be investigated including older drugs such as carbamazepine, phenytoin, phenobarbitone and sodium valproate and newer agents such as lamotrigine, *e.g.* Lamictal, and vigabatrin, *e.g.* Sabril, and topiramate, *e.g.* Topamax. Older generation anticonvulsants, particularly phenobarbitone and carbamazepine can cause severe toxicity in overdose.

23.6.4 Antipsychotic Drugs

A range of potent "neuroleptic" drugs may be used in the control of symptoms of psychoses or control of aggressive behaviour. This includes older drugs such as chlorpromazine, thioridazine and haloperidol. However, many newer drugs are more frequently prescribed including zuclopenthixol, *e.g.* Clopixol; clozapine, *e.g.* Clozaril; olanzapine, *e.g.* Zyprexa; and quetiapine, *e.g.* Seroquel.

23.6.5 Hypnotics

Older hypnotics such as the barbiturates, *i.e.* butobarbital, amobarbital, quinalbarbital, are no longer recommended or generally prescribed for insomnia, but may still be found in some cases of poisoning. Newer agents, which include some benzodiazepines, *e.g.* temazepam, lorazepam, flunitrazepam, may be seen or even newer benzodiazepines like hypnotics such as zaleplon, *e.g.* Sonata; zolpidem, *e.g.* Stilnoct; and zopiclone, *e.g.* Zimovane.

Clinical or forensic laboratory investigations in the investigation of suspected poisoning or fatality may be extended to the detection of any prescribed or unprescribed medication that a patient may have had access to. Detection of such drugs in specimens of biological fluids can be helpful where diagnosis in uncertain, or cases involving children. In the investigation of drug-related fatalities, it may be important to look for the presence of a wide range of drugs and other poisons, and also measure their concentration, or that of an active metabolite, in blood in order to determine their significance in relation to the likely cause of death. However, it is always important to take into account the history of the patient, clinical signs and symptoms, and also findings at post-mortem examination in the case of a fatal outcome.

BIBLIOGRAPHY

R.A. Braithwaite, 'Metals and anions', in *Clark's Analysis of Drugs and Poisons*, 3rd edn, Vol 1, A.C. Moffat, M.D. Osselton and B. Widdop (eds), Pharmaceutical Press, London, 2004, 259–278.

D.R. Camidge, R.J. Wood and D.M. Bateman, The epidemiology of self-poisoning in the UK, *Br. J. Clin. Pharmacol.*, 2003, **56**, 613–619.

M. Eddleston, Patterns and problems of deliberate self-poisoning in the developing world, *Q. J. Med.*, 2000, **93**, 715–731.

R.J. Flanagan, 'Volatile substances', in *Clark's Analysis of Drugs and Poisons*, 3rd edn, Vol 1, A.C. Moffat, M.D. Osselton and B. Widdop (eds), Pharmaceutical Press, London, 2004, 2002–2025.

R.J. Flanagan, R.A. Braithwaite, S.S. Brown, B. Widdop and F.A. de Wolff, *Basic Analytical Toxicology*, World Health Organisation, Geneva, 1995.

M. Kala, 'Pesticides', in *Clark's Analysis of Drugs and Poisons*, 3rd edn, Vol 1, A.C. Moffat, M.D. Osselton and B. Widdop (eds), Pharmaceutical Press, London, 2004, 202–225.

J.A. Vale and A.T. Proudfoot, *Acute Poisoning. Concise Oxford Text Book of Medicine*, Oxford University Press, Oxford, 2000.

I. Watson and A. Proudfoot, *Poisoning and Laboratory Medicine*, ACB Venture Publications, London, 2002.

Chapter 24

Pharmaceutical Toxicology

ROBIN A. BRAITHWAITE

24.1 INTRODUCTION

The analysis of drugs and their metabolites in biological fluids and pharmaceutical products has an essential role in new drug development and the registration of a new drug for human or animal healthcare. The analytical work that is undertaken may be applied initially to the study of the pharmacokinetics and metabolism of a candidate substance in an animal model or *in vitro* system, such as isolated cell cultures. Later work requires the analysis of parent drug and active metabolites in human volunteer studies and subsequent clinical trials in patient groups. A good understanding of the metabolism of a drug and pharmacokinetic profile in appropriate groups of patients is essential for a complete understanding of the drug's clinical activity and possible toxicity. In addition, the development of different formulations requires that analytical work be undertaken to ensure that the final product meets the required specifications for bioavailability and safety. Measurement of plasma drug concentrations is also important in the continuing investigation of a drug's therapeutic action and toxicity once the product has been introduced into clinical practice.

24.2 SPECIMEN COLLECTION AND APPLICATION TO LABORATORY TECHNIQUES

Sensitive and specific methods of analysis are an essential requirement for the measurement of all drugs and their metabolites in biological fluids and tissues. The most common approach is that of gas chromatography–mass spectrometry and liquid chromatography–mass spectrometry.

In most cases, blood is the most appropriate specimen of biological fluid to be collected for the analysis of drugs and their metabolites. Generally, venous anticoagulated blood (heparinised or EDTA) is the most appropriate specimen collection. Plasma should be separated and refrigerated or stored deep frozen prior to analysis. Urine specimens can be useful for the identification of both active and inactive metabolites, particularly drug conjugates. Large urine collections are used for the isolation and purification of particular drug metabolites. In every patient, it is important to indicate the date and time of the specimen collection in relation to the time of the last drug administration. In the case of post-mortem investigations, it is important to indicate the exact site of specimen collection; it is generally recommended to take samples from only peripheral sites *e.g.* femoral vein. Other specimens may also be used at post-mortem, including urine, stomach contents and vitreous fluid, and in some cases tissues, muscle or liver for example, if the body is badly decomposed.

24.3 PHARMACOKINETICS AND PHARMACODYNAMICS

The study and mathematical description of the process of absorption, distribution and elimination of drugs is known as pharmacokinetics, and is described in detail in Chapter 3. In essence this is what the body does to drugs (or any xenobiotic) following its entry into the body. However, the pharmacological or toxicological action that takes place as a drug and its various metabolites interact with various receptors, enzymes or specific cellular processes is known as pharmacodynamics; in essence, this is what the drug does to the body.

Whenever the potency or duration of action of a drug in the body is considered, it is important to consider both the pharmacokinetic and pharmacodynamic processes involved. Moreover, other factors such as age, sex, disease state and genetic predisposition are also important to consider and may have a major influence on the pharmacokinetics and pharmacodynamics of drugs.

Absorption of any drug will depend on its chemical composition, particular formulation and route(s) of administration. The pharmacokinetics of some drugs may also vary according to dosage. Drugs may be administered by a variety of routes depending on the drug and clinical needs of the patient.

Common routes include oral and sub-lingual (mouth), inhalation (lung), subcutaneous, intravenous and dermal (skin). The rate of absorption of the drug will greatly depend on the route of administration. If a drug is given by the intravenous route, the effect of drug will be apparent within seconds, *e.g.* use of a drug in the induction of anaesthesia. In contrast, if a drug is swallowed, dissolution of the drug and its absorption by the stomach or small intestines may take place over many minutes or some hours, particularly if a sustained release formulation is used. In such cases, there will be a lag-phase before a drug enters the circulation and exerts its effect. In addition, following absorption of a drug by the gut and its passage to the liver via the hepato-portal vein, a proportion of the drug may be broken down, during its passage through the liver, known as the first pass effect. This is particularly important for some drugs, which are known not to be particularly effective by the oral route, *e.g.* morphine.

Once absorbed, drugs may be eliminated by two main processes, excretion by the kidney and metabolism by the liver. With some drugs, urinary excretion of unchanged (active) drug is an important factor in determining their duration of activity. In cases where the functioning of the kidney may be reduced, *e.g.* in the elderly, or impaired as in renal failure, the elimination of a particular drug may be greatly reduced and its action prolonged, *e.g.* digoxin. This may also lead to accumulation of the drug and any active metabolites in the body. In contrast, some drugs may be largely eliminated by metabolism in the liver, by the cytochrome P450 group of enzymes. This may take place in two phases; phase I metabolism is where a functional group is introduced into the drug molecule by oxidation, reduction or hydrolysis. For example, a methyl group may be removed by nitrogen or oxygen demethylation or a hydroxyl group may be added to the molecule by hydroxylation. In general, this process is to increase the polarity of the drug molecule to facilitate further metabolic processes. In phase II of this process, the drug molecule or its metabolite will be conjugated with an endogenous substrate, typically a glucuronide, to make the molecule even more water soluble and more easily excreted in urine. In general, in phase I metabolism, the pharmacological activity of a molecule may be reduced, modified or even increased in a few selected cases. In phase II metabolism, conjugates are largely inactive and easily excreted by the kidney, except in situations of renal impairment.

The rates of elimination of drugs by metabolism are known to vary between individuals. This is known to be under genetic control and some individuals may have particular genetic mutations that lead to either a reduced or increased quantity of a particular cytochrome P450 isoenzyme involved in different pathways of drug metabolism. Techniques are now available to study the genetic aspects of drug metabolism and genotyping of individual patients.

Pharmacokinetic studies can be carried out to understand the pharmacokinetics of drugs in different populations of patients and volunteers. This is an important aspect of drug development and the registration process. Typical data that can be collected include peak and time of plasma drug concentration following dosage and routes of administration, *e.g.* comparison of standard and sustained release formulation. If multiple blood specimens are taken following dosage, the plasma drug *vs.* time concentration profile may be studied. This shows the extent of the absorption and elimination of a drug with time (its bioavailability). By comparing the concentration/time curve for the intravenous route (100% availability) with other routes, *e.g.* oral route, the relative bioavailability of the drug following various routes may be determined. This may be calculated from the area under the curve (AUC) from the time of drug administration (t_0) to infinity (t_∞).

Thus, the percentage bioavailability (F) of an oral formulation may be calculated from

$$F = \frac{AUC_{oral}}{AUC_{i.v.}} \times 100$$

Following entry into the circulation, the concentration of drug observed will be determined by its relative distribution between blood and tissues. This is determined by the chemical characteristics of the drug, particularly its solubility in fatty tissues, *i.e.* lipid solubility. The extent of this relative distribution is known as the distribution volume (V_d), which has units of litres or litres per kilogram of body weight. For simple one-compartment models, this can be described by the equation:

$$V_d = \frac{Dose}{C_0}$$

where C_0 is the concentration in blood at time zero.

Distribution volume is a useful concept that has implications for the assessment of drug elimination half-lives ($t_{1/2}$). The elimination half-life is the time taken for the drug concentration to fall by one-half. Drug elimination rates depend on distribution volume (V_d) and clearance of the drug (CL). However, different models may be used to describe the elimination process, in general either a single or two compartment model may be used.

In the case of a one compartment model, the simplest, the rate of elimination (K_{el}) of a drug from the body is proportional to its clearance, but inversely proportional to the distribution volume:

$$\text{Rate of elimination } (K_{el}) = \frac{\text{CL}}{V_d}$$

Clearance may be calculated from the area under the plasma drug concentration/time curve.

$$\text{CL} = \frac{\text{Dose}}{\text{AUC}}$$

This is usually calculated from an intravenous dose

$$\text{CL} = \frac{\text{Dose}_{i.v.}}{\text{AUC}_{i.v.}}$$

In the case of an oral dose, where F is the function of the dose that is bioavailable.

$$\text{CL} = \frac{\text{Dose} \times \text{F}}{\text{AUC}_{oral}}$$

However, in some situations clearance of a drug following an oral dose may be usefully expressed as:

$$\text{Apparent clearance (oral)} = \frac{\text{Dose}}{\text{AUC}_{oral}}$$

The rate of elimination of a compound may be described by detailed blood concentration measurements taken over time. Most compounds follow a first-order elimination process in that the rate of change of drug concentration with time is proportional to the concentration of drug.

If the log of the concentration of drug is plotted against time, then this may be expressed by a first-order equation (*i.e.* $y = kx + C$), where k (the elimination rate constant) is the slope of the curve and C is the intercept at time zero.

$$\text{Half-life } (t_{1/2}) = \frac{(\ln 2)}{K_{el}} = \frac{0.693}{K_{el}}$$

A few drugs, *e.g.* alcohol, largely undergo elimination at a constant zero-order rate, *i.e.* elimination is *not* proportional to concentration, but at a fixed rate, *e.g.* $18 \, \text{mg} \, \text{dL}^{-1} \text{h}^{-1}$.

The half-life is an important factor to consider when drugs are given by repeated doses. Those drugs with a long half-life, *i.e.* longer than the dosage interval, will accumulate until a so-called steady-state concentration is reached.

$$\text{Steady-state concentration } (C_{ss}) = \frac{\text{Dose} \times F}{\Delta T \times \text{CL}}$$

where ΔT is the dosage interval and CL is clearance.

Thus, steady-state concentrations are inversely proportional to drug clearance. Thus if clearance is reduced, steady-state concentrations will increase. This is particularly important if there is inhibition in drug metabolism, due to dose-dependent clearance or a metabolite–drug interaction.

Pharmacokinetics is important when considering the desirable duration of action of a drug. Thus, if the oral bioavailability is poor, the drug will have to be given by alternative routes. If the drug has a slow clearance, or long elimination half-life, these are undesirable properties for a hypnotic drug and will cause 'hang-over' effects the next day.

Studies of the pharmacodynamics of drugs are generally less well understood. The pharmacological effects of a drug, including side-effects, may change over time as drug receptors undergo modification. An example of this is the development of drug tolerance to opioids following repeated dosing. Drug receptors may also be under genetic control and the effect of a drug may depend on the severity of a particular disease state and the many environmental and social factors that influence actual effects and clinical efficacy in patients.

24.4   SELECTED PHARMACEUTICALS AND THEIR CLINICAL USE

The bulk of this chapter will consider those compounds and groups of compounds that have a pharmacological use and consider their toxic effects of overdose and misuse. Such a list cannot be all inclusive in a book such as this, but it is a compilation of those that the student is most likely to encounter.

24.5   ANAESTHETIC AGENTS

A wide range of drugs may be used in anaesthesia and generally they are divided into two broad areas of application, general anaesthesia and local anaesthesia.

24.5.1   General Anaesthesia

A number of different drugs are commonly used at the same time during surgical anaesthesia, particularly during a lengthy procedure. These drugs may include intravenous anaesthetic agents such as thiopental, ketamine and propofol; inhalation gases or volatile agents such as halothane, enflurane and nitrous oxide; sedating agents such as diazepam and midazolam; opioid analgesics such as alfentanil, fentanil and morphine; neuromuscular blocking agents that cause muscle relaxation such as atracurium and vecuronium. Patients undergoing general anaesthesia are generally carefully screened for any obvious risk factors prior to their operation. Patients on pre-existing medication may sometimes be asked to stop their normal medication prior to surgery if this is thought to be a significant risk of interaction or harmful side-effect during anaesthesia.

24.5.2   Local Anaesthesia

A wide range of drugs may be used in local anaesthesia, which generally act by reversibly blocking nerve conduction. Commonly used agents include lidocaine (lignocaine) and bupivacaine. Local anaesthesia carries a much lower general risk to the patient in comparison with general anaesthesia. Local anaesthesia is widely used in performing minor dental or surgical procedures.

24.6 ANTIBIOTIC AGENTS

There is an extremely broad range of drugs covering antibacterial, antifungal, antiviral, antiprotozoal and anthelmintics agents, and it is difficult to describe the extremely diverse range of substances with useful antibiotic action.

24.6.1 Antibacterial Agents

A wide variety of agents are available, however, the choice of the most appropriate antibacterial agent is extremely important. This will depend on the nature of the infection and organ system most affected, *e.g.* gastrointestinal system, urinary tract, *etc.*, and likely sensitivity of an organism to an antibiotic agent. Some agents may only have activities towards particular types of organisms, others may have a broad spectrum of activities. Major classes of drug used include the penicillins, cephalosporins, tetracyclines, aminoglycosides, macrolides and sulphonamides.

24.6.2 Antifungal Drugs

Specialist use of antifungal drugs is required when there is a systemic fungal infection, particularly in an immunologically compromised patient, *e.g.* human immunodeficiency virus (HIV) infection or post-transplantation. Some drugs may be used for systemic infection, *e.g.* amphotericin and fluconazole. However, there are a number of drugs widely used for the treatment of local fungal infections, *e.g.* Nystatin.

24.6.3 Antiviral Agents

A number of important antiviral drugs including nucleoside reverse transcriptase inhibitors are available, particularly for the treatment of those patients who are suffering from an immunodeficiency disorder, *e.g.* infection by HIV. Important drugs include zidovudine, abacavir and nevirapine. Because of the increasing importance of controlling HIV, there is a great deal of research activity in developing more effective and cheaper agents in this field.

An additional important usage is in the treatment of herpes viruses, *e.g.* cold sores and genital herpes, and to treat conditions in as skin, eye and genitals. Acyclovir is a well-used drug in this respect.

24.7 ANTICONVULSANTS

The main purpose of the use of anticonvulsant drugs is the prevention or reduction of seizures, epileptic fits. This may be possible with the use of one or more drugs, however, the key objective is to control seizures, with a minimum of side-effects, using the lowest possible dose. Ideally, only a single drug should be used, and additional drugs are added only if the drug therapy with a single agent does not provide adequate control. The older generation of drugs include phenobarbital, phenytoin, carbamazepine and valproate, but these are still widely used. Newer agents include lamotrigine and vigabatrin. Some drugs used in the treatment of epilepsy may cause interaction with other drugs, which can be a serious problem if overlooked. Important types of interaction include induction of drug-metabolising enzymes and displacement of drugs from plasma protein binding sites. Therapeutic drug monitoring, the measurement of plasma drug concentration, of several of the anticonvulsants has been shown to be useful in improving patient compliance and individualising dosages to meet the requirement of the particular patient.

24.8   ANTIHISTAMINES

Antihistamines are generally used in the treatment of allergies, most commonly those with nasal allergies, *e.g.* hay fever. They are also used in reducing symptoms caused by allergic reaction to insect bites or by some types of drug-related allergy. Some members of this group of drugs, particularly older agents such as promethazine, chlorpheniramine and diphenhydramine, are relatively sedating due to their pharmacological action and uptake by the brain, so may cause drowsiness and problems with skilled tasks such as driving. However, tolerance may develop following usage over time. In recent years, a range of non-sedating antihistamines has been developed, particularly useful in the control of symptoms of hay fever; these drugs include acrivastine and loratadine. Some products can now be purchased over the counter. Some of the older antihistamines that have sedating properties, *e.g.* diphenhydramine, have recently been marketed as hypnotics and on direct sale to the public as these are considered to be relatively safe in overdose.

24.9   ANTIMALARIALS

The treatment and prophylaxis of malaria is an important worldwide problem, particularly with mass travel to parts of the world that are not free of the malaria parasite. In addition, malaria is known to be a major cause of death in developing countries. There are specific recommendations for the choice of antimalarial drugs depending on the geographical area and the likely resistance to particular drugs. Older antimalarial drugs such as quinine and chlorquine are still in use, but some malaria parasites are now resistant to chloroquine. Newer alternatives include mefloquine (Larium).

24.10   BARBITURATES

The barbiturates comprise a large group of drugs that have been widely used in medicine for many decades, particularly as general sedatives and hypnotics. However, apart from specific compounds their use has become very restricted in recent years. These drugs cause depression of the central nervous system (CNS) and some intermediate-acting barbiturates may still be used as hypnotics, but their use is discouraged in all but the severest cases, such as those patients already taking barbiturates over many years. Commonly known barbiturate hypnotics include amobarbital, butobarbital and secobarbital, or a mixture of amobarbital and secobarbital (Tuinal). Such drugs may cause dependence, and tolerance to their sedative effects may easily develop. Abrupt withdrawal may cause severe symptoms, including rebound insomnia, anxiety and convulsions. There is also the potential for interaction with other sedative type drugs and alcohol, particularly where there is risk of abuse or overdosage. Excessive dosages may cause drowsiness, confusion and loss of consciousness, also respiratory depression. Use of such drugs is avoided in the elderly or those with a history of drug and alcohol abuse. Overdosage can easily lead to fatalities.

Certain barbiturate derivatives continue to have clinical uses in other areas. For example, phenobarbital has been used as an anticonvulsant drug for many years, but is much less used currently. Chronic use of this drug may cause the induction of hepatic microsomal enzymes, which may reduce the effectiveness of other drugs that may be concurrently prescribed. One barbiturate that continues to have a very useful clinical application is thiopental, which is used in the induction of anaesthesia, the treatment of patients with bouts of continuous seizures (status epilepticus) and also in cases of cerebral oedema to reduce brain swelling. However, prolonged infusion of thiopental can have risks, particularly when given over several days and may cause severe depression of the CNS and respiratory depression.

24.11   BENZODIAZEPINES AND OTHER HYPNOTICS

Often problems of insomnia may be caused by anxiety, so a number of drugs that are widely prescribed as hypnotics, because of their sedating effects, also have anxiolytic action. This particularly applies to the benzodiazepine group of drugs. However, hypnotic drugs that have a prolonged duration of action due to a slow elimination, or produce active metabolites, may cause hangover effects and problems during the daytime, including impairment of skilled tasks such as driving or operating complex equipment. As a consequence, a number of drugs that have historically been used as hypnotics such as barbiturates are no longer recommended for such purposes. The ideal hypnotic drug is one that is rapidly absorbed and eliminated over a short period of time, without producing accumulation in the body, or producing active metabolites. In addition, there is continuing concern over the risk of dependence of use of such drugs in patients over prolonged periods of time for the treatment of insomnia.

Generally, most hypnotics are intended for short-term use, *i.e.* a few weeks, and are discontinued as soon as it is feasible. A number of benzodiazepine-type drugs have historically been used as hypnotics, *e.g.* nitrazepam (Magodon), flurazepam (Dalmane) and flunitrazepam (Rohypnol). However, these drugs tend to have a longer duration of action and their use as hypnotics is not recommended now. Also flunitrazepam (Rohypnol) has recently been discontinued because of its abuse potential and its possible association with drug-facilitated sexual assaults (DFSA). Those benzodiazepines that continue to be recommended for use as hypnotics include lorazepam and temazepam. These drugs have minimal residual side-effects when used as a hypnotics, but in the case of temazepam, it has been associated with serious illicit abuse over many years, including intravenous administration in heroin abusers.

Because as a class, the benzodiazepines have been strongly associated with abuse and dependence, a number of alternatives have been developed. The benzodiazepines exert their action on specific benzodiazepine receptor sites in the body. A number of these newer drugs are structurally different from the benzodiazepines and have a short duration of action, but act at the same or similar receptor sites as the benzodiazepines. These newer drugs include zaleplon (Sonata), zolpidem (Stilnoct) and zopiclone (Zimovane). However, these newer drugs are also sometimes associated with their abuse. A number of older drugs may be used as hypnotics that include chloral hydrate and clomethiazole (Heminevrin). All of these drugs may cause dependence and a potential for hangover effects at higher dosages, including drowsiness, confusion, lack of coordination and slowed reaction time. More serious effects may be seen following overdosage, including loss of conscious, but rarely death, unless taken with other drugs or alcohol. A number of other benzodiazepines are commonly used as anxiolytics, the most well known being diazepam (Valium) and chlordiazepoxide (Librium). These drugs, particularly diazepam, are very effective in the short-term treatment of anxiety and agitation, and may be given by different routes of administration, *i.e.* oral, intravenous, intramuscular or rectal. Diazepam also has widespread use as an anticonvulsant in the emergency treatment of fitting or convulsions, particularly in a hospital setting.

One other benzodiazepine, midazolam, because of its particular high lipid solubility, is widely used as an anaesthetic agent, often in combination with other drugs in patients treated in intensive care in hospitals.

24.12   CARDIOLVASCULAR AGENTS (HEART AND CIRCULATION)

An extremely diverse range of drugs may be used in the treatment of disorders of the heart and circulation, including the control of blood pressure, and blood clotting. These agents may come from a diverse range of chemical and pharmacological entities.

24.12.1   Cardiac Glycosides

Some of the oldest agents used to treat heart failure are digoxin and digitoxin. These drugs are derived from the foxglove plant. Digoxin is the most widely used cardiac glycoside; however, it has an extremely long elimination half-life, which can sometimes increase in patients with poor renal function, such as the elderly. Digoxin toxicity can sometimes be a problem and can be confirmed by the measurement of the plasma digoxin concentration, but can also be influenced by the presence of low serum potassium concentration, hypokalemia. A digoxin-specific antibody, Digibind, can be used in the treatment of severe digoxin toxicity or overdosage.

24.12.2   Diuretics

A wide range of different types of diuretics may be used to treat the accumulation of excess fluid, oedema, that can sometimes be present in certain types of heart failure. Use of certain types of diuretics can sometimes cause loss of important electrolytes such as potassium and have a deleterious effect on the heart. This is seen particularly with thiazide and loop-type diuretics.

Thiazides and related compounds are one of the best-known diuretics and act by inhibiting the re-absorption of sodium ions at the beginning of the distal tubule in the kidney. Common thiazide diuretics include bendroflumethiazide and indapamide. Such drugs are widely used in the management of hypertension. Another common group of diuretics are known as loop diuretics and act by inhibiting re-absorption from the ascending limb of the loop of Henlé in the kidney. Common 'loop'-type diuretics include frusemide and bumetanide. Some diuretics are known as 'potassium-sparing diuretics' and will cause retention rather than loss of potassium, *e.g.* amiloride. A number of combination products containing potassium sparing with other types of diuretics are also widely used.

24.12.3 Antiarrhythmic Agents

Different types of heart rhythm disturbance, arrhythmia, require the use of different classes of drug and there is a well-established clinical classification system. These agents are classified according to their effect on the electrical activity of the heart. There are four main classes of these drugs (classes I–IV) also some sub-classes within group I. Commonly used arrhythmic agents include amiodarone, disopyramide and lignocaine.

24.12.4 Beta-blocking Drugs

This important group of drugs are used in the treatment of a range of conditions including hypertension, angina and myocardial infarction; commonly used beta-blockers include atenolol, metoprolol and propranolol. Some beta-blockers are relatively lipid soluble and hence enter the brain easily and are more likely to cause sleep problems. Water-soluble beta-blockers may therefore be preferable in certain types of situation.

24.13 CYTOTOXIC/ANTICANCER DRUGS

This group of drugs is largely used in the treatment of malignant disorders such as cancer. Many such drugs, because of the nature of their action, may cause severe side-effects due to damage caused to healthy cells. Cancer chemotherapy is generally carried out only by cancer specialist (oncologists) working in centres having sufficient expertise in the use of such drugs. Complex combinations, or cocktails, of drugs may be used in some patients, as these may be more effective than therapy with a single agent. There are a number of different categories of cytotoxic drugs in common use, which include the following.

24.13.1 Alkylating Drugs

These are some of the most widely used drugs in cancer chemotherapy. They act by damaging DNA in the cell nucleus thereby reducing cell replication. The most common drugs in this class include cyclophosphamide, chlorambucil and melphalan.

24.13.2 Cytotoxic Antibiotics

This other widely used group of drugs includes bleomycin, doxorubicin and epirubicin. These may be used in a wide variety of different cancers.

24.13.3   Antimetabolites

These drugs act by interfering in cell division. The main drugs in this class include methotrexate, fluorouracil and mecaptopurine.

24.13.4   Other Antineoplastic Drugs

A wide range of other drugs may be used in treating cancers including platinum compounds, such as cisplatin and carboplatin, and taxanes, such as paclitaxel and docetaxel. These are widely used in the treatment of ovarian cancers; sometimes both groups of drugs may be used in combination.

A wide range of 'novel' agents may sometimes be used experimentally in patients with terminal illness or where existing treatment has been ineffective. Many novel compounds may be derived from 'natural products', and is currently a very active area of new drug discovery.

24.14   IMMUNOSUPPRESSANT AGENTS

Corticosteroid immunosuppressants such as prednisolone are often used in the treatment of cancers, particularly in certain types of 'blood' cancers such as leukaemia and lymphoma. These drugs are also used in organ transplantation, particularly in the treatment of acute episodes of organ rejection.

A wide range of other potent immunosuppressives are now essential in preventing graft rejection following transplantation. The most commonly used agents include cyclosporin, tacrolimus and sirolimus. These agents are extremely complex molecules and are derived from natural products. Therapeutic drug monitoring is now widely used in adjusting dosage levels to within an appropriate therapeutic range. This is an important factor in reducing the chance of organ rejection following transplantation and ensuring long-term survival of the organ graft and of the patient. This is particularly important in heart and liver transplantation. Immunosuppressant drugs are also used in the treatment of a wide range of other diseases that have an immunological origin.

24.15   INSULIN AND ORAL HYPOGLYCAEMIC AGENTS

Insulin and a number of orally administered drugs are widely used in the management of diabetes. The incidence of diabetes is growing rapidly in developed as well as some underdeveloped countries. It is a chronic metabolic disorder characterised by the presence of a raised blood glucose concentration, hyperglycaemia. This occurs because of a lack of insulin production by the pancreas or resistance to its action at the cellular level. There are two main types of diabetes, type 1 diabetes and type 2 diabetes.

24.15.1   Type 1 Diabetes

This is defined as insulin-dependent diabetes mellitus (IDDM) and usually has a rapid onset of action due to a lack of insulin production caused by destruction of insulin-producing cells in the pancreas. It is generally accepted that, this is caused by an autoimmune-type disorder. Patients diagnosed with IDDM require regular insulin administration to replace all of the body's own insulin previously secreted by the pancreatic beta cells.

24.15.2 Type 2 Diabetes

This is defined as non-insulin-dependent diabetes mellitus (NIDDM) and is caused by a reduced or failing production of insulin, or in some patients a loss in its ability to control adequately the blood glucose concentrations. Generally, this may effect a much broader population of patients, who may present with a wide range of symptoms. Some patient may have had the disease for some years, without the diagnosis having being made; common symptoms include thirst, polyuria and weight loss. The growing incidence of diabetes is strongly linked to diet, obesity and lack of exercise. In some cases, the disease may be controlled by a change in diet and lifestyle. It can also be 'controlled' by the use of oral drugs also by insulin itself, or a combination of treatment and lifestyle changes.

24.15.3 Insulin

Insulin is synthesised in the beta cells of the islets of Langerhans in the pancreas. It is formed from a precursor protein known as pro-insulin, which is converted into one molecule of insulin and one molecule of C-peptide. Insulin itself is a 51 amino acid polypeptide chain linked by two disulphide bridges. A wide range of different insulin preparations is available. These may be of animal pancreatic origin, mainly pig, or more recently by recombinant DNA biosynthesis of the human insulin sequence. Insulin is generally given by subcutaneous injection often using a type of injection device that is able to control and meter the dose given. Patients may monitor their own blood glucose concentrations using a glucose meter with the aim of keeping the concentration within a minimum and maximum range agreed by their doctor. A better marker of glucose control is the laboratory measurement of glycosylated haemoglobin (Hb_{A1c}). The major objective in using insulin in the treatment of diabetes is the control of blood glucose. The measurement of plasma insulin and C-peptide can be useful in the investigation of unexplained hypoglycaemia in patients on insulin. In addition, the ratio of insulin to C-peptide is used in the investigation of suspected insulin overdosage, particularly where there are suspicious circumstances.

24.15.4 Oral Hypoglycaemic Drugs

A number of oral drugs are used in the treatment of non-insulin-dependent diabetes. These are generally given to patients whose glucose control is not controlled well enough by diet or exercise alone, but do not require insulin. One group of oral hypoglycaemic drugs that have been prescribed for many years is the sulphonylureas, which act by enhancing insulin secretion from failing beta cells in the pancreas. Some of the older sulphonyl drugs such as chlorpropamide and glibenclamide are associated with a higher risk of adverse effects. Alternatives such as gliclazide, glimepiride, glipizide or tolbutamide are now recommended.

Other drugs used in the control of NIDDM may have a different mode of action, such as metformin, which is a biguanide-type compounds. Newer agents such as nateglinide (Starlix) and repaglinide (NovoNorm) stimulate insulin release from functioning cells. Other drugs such as piaglitazone (Actos) and rosiglitaxone (Avandia) are derivatives of thiazolidinedione. The measurement of oral hypoglycaemic drugs in blood and urine is sometimes useful in the investigation of adverse reaction or unexplained hypoglycaemia or other metabolic problems.

The consequences of poor glucose control can be very serious, resulting in a range of medical complications such as retinopathy, leading to blindness and peripheral neuropathy (nerve damage).

24.16   LIPID-LOWERING DRUGS

Lipoproteins are very complex spherical structures containing a number of different components including triglycerides, cholesteryl esters and phospholipids. Although cholesterol is sometimes viewed as a harmful substance, it is synthesised by the liver and is an essential precursor of cell membranes, bile acids and steroid hormones. There are three major classes of lipoproteins, which are known as:

Very Low-Density Lipoproteins (VLDL)
Low-Density Lipoproteins (LDL)
High-Density Lipoproteins (HDL)

In the body, LDL is the main carrier of cholesterol and is seen as 'bad cholesterol' whereas HDL is seen as 'good cholesterol' and has a protective function in transporting cholesterol to the liver. However, this is very much an over simplification of the role of cholesterol and lipoproteins as a risk factor in heart and circulatory disease. In the case of primary disorders of lipid metabolism, dislipidemia, measurement of various lipids and their relative ratio in plasma plays an important role in the investigation of such disorders. As a consequence, reducing the concentration of (LDL) cholesterol and/or raising the concentration of HDL cholesterol in blood plasma is generally accepted to reduce the risk and progression of atherosclerotic heart disease. However, smoking, diet, weight and exercise are also all important risk factors.

Drugs, particularly the statins have been shown to be very effective in reducing cholesterol, particularly when combined with changes in diet and lifestyle. Well-known statins include simvastatin and pravastatin. Side-effects of such drugs are relatively rare. These drugs act by inhibiting an enzyme involved in cholesterol synthesis, particularly in the liver. Some statins are now available as over the counter medicines.

24.17   LITHIUM SALTS

Lithium salts have been used for many years for the prophylaxis and treatment of manic-depressive illness, bipolar-affective disorder. The drug, however, has a narrow therapeutic range and it is important to monitor the plasma concentration of lithium to achieve an effective concentration with a minimum risk of side-effects.

Dosages are generally adjusted to achieve a plasma lithium concentration of $0.4–1.0\,\mathrm{mmol\,L^{-1}}$ in a specimen of blood taken 12 h following the last dose. Generally, plasma lithium concentrations are checked every few months in patients to confirm that they are maintained within an optimum therapeutic range, also to assess any side-effects that may develop. There are many different formulations of lithium salts that may be prescribed, most of these are based on lithium carbonate or citrate. But the bioavailability of different formulations may vary and caution is required when there is a change of formulation or brand of drug prescribed to a patient. The side-effects of lithium are well documented. Milder, commonly reported side-effects include gastrointestinal disturbances, fine tremor and weight gain. There may also be a risk, particularly in women, of developing hypothyroidism and symptoms such as lethargy and feeling cold should be noted each time the patient is seen in clinic. Thyroid function should be checked at least once a year. More severe side-effects include blurred vision, diarrhoea, vomiting, muscle weakness and renal impairment. Severe symptoms are generally associated with lithium above $1.5\,\mathrm{mmol\,L^{-1}}$ and overdosage is associated with concentrations above $2.0\,\mathrm{mmol\,L^{-1}}$, when emergency treatment is required.

24.18 ANTIPSYCHOTIC DRUGS (NEUROLEPTICS)

A wide range of drugs are used in the treatment of patients with psychotic-like symptoms or an established mental illness such as schizophrenia. This chronic, very debilitating disorder affects about 1% of the population and usually begins in late adolescence and may continue throughout life. Characteristic symptoms of the disorder, whose cause is unknown, include delusions, hallucinations, disorganised speech and bizarre behaviour. The first generation of drugs that were introduced into treatment some 50 years ago was the phenothiazines, *e.g.* chlorpromazine, thioridazine and fluphenazine. These drugs acted by reducing transmission of the neurotransmitter dopamine and helped support the dopamine theory of schizophrenia and its pharmacological treatment. This older generation of drugs, although partially affective in reducing symptoms in many patients, is associated with the development of significant side-effects and toxicity. Important and debilitating side-effects include parkinsonian symptoms, *e.g.* tremor, and a wide range in abnormal body and facial movement disorders, *e.g.* dystonia, akathisia and tardive dyskinesia. Parkinsonian-like side-effects can be reduced by use of 'antimuscarinic' drugs that influence the excess cholinergic activity in the brain resulting from a reduction in dopamine. These drugs include orphenadrine and procyclidine.

In recent years, a newer generation of drugs has been developed that generally has a more specific mode of action on the dopaminergic neurotransmitter system. These newer drugs include clozapine (Clozaril), rispiridone (Risperdal), olanzapine (Zyprexa) and quetiapine (Seroquel). Apart from clozapine, the side-effects of these newer agents are far fewer and less severe than first generation of drugs and are now recommended in the first line treatment of newly diagnosed schizophrenic patients. Clozapine, although an extremely effective antipsychotic drug, has an associated risk of agranulocytosis and myocarditics, and patients receiving this drug require careful blood monitoring.

24.19   NON-STEROIDAL ANTI-INFLAMMATORY DRUGS

This large and widely prescribed group of drugs has both analgesic and anti-inflammatory properties. This makes them very useful in the management of painful inflammatory disorders such as rheumatoid arthritis, or common but less well-defined disorders, such as back pain. However, in some non-inflammatory but painful conditions, paracetamol alone may be more appropriate. The main side-effects associated with this group of drugs include gastrointestinal bleeding and ulceration, particularly in the elderly and those taking repeated dosages.

The common older members of this group of drugs include Ibuprofen, Naproxen and Diclofenac. The mode of action is related to the inhibition of cyclo-oxygenase activity (COX). A number of newer non-steroidal anti-inflammatory drugs (NSAIDs) that have recently been introduced are more specific COX-2 inhibitors, *e.g.* Celecoxeb and Rofeoxib.

24.20   OPIOID ANALGESICS

Opioid, or opiate-like, analgesic drugs are used in the management of moderate to severe pain. However, it is well recognised that such drugs may produce tolerance following repeated administration, also dependence. Opioid drugs exert their action by interacting with specific receptors in the body. Three major classes of opioid receptor have been described as μ (mu), κ (kappa) and δ (delta), and also a number of sub-classes of these receptors are known. The most important analgesic mode of action concerns the interaction with μ receptors in the brain, but interaction with other receptors in the brain and gut is also important, particularly at higher doses. Classically, opioid drugs were all derived from opium, which was obtained from the opium poppy (Papaver Sumniferum L). Morphine is the main opiate analgesic drug present in opium, and a number of important 'opiate' drugs are structurally closely related to morphine such as codeine, diamorphine (heroin) and dihydrocodeine (DHC).

Morphine, particularly in the form of its sulphate salt, is still widely used in the treatment of severe pain. It is often given orally by tablet or solution, or by intravenous infusion. The diacetyl derivative of morphine (diamorphine), also known as heroin, is widely used in the treatment of severe pain. It is more potent than morphine, has a greater water solubility and can be given in smaller volumes, particularly by subcutaneous infusion. Other derivatives of morphine are also used in the treatment of mild to moderate pain and include codeine and DHC. These drugs may be combined with paracetamol and are available as over the counter medicines, *e.g.* co-codamol, co-dydramol.

The main side-effects of opioid-type drugs include nausea, vomiting, constipation and drowsiness. High doses may cause respiratory depression and hypotension. In cases of severe respiratory depression or overdosage, the effects of opioid drugs can be revised by the use of specific opioid antagonists (antidotes) such as naloxone (Narcan).

A wide range of synthetic or semi-synthetic opioid drugs has been developed over the last 60 years. The most important of these drugs are dextropropoxyphene often combined with paracetamol (Coproxamol), methadone, fentanyl, pethidine, tramadol and buprenorphine. Some of these drugs, particularly methadone and buprenorphine are also prescribed for the management of heroin abuse. A large number of patients may be maintained on these drugs, particularly methadone, which may be prescribed as a liquid formulation, sometimes to be taken in controlled situations, *e.g.* at a pharmacy.

24.21 PARACETAMOL, ASPIRIN AND OTHER NON-OPIOID ANALGESICS

A number of non-opioid-type drugs are widely available over the counter for the relief of mild to moderate pain or mild fever. The oldest known drug is aspirin (acetyl salicylic acid). However, chronic use of aspirin may cause gastric irritation and bleeding, although this problem may be reduced if the drug is taken after meals. Aspirin can also cause significant interaction with other drugs, particularly anticoagulants such as warfarin. However, aspirin in lower doses is widely prescribed as an effective anticoagulant in the prophylaxis of thrombotic episodes, particularly in the elderly. Use of aspirin in children is now restricted because of its association with a higher risk of Rey's Syndrome. Overdosage with aspirin may cause characteristic signs and symptoms such as pyrexia, sweating and tinnitus. In more severe cases of overdosage, this may progress to loss of consciousness, coma and severe metabolic acidosis. Urgent medical intervention is essential.

Paracetamol, known as acetaminophen in USA, is probably the most commonly used non-opioid analgesic drug for the treatment of mild to moderate pain or mild fever. It is a relatively safe drug when used within recommended dosages; it is also suitable for children. However, there is caution in the use of the drug in patients with liver or kidney impairment. In adults the recommended oral dose is 0.5–1 g every 4–6 h to a maximum of 4 g daily. Most formulations contain 500 mg of the drug. In addition, there is a large number of products containing paracetamol, which also contain an opioid, such as codeine, DHC or dextropropoxyphene, by prescription only. There can be a risk of accidental overdosage in adults who may use paracetamol-containing medicines with different brand names, but remain unaware that they may be exceeding the recommended daily dosage of paracetamol. Use of adult products in children may also lead to overdosage including liver damage and fatalities. It is known that as little as 10–15 g of paracetamol (20–30 tablets) consumed within 24 h may cause severe liver damage. The main result of overdosage in adults or children is with liver and sometimes kidney damage. In cases of self-poisoning, there may be few early signs and symptoms of toxicity, apart from nausea, vomiting and loss of appetite. However, overdose paracetamol is a powerful hepatotoxin that can lead to severe hepatic damage and fulminant liver failure and death due to hepatic failure within 5–10 days. In patients admitted to hospital emergency departments, the measurement of plasma paracetamol concentration is an essential part of the initial patient investigation along with blood clotting and liver function tests, to determine the severity of the overdose. Widely available nomograms are used to decide the need for urgent antidotal therapy according to concentration of drug and time of ingestion. The most commonly used antidote in the UK is *N*-acetylcysteine (Parvolex), which is given as an intravenous infusion. The antidote is most effective in reducing the risk of liver damage when given within 10–12 h of the ingestion of the overdose. In patients who present 'late' to hospital, or have certain risk factors, *e.g.* poor nutritional state or induced liver drugs metabolising enzymes, the risk of liver damage is much higher.

24.22   ANTIDEPRESSANTS

A wide range of drugs have been developed over the last 40 years for the treatment of depressive disorders. There are 4 main classes of antidepressant drugs, which include tricyclic and related antidepressants (TCAs), monoamine oxidase inhibitors (MAOIs), selective serotonin re-uptake inhibitors (SSRIs) and miscellaneous drugs with differing or related modes of action to that of the TCAs or SSRIs.

24.22.1   Tricyclic Antidepressants

The first major class of drugs to be developed for the specific treatment of depression was the tricyclic antidepressant group of drugs, which are all based on a 3-ring structure. The first effective drug was imipramine (Tofranil) followed by amitriptyline (Triptofan). Other well-known tricyclic drugs included clomipramine (Anafranil) and dosulepin (Prothiaden). But many other TCAs are available for prescription. The *N*-desmethyl active metabolites of some of these drugs were developed as a later generation of tricyclic antidepressants, such as nortriptyline (Notival) derived from amitriptyline. But these were found to be no more effective than their parent drugs. The main mode of action of these drugs is believed to be associated with their action on noradrenergic and serotonergic neurotransmitter re-uptake processes within neurones. This group of drugs has been shown to be effective in the treatment of severe depression, but there is always a high incidence of side-effects such as dry mouth, sedation, blurred vision, constipation and ECG changes. In addition, such drugs are extremely toxic when taken in overdose, causing coma, respiratory failure and convulsions. A large number of deaths have been associated with suicidal ingestion of these drugs, particularly dosulepin and amitriptyline. Use of such drugs in patients with a history of overdosage or suicidal ideation is not recommended. As a consequence, the use of this group of drugs has greatly declined in recent years.

24.22.2   Monoamine Uptake Inhibitors

Another 'older generation' of antidepressant drugs are those known as MAOIs, which have a mode of action different from the tricyclic antidepressants. They act by inhibiting monoamine oxidase enzymes within the neurone to increase the amount of neurotransmitter, noradrenaline, that is thought to be linked to the aetiology of depression. The older generation of MAOIs includes phenelzine (Nardil), and tranylcypromine and isocarboxazid. Because these inhibitors of MAO are not reversible, there is a greater risk of side-effects and adverse reactions. The most commonly reported side-effects include postural hypotension, drowsiness, headache, dry mouth, constipation and other GI disturbances. Important adverse reactions may be caused by the consumption of foods, *e.g.* cheese or beverages (red wine, which contains certain amines such as tyramine). Some other drugs, including tricyclic antidepressants, may also cause adverse (cross) reactions, which can result in a severe hypertensive crisis. There may also be problems in patients who suffer from pre-existing disease states such as diabetics or cardiovascular disease. More recently a reversible MAOI, Moclobemide, has been introduced, which is claimed to have a few side-effects compared with earlier MAOI drugs.

24.22.3 Selective Serotonin Re-Uptake Inhibitors

An important group of a 'new generation' of drugs has been introduced over the last 20 years for the treatment of depressive disorders. All these drugs act on the serotonin neurotransmitter system, which is thought to be associated with the aetiology of depressive disorders. This group of drugs is therefore known as the SSRI group of antidepressants. The first of this new generation of drugs to be introduced was fluoxetine (Prozac) followed by a number of other drugs with a similar pharmacological action such as citalopram (Cipramil), paroxetine (Seroxat) and sertraline (Lustral). These drugs have had the added benefit that they were less sedating than the tricyclic group of antidepressants, also less likely to cause death following overdosage. However, in recent years, concern has been expressed regarding the efficacy of the SSRIs that they may not be particularly effective in some patients, or may even worsen some symptoms of depression such as suicidal thoughts. Interestingly, a number of newer drugs have been introduced recently that have a specific action on the noradrenergic neurotransmitter system, *e.g.* reboxetine (Edronax); and others that have an action on both the serotonin and noradrenergic systems, *e.g.* venlafaxine (Effexor). This trend indicates the complexity of understanding the mode of therapeutic action of antidepressant agents and the balance between efficacy and toxicity when developing new and safer drugs.

BIBLIOGRAPHY

British National Formulary (BNF), British Medical Association and Royal Pharmaceutical Society of Great Britain, London, 2005.
A.C. Moffat, M.D. Osselton and B. Widdop (eds), *Clarke's Analysis of Drugs and Poisons*, 3rd edn, Vols 1 and 2, The Pharmaceutical Press, London, 2004.
J.G. Hardman and L.E. Limbird (eds), *Goodman and Gilman's The Pharmacological Basis of Therapeutics*, 10th edn, McGraw-Hill, New York, 2001.
S.C. Sweetman (ed), *Martindale The Complete Drug Reference*, 32nd edn, The Pharmaceutical Press, London, 2004.

Chapter 25

Safe Handling of Chemicals

HOWARD G.J. WORTH

25.1 INTRODUCTION

The introduction to this textbook states that toxicology is the fundamental science of poisons. In the broadest sense of this definition the adverse action of any chemical on living tissues may be regarded as toxicological. This ranges from the corrosive effect of the spillage of a strong mineral acid; to the injurious effect of the exposure to radionuclides; to the exposure to biological fluids containing potentially danger- ous pathogens; to the environmental and occupational exposure to chemicals. The chemist is often seen as the person who can give help and guidance on the handling of chemicals, on the toxicological effects associated with them, and advice on how to deal with an incident when it occurs. It is not surprising that chemists are regarded as the professional group that should be able to give such help and advice. However, this is frequently not recognised in the curriculum for the training of chemists and indeed, apart from what they pick up as part of their educational progress, there is often no formal teaching of toxicology. This makes chemists vulnerable, as there is considerable legislation in many counties concerned with the toxicity of the handling of chemicals. For this reason, a proposed curriculum of fundamental toxicology for chemists is included in Appendix A.

25.2 LEGISLATION

In recent years many countries have passed legislation concerned with safe practice in the working place. Inevitably, this has implications for the handling of chemicals. The considerable use of chemicals in the domestic and non-technical environment means that their safe handling is no longer just a concern of those employed in the chemical industry. Domestic cleaners, solvents and detergents, weed killers and pesticides, and proprietary medicines are examples of chemicals available to the public, whose safe handling may well be the subject of legislation or other related documentation.

In the UK in 1974, the Health and Safety at Work Act was passed, which required the appointment of safety representatives, the establishment of safety committees and the inspection of the workplace by representatives and/or members of the safety committee. This has implications for handling of chemicals where these are used in the workplace. The Act also indicated that every employee is responsible for ensuring, through his or her action, that the workplace is a safe environment. This has implications for the most junior chemistry graduate in a laboratory who will be held responsible, by the Act, for his or her safe use and handling of chemicals.

As a consequence, documentation was produced concerned with specific aspects of handling potentially toxic material, for example, the Department of Health and Social Security produced a series of documents culminating in the 'Code of Practice for the Prevention of Infection in Clinical Laboratories and Post Mortem Rooms' in 1978. This has now been updated by the Health and Safety Executive as 'Safe Working and the Prevention of Infection in Clinical Laboratories and Similar Facilities' (2003) and recently republished. Other more recent regulations for the Control of Substances Hazardous to Health (COSHH), require laboratories and institutions to compile and return inventories of chemicals used. This must include statements concerning their toxicity, safe handling procedures and action that should be taken in the event of an accident or spillage. Guidelines are now produced by many chemical reagent supply companies, which are invaluable compendia of the hazards related to individual chemicals (for example, The Sigma – Aldrich Library of Regulatory and Safety Data, 1992).

Such legislation is not restricted to the UK; most European and North American countries have similar requirements. Eventually all nations will have similar laws. The International Programme on Chemical Safety (IPCS, 1990), which is a joint programme of the United Nations Environment Programme (UNEP), the International Labour Office (ILO) and the World Health Organization (WHO) was set up to facilitate this process. One of its many useful outputs is the set of International Chemical Safety Cards, which it produces in conjunction with the Commission of the European Communities (CEC, 1996). These cards and other useful sources on information can be found on the 'Information from the Intergovernmental Organizations' website www.inchem.org/.

25.3 TOXICOLOGICAL REACTION

It is impossible to deal with every conceivable chemical toxicological reaction in one chapter, and indeed some areas are dealt with in more detail in other chapters. This chapter will highlight some of the more common areas of possible reaction and give some examples of these.

25.3.1 Corrosion

Corrosive chemicals have a wholly destructive action on tissues. This is often a hydrolytic reaction through the high water content of the cell leading to structural and chemical destruction, which manifests itself in burns of varying degrees of severity according to penetration and the degree of contact. This occurs most commonly through skin reaction but not exclusively so; corrosive chemicals could be ingested, or even inhaled. An obvious example is the reaction of a strong mineral acid or base. Acid salts such as phosphorus(V) chloride are similarly reactive but neutral salts are not. Contact with sodium chloride is not injurious, although it is corrosive in the sense that it will react with other compounds, metals, for example, usually over a prolonged period of time. But ingestion of large quantities of sodium chloride would be toxic because of its absorbance into cells, which would not occur by normal skin contact. The toxic effect of corrosive chemicals is not confined to inorganic compounds, although corrosive organic compounds would usually have a relatively low or high pK value as with ethanoic acid or tertiary ammonium compounds, or have an inorganic component as in benzoyl chloride.

25.3.2   Organic Compounds

The range of compounds that might be considered under this heading is enormous and wide ranging. When toxicity arises primarily due to the organic nature of the compound, it is often a consequence of its volatility and associated properties as a solvent and its flammability. These terms are often used to indicate toxicity, but can be misleading as there are many toxic organic compounds that are not solvents, and some organic solvents that are toxic but are not flammable. Secondly, flammability clearly indicates a hazard but is not necessarily an indication of toxicity. The confusion arises because most organic solvents are volatile and toxic. Almost by definition they are good solvents because they are volatile, and since they are volatile they are easily absorbed, through the skin and by inhalation. The student needs to be clear about these differences, and they are best demonstrated by example.

Benzene derivatives and ketones form the basis of many industrial-solvent mixtures, used, for example, in adhesive preparations whose misuse is frequently referred to as 'glue sniffing'. Their toxicity is well recognised through inhalation and they are good solvents, volatile and flammable. The simple alcohols are good solvents, they are flammable and toxic but their toxicity is usually due to ingestion, not inhalation. Other solvents are volatile and toxic but not flammable, for example, the halogenated hydrocarbons such as chloroform and carbon tetrachloride. Carbon tetrachloride is so toxic (see Chapter 15) that it has been withdrawn largely as a commercial chemical, and totally as a domestic dry cleaning agent. Similarly, heterocyclics, pyridine and its derivatives, are good solvents, volatile and highly toxic, but non-flammable. The modes of action of these various groups of compounds may be very different, the alcohols have a long-term chronic effect on liver cells, whereas halogenated hydrocarbons, which are also extremely hepatotoxic, cause acute liver failure. Others such as benzene and related derivatives and aromatic amines have carcinogenic properties; this makes them toxic by ingestion as well as inhalation. It is interesting to note that toluene-based compounds are much less toxic in this respect than benzene derivatives, and toluene is therefore preferred as a solvent. This is because it is metabolised to benzoic acid, which is water-soluble and may therefore excreted through the kidneys. It is, nonetheless, highly flammable.

Finally, extreme examples of rapid toxicity are those that directly affect the nervous system or the mitochondrial respiratory pathway, for example, many organophosphorus compounds and cyanides (see Chapter 17).

25.3.3   Biological Materials

Chemistry students frequently study modules in biochemistry and other biological sciences, and later embark on a career in biological sciences. This involves the handling of biological fluids and substances, and the risk of exposure to dangerous pathogens.

In the handling of human material, the most common, potentially serious pathogens are hepatitis viruses and human immunodeficiency virus (HIV). The pathological activity of body fluids and other materials is dependent upon the likelihood of containing the virus, but all should be treated as potentially hazardous. The most likely hazard to laboratory workers handling pathogens is by the introduction of the fluid into the blood stream, which can occur through contact with cuts, abrasions and other breakages in the skin's surface, or through needle stick injuries. The likelihood is extremely low, but the risk is high because of the poor prognosis. For this reason, the chemist must know how to handle biological fluids safely and be aware of the use of protective clothing.

25.3.4 Allergens

Individuals can have an allergic reaction to compounds that are otherwise harmless. This involves the stimulation of the immune system to produce an apparently unnecessary defensive mechanism. It may occur as a result of sensitisation to a particular compound due to an earlier exposure. The possible reaction, both in terms of the allergen and the seriousness of the reaction, is enormously variable. Almost any compound can be responsible, for example, adverse reactions are well recognised with diazomethane, formaldehyde, isocyanates, phenols, nickel salts and chromates (VI).

Common allergies are also recognised in certain foodstuffs, strawberries and nuts for instance, and in pharmaceutical preparations such as penicillin. Reactions are widespread, ranging from an irritating, benign rash to a life threatening anaphylactic reaction. Individuals with a reaction to penicillin should have this recorded in their medical notes and a doctor, if thinking of prescribing penicillin, will ask the patient if he or she has a known reaction. This is an indication of its potential severity. Strawberries may cause a transitory embarrassing rash whereas nuts, particularly peanuts, may result in a fatal reaction. Some everyday products such as latex may cause a reaction. Latex has wide ranging uses especially in patient care, and may be a hazard to both patient and carer.

Clearly reactions are wide ranging in intensity and very enormously from one individual to another. It is therefore impossible to lay down any specific requirements for safe handling of chemicals that may cause an allergic reaction. If good handling procedures are employed, the risk is minimal. However, if an individual is aware of being sensitive to a particular compound then it should be handled with special care or avoided completely.

25.3.5 Pharmaceuticals

Many chemists work in the pharmaceutical industry and consequently handle chemicals with a pharmacological activity. Most pharmaceuticals are not toxic unless taken in quantities that would indicate malicious or suicidal overdosing rather than accidental ingestion. Consequently, if safe chemical handling procedures are adopted, pharmaceutical compounds should not require special mention. There are many formularies and compendia, which list the side effects of drugs, including overdosing, as well as their pharmaceutical usage, for example, the Association of the British Pharmaceutical Industry (ABPI, 1999) Compendium of Data Sheets and Summaries of Product Characteristics.

However, there is a security aspect to the safe handling of pharmaceutical compounds that have an unlawful market. 'Hard line' drugs such as opiates fall into this category, as may others that are not available on the open market, and particularly those that are no longer readily available through medical practitioners, for example, barbiturates and benzodiazepines. High security cupboards are therefore recommended for the storage of such compounds. In many countries legislation requires such security for certain prescribed drugs.

25.3.6 Radionuclides

The use of radionuclides has been discouraged over recent years on health and safety grounds, and in many instances alternate methods have been introduced. This is particularly true in the analytical field of immunoassay where radioactive isotopes have largely been replaced by alternative labels. Nonetheless, radionuclides are still used and handled by chemists in many different areas. Their use and handling is controlled by legislation in most countries, and this is often complex and variable because the hazards involved are dependent upon the type of emission and nature of the labelled compound. For example, a radioactive isotope in a compound that is metabolised if ingested, can end up in many other compounds in the body, in some instances, concentrated in particular organs, which constitutes a much greater hazard than the ingestion of the same isotope in a compound that is not metabolised.

For these reasons legislation requires purpose-designed laboratory areas put aside solely for the handling of isotopes, and the safety level graded according to the types of isotopes and compounds that are going to be handled. Reference must be made to local legislation for details of local requirements. There may also be a government inspection process before the handling of radionuclides is permitted.

25.4 GOOD LABORATORY PRACTICE

The chemist who is trained in good laboratory practice will be practicing the safe handling of chemicals. It is impossible to lay down criteria that will guarantee safe handling because it is not possible to predict the eventualities and situations that could arise in a chemical laboratory. However, safe practice, whether it be in a chemistry laboratory or any other workplace, consists of a combination of common sense, experience and a technical understanding and appreciation of the procedures that are being carried out.

25.4.1 Storage

A well-organised laboratory will have a continuous supply of the chemicals that are required for its work, which means that there must be a comprehensive storage system. The number and quantities of chemicals retained at the bench should be minimal, the bulk being kept in a controlled and safe store. Bench chemicals should never be stored on the actual bench, but on shelving where they are easily assessable without the operator having to stretch unduly to reach them or to stretch over equipment on the bench, thus running the risk of protective clothing being caught in moving apparatus or coming into contact with chemicals in use in the bench.

Commercially purchased chemicals should be stored in the bottles or containers in which they are received from the manufacturer, thus preserving the correct manufacture's labelling. Reagents prepared within the laboratory should be clearly labelled and contained in an appropriate receptacle. There should never be more than one container of a given chemical or reagent stored on the working bench at any time.

When flammable organic solvents are used, not more than 500 mL should be stored at the working bench. Compounds that are particularly hazardous need special consideration. For example, working quantities of flammable solvents should be stored in a steel storage cabinet, not on the bench shelves, and compounds of known high toxicity must be stored in a fume cupboard. The fume cupboard should be designated for storage, and not used as a working cupboard.

Chemicals that are not intrinsically hazardous, but are highly toxic if ingested, need to be stored in a special high-security cabinet. This includes extreme, acute poisons, and some pharmaceutical drug preparations. Indeed, the handling of some substances may require government authority. In the UK, for example, a Home Office licence is required for the storage of certain categories of drugs, in which case a high-security cabinet is required, and an inventory of the purchase and use of such compounds must be maintained.

It has already been mentioned that the bulk of chemicals should not be stored in the laboratory, but in a safe store. For most, a dry store with assessable shelving is adequate. An inventory of the store must be maintained and a sequence of chemicals instituted. The obvious sequence is an alphabetical one, but this is not always recommended as it may result in the proximity of chemicals, which, if spilt, would result in a potentially dangerous reaction. However, alphabetical storage is quite satisfactory, so long as it is overridden where known hazardous combinations might arise. The store area needs to be well ventilated and equipped with appropriate safety facilities, fire extinguishers, smoke detectors, drip trays, *etc.*

A special fireproof store must be used for the storage of flammable solvents and strong acids. This must be a separate building, located near to, but outside, the laboratory complex and must comply with local safety regulation requirements. It must be able to maintain a major spillage and be equipped with appropriate safety facilities, *i.e.* smoke and fire detectors, fire extinguishers and suitable ventilation, *etc.*

All store areas must be specifically designated, and should used for no other purpose.

25.4.2 Reagent Preparation

Laboratories buy many of their reagents already prepared from commercial suppliers, but some, particularly those used in a research environment, are prepared in the laboratory. It is therefore important that the student learns how to prepare reagents correctly and safely. In mixing chemicals, the student must be aware of possible reactions of mixing. These may be potentially hazardous as in the simple dilution of concentrated sulphuric acid with water where sufficient heat may be generated to cause the resultant solution to boil. In this case, the concentrated acid must be added to the water and not *vice versa*, thus if there is a spillage as a consequence of overheating, it is a dilute acid solution that is spilt, not a concentrated one. All bench techniques involved in reagent preparation must be carried out carefully to avoid chemical spillage whether it be on the balance pan during weighing or elsewhere during transfer. The reagent, when prepared, must be stored in a suitable receptacle and must be labelled correctly. The labelling must include details of all constituent components and their quantities. The date upon which the reagent was prepared and the name of the person who prepared it must be recorded.

Reagents must be prepared and stored in appropriate containers. In most cases, a container will be a glass bottle with a glass stopper. However, there are exceptions and these must be known and observed. For instance, solutions containing alkali at a pH greater than 12.0 should be stored in polythene bottles with inert plastic stoppers or screw tops. Glass bottles should not be used and under no circumstance glass bottles with glass stoppers be used. Strong hydroxides will react slowly with silicates in the glass and atmospheric carbon dioxide forming a thin but hard layer of silicates and carbonates around the stopper making it extremely difficult to remove. There are more laboratory injuries through breakage of glassware than any other single cause, and the removal of glass stoppers is the most common cause of accidents with glassware.

Picric acid solution is another example of a common possible hazard. The solution is stable, but the solid is potentially explosive. A thin layer of dried picric acid around the neck of a glass-stoppered bottle may be quite dangerous.

Organic solvents should not be stored in plastic bottles. Although many plastics such as polythene are inert to most solvents, there is a long-term effect, which will cause the vessel to split suddenly.

Reagents that are light sensitive need to be stored in dark bottles, and those that are very sensitive should be stored in dark bottles in a closed cupboard. Reagents that are labile at room temperature need to be stored in a refrigerator or even a freezer. However, flammable liquids must never be stored in a domestic type refrigerator or freezer because of the risk of ignition through electrical discharge when the thermostat switches on or off. Only cooler units specifically designed to contain flammable material should be used. These are a few examples of many reagent vessel storage problems; but the student will learn most effectively by experience.

25.4.3 Biological Fluids

The potential danger of biological fluids has already been discussed from which it is clear that the hazard is dependent upon what, if any, infective agent is present. Every fluid should therefore be regarded as potentially infected and be handled accordingly. Good practice is largely a matter of hygiene and the appropriate use of protective clothing. These are dealt in Section 25.5. In addition, there should be a laboratory policy for the swabbing of benches and other areas with a suitable disinfectant at frequent defined intervals. Under no circumstances should biological fluids be apportioned by mouth pipetting. Indeed mouth pipetting of any liquid should not be permitted. Pipette fillers should be used at all times and if they become contaminated, a decontamination procedure must be adopted, and if necessary, the filler discarded.

25.4.4 Radionuclides

The hazard and possible toxic effects of radioisotopes are dependent upon the radiation emitted and the metabolism of the labelled compound, if ingested. This has been referred to earlier, as has the fact that many countries enforce strict legal requirements on the use and handling of isotopes. In addition to complying with these requirements, safe handling should be achieved by adhering to the same requirements as those for handling biological fluids.

25.4.5 Gases

Many laboratory procedures require the use of gases supplied commercially in pressurised bottles. These present two hazards, those associated with the chemical hazard of the gas, and those associated with a gas under pressure, even though it may be chemically inert. It must be remembered that a serious leak of an inert gas can cause asphyxiation by replacement of the atmospheric oxygen.

A fully charged cylinder may have a pressure as high as 15 MPa, *i.e.* in excess of 100 atmospheres. The mishandling of gas cylinders therefore represents a considerable hazard. Cylinders should not be stored or operated from within the laboratory, but kept in a purpose built store with good ventilation and plumbing. If it is necessary to store a cylinder in the laboratory, it must be contained in a purpose built stand and clamped securely to the bench. Transportation of cylinders should only be carried out using a purpose designed cylinder carrier.

Ideally, cylinders should be operated from the store, the one in use being plumbed into the laboratory through a permanent piping system. Expert installers must be consulted in order to ensure the correct material is used for each gas (this is not necessarily the same for each gas). Reducing valves must be used at all times with high-pressure cylinders, and the operator must ensure the correct valve is used and be aware of the differences between valves. Generally, flammable gases are fitted with left-hand thread valves whereas other gases are fitted with conventional right-hand thread valves.

Acetylene is probably the most explosive and dangerous chemical that is commonly used in chemical laboratories and there are a number of special precautions required in handling this gas, both in relation to the cylinder and plumbing. Chemists using acetylene must familiarise themselves with these requirements and satisfy themselves that they are being adhered to.

Hydrogen is almost as dangerous because of its explosive nature and its very low molecular mass. It is the smallest known molecule and as a consequence is highly diffusible. This means that leaks can easily occur and the gas 'creeps', accumulating in potentially explosive mixtures at locations some distance from the source of the leak. This is a hazard of which the chemist must be well aware.

When cylinders are being moved, whether full or empty, they must be handled with care. They should not be rolled or manhandled, they should not be moved with the reducing valve in place because of the risk of damage, nor should the main valve be left open when the cylinder has been emptied. Finally, grease should never be used on any cylinder for lubrication unless it is specified for use with that cylinder. Flammable and non-flammable gases should be stored separately, and separate from oxygen.

25.4.6 Equipment

The safe use of equipment is related to the safe handling of chemicals. If the equipment is not used correctly, then there is a greater likelihood of accidents with undue exposure to chemicals and their toxic and hazardous properties. Operational equipment should be maintained to a high level of electrical and mechanical safety, if necessary having a service contract with the supplier. A sequence of check procedures must be established and adhered to. General laboratory tidiness is part of good equipment maintenance. Loose paper left lying in inappropriate places may result in the blocking of air vents in large machines causing overheating. Reagent bottles or small equipment left in the wrong place may cause an obstruction resulting in a chemical spillage, possibly over equipment.

Finally, there are specific hazards related to particular pieces of equipment; for example, equipment that requires gas supplies from high-pressure bottles, some of which may be flammable and potentially explosive. Gas chromatographs not only require a flammable gas supply, but also have ovens set at relatively high temperatures. Centrifuges, because of their high speed of operation, pose a particular hazard. In many countries their use is controlled by legislation, particularly if they are being used for the processing of biological fluids. It may be necessary to fit a bench centrifuge with a restraining bar to prevent movement in the event of it not being balanced correctly. These are examples of specific equipment hazards and the student needs to be familiar with the particular hazards associated with the equipment he or she is using, and must adopt a good practice approach to all equipment.

25.5 HEALTH AND SAFETY

Good laboratory practice and the safe handling of chemicals are related to good health and safety practice which includes standards of personal hygiene and the appropriate use of protective clothing.

The laboratory is a workplace where potentially dangerous chemicals are in constant use and must therefore remain solely a workplace. Other related work, such as paperwork, should be carried out in an adjacent but specified area where chemicals are not in use. Other activities are totally forbidden, for example, eating, drinking, smoking, *etc.*

Hands should be washed before leaving the laboratory area because of the risk of contamination from compounds handled in the laboratory.

Protective clothing is equally important. A white coat, or equivalent, should be worn in the laboratory at all times and this should not be removed from the laboratory area, except for laundering. This is particularly important when dealing with biologically hazardous substances in which case the microbiological high-neck type laboratory coat may be appropriate. When handling biological substances it is also important to ensure that areas of broken skin that may come into contact with such substances should be covered. Gloves and goggles or facemasks should be worn where the risk is high. Mouth pipetting is totally forbidden.

For rapid and potentially dangerous or explosive reactions, a safety shield should be placed between the equipment and the operator, with the whole system placed in a fume cupboard. Fume cupboards should also be used for processes that involve the use or release of potentially harmful or volatile compounds. Fume cupboards are only effective if used properly. This means that the fan must be operational and the front sash window closed to at least the maximum operational position. This is either marked on the side of the cupboard or the information is available from the installers. While working in the cabinet, the window should be as low as possible, leaving only a minimal reasonable operating space between the floor of the cabinet and the bottom of the window. No fume cupboard will operate effectively with the window fully open. When not in use the window should be kept in the closed position, which leaves a gap at the bottom for airflow. The lower the window, the more effective the cabinet is. From time to time the airflow through the cabinet must be checked with an airflow meter to ensure that the minimum, specified air flow is achieved.

25.6 POST INCIDENT PROCEDURES

Through good practice and safe handling of chemicals it is hoped that untoward incidences will not occur. However, it must be recognised that there will be accidents even with the operation of safe procedures, and the student must be aware of how to deal with these. Procedures may be divided into two categories, those dealing with an affected individual and those concerned directly with the incident.

An affected individual should be removed from the immediate site of the incident. Qualified first aid or medical assistance should be sought urgently if this is considered necessary. If there is skin contact with corrosive materials the affected areas should be doused with large volumes of cold running water. Similarly, if there is an eye contact, the eye and surrounding tissue should be washed thoroughly. If in doubt, large quantities of cold running water should be administered until more expert advice is available. For larger affected areas the individual should be placed under an emergency shower, which must be available in all chemistry laboratory areas. Clothing should be removed and reservations due to modesty overridden in favour of safety and treatment. In cases of ingestion, forced vomiting or the administration of an antidote must be considered. Antidotes, if they exist, should be available where chemicals of specific high toxicity are being used. For instance, if a laboratory uses cyanides, the antidote must be available and the laboratory staff must be trained in its use.

The most common chemical accident is spillage, either as a consequence of the breakage of a reagent or chemical container, or as the result of a fracture to a vessel in which a chemical reaction is taking place. In either case the area should be evacuated and a quick assessment made as to whether the spillage can be dealt with without further risk or injury. The spillage should be contained and mopped up. This may mean the use of large quantities of water if the contaminant needs to be diluted, or if it contains a strong acid, the use of a weak base such as solid sodium carbonate may be more suitable. Equally, if a strong alkali is present, a weak acid, such as dilute ethanoic acid, may be used. Water must not be used to remove organic solutions that are not miscible with water. In such situations, an absorbing reagent such as vermiculite should be used.

If there is any possibility of further exposure of individuals, then the area should be evacuated, sealed off and the assistance of professional services be sought, such as the fire service. The release of obnoxious fumes, or fire, should *only* be dealt with if there is no risk to individuals. The intensity of fumes is reduced by opening as many windows as possible to create ventilation and a through draft, and by removing chemicals or the reaction from the source that is causing the fumes, for example, removing the reaction vessel from the heat source, switching off equipment, *etc.*

Small fires may be dealt with by applying the appropriate fire extinguisher, if there is no risk to laboratory personnel. All laboratory workers should be aware of the location of fire extinguishers, which type is appropriate for which type of fire, and apply accordingly. Similarly, laboratory personnel should be aware of the location of other safety aids, such as first aid cabinets, fire blankets, fire alarm points and respiratory and other protective clothing and equipment.

25.7   PROTOCOLS AND PROCEDURES

There is an increase in legislation in many countries relating to good laboratory practice and the safe handling of chemicals. This leads to a requirement of compliance with safety documents, the establishment of protocols for carrying out routine procedures, and the listing of chemicals with their particular hazards and toxic properties. Examples of this in the UK have already been mentioned, such as the COSHH requirements and the Department of Health's documentation on the handling of biological material. Increasingly, protocols are required for laboratory registration programmes such as accreditation, good working practices and total quality management. Many professional chemists may find these legislative requirements imposing upon their professional acumen. Nonetheless, they produce a general awareness of safety requirements, and help to establish acceptable minimal standards.

It is also recognised that there is a need to produce Statements of Operational Procedures. These detail the work (analytical) that is being carried out within a laboratory. Each step of a procedure is described, giving reagents and quantities that are required, incubation times, *etc.* In addition these procedures should include any relevant details concerning the hazards or toxicity of any of the reagents involved, and if necessary procedural details in the case of a spillage or an accident. Not only do such details help to maintain safety standards, but also they improve analytical competence by ensuring that different operators follow the same protocol.

Chemists involved in research may find it difficult to describe their work in this way, as these procedures are more applicable to routine processes. Nonetheless, documentation such as this will enhance the safety of operational procedures, and the student needs to be aware of their existence, and when to apply them. Knowledge of the existence of such protocols is important because the student will undoubtedly meet them as his or her career develops.

BIBLIOGRAPHY

ABPI, *ABPI Compendium of Data Sheets and Summaries of Product Characteristics 1999–2000*, Datapharm Publications Limited, London, 1999.

P.H. Bach, S.S. Brown, J.A. Haines, R.M. Joyce, B.C. McKusick and H.G.J. Worth (eds), in *Chemical Safety Matters*, Cambridge University Press, Cambridge, 1992.

CEC, *European Community Environmental Legislation*, Vols 1–7, Office for Official Publications of the European Communities, Luxembourg, 1996.

A.K. Farr (ed), *CRC Handbook of Laboratory Safety*, 5th edn, CRC Press, Boca Raton, 2000.

HMSO, *Code of Practice for the Prevention of Infection in Clinical Laboratories and Post Mortem Rooms*, HMSO, London, 1978.

HSE, *Safe Working and the Prevention of Infection in Clinical Laboratories and Similar Facilities*, 2nd edn, HSE, London, 2003.

IPCS/CEC, *Industrial Health and Safety. International Chemical Safety Cards*, Commission of the European Communities, Luxembourg, 1990.

R.E. Lenga and K.L. Votoupal (eds), *The Sigma – Aldrich Library of Regulatory and Safety Data*, Sigma–Aldrich Corporation Inc, Milwaukee, WI, 1992.

S.G. Luxon (ed), *Hazards in the Chemical Laboratory*, 5th edn, Royal Society of Chemistry, London, 1992.

Merck Ltd, *Health and Safety Data Sheets*. Merck Ltd, Poole, Dorset, 1995.

USEFUL WEBSITES

Chemical Safety Information from Intergovernmental Organizations www.inchem.org/
Control of Substances Hazardous to Health www.hse.gov.uk/coshh
International Programme on Chemical Safety www.who.int/pcs

Appendix A: A Curriculum for Fundamental Toxicology

Increasingly, chemists and others are faced with legislation requiring assessment of hazard and risk associated with the production, use and disposal of chemicals. In addition, the general public are concerned about the dangers that they hear may result from the widespread use of chemicals. They look to chemists for explanations and assume that chemists understand such matters. When they find that chemists are ignorant of the potential of chemicals to cause harm, their confidence in the profession is lost and chemophobia may result. Thus, it has become essential to introduce toxicology into chemistry course. In order to facilitate this, the International Union of Pure and Applied Chemistry (IUPAC) Commission on Toxicology and the IUPAC Committee on the Teaching of Chemistry in 1993 began to draft a curriculum in fundamental toxicology for chemists and this textbook to support it. The curriculum is detailed below.

It will be seen that the chapter titles in this book are echoed in the suggested curriculum. The complete curriculum shown would provide a thorough basis in toxicology, but the 60 h suggested will be hard to find in any existing degree structure where the continuing increase in knowledge puts time at a premium. We therefore suggest that time be sought for a minimum curriculum consisting of Sections 1, 2, 5 and 7 taking about 30 student contact hours. This would correspond approximately to the traditional term at a British university or a half-semester in the American system. Alternatively, the minimum curriculum could be taught as a 1 week intensive full-time course. The full curriculum could similarly be taught intensively over 2 weeks.

Ultimately, any curriculum will be decided by the faculty members who teach it. Our primary aim in producing this book is to provide the building blocks from which faculty can construct their own curricula appropriate to their own courses. Thus, although we have had the curriculum below in mind in planning this book, we regard it simply as a starting point and we hope that it will evolve in practice to different curricula tailored to the local needs of innovative college and university departments and their graduates.

Since this curriculum was designed, supplementary presentations have been prepared by the authors, approved by IUPAC and made available on the IUPAC website at http://www.iupac.org/publications/cd/essential_toxicology. These presentations are downloadable in PDF format. In themselves, they constitute a course in essential toxicology, but they are also intended to be a source of material for anyone who may start a course in toxicology. The presentations may be used for independent learning by those who are sufficiently motivated.

THE DETAILED CURRICULUM

Section 1 Introduction to Toxicology
Hazard, toxicity. And fundamental concepts. Toxicity testing and epidemiology as sources of information
Toxicokinetics
Exposure, absorption, distribution, storage, metabolism and excretion
Toxicodynamics
Mechanisms of toxicant production of harmful effects

Section 2 Using Toxicity Data
Toxicity data interpretation
Risk assessment and risk management
Monitoring exposure

Section 3 Specialised Aspects of Toxicology
Mutagenicity
Carcinogenicity
Reproductive toxicity
Immunotoxicity
Toxicogenomics
Immunotoxicology
Clinical toxicology

Section 4 Organ Toxicology
Skin toxicology
Respiratory toxicologist
Hepatotoxicity
Nephrotoxicity
Neurotoxicity

Section 5 Complex Aspects of Toxicology
Behavioural toxicology
Environmental fate and ecotoxicity

Section 6 Special Toxicants
Radionuclides
Biocides
Pharmaceuticals

Section 7 Safe Handling of Chemicals

Summary and Review

Proposed Timetable

Section 1 Introduction to toxicology	12 h
Section 2 Using toxicity data	8 h
Section 3 Specialised aspects of toxicology	14 h
Section 4 Organ toxicology	10 h
Section 5 Complex aspects of toxicology	7 h
Section 6 Special toxicants	6 h
Section 7 Safe handling of chemicals	2 h
Summary and Review	1 h
Total	60 h

Appendix B: Glossary of Terms Used in Toxicology

This glossary is an abridged version of a compilation of the glossaries of terms used in toxicology[1] and terms used in toxicokinetics[2] approved by IUPAC and published in 1993 and 2004 in the following papers:

1. J. H. Duffus, *Pure Appl. Chem.*, 1993, **65**, 2003–2122.
2. M. Nordberg, J. Duffus and D. M. Templeton, *Pure Appl. Chem.*, 2004, **76**, 1033–1082.

absolute lethal concentration (LC_{100})
Lowest *concentration* of a substance in an environmental medium which kills 100% of test organisms or species under defined conditions.
Note: This value is dependent on the number of organisms used in its assessment.

absolute lethal dose (LD_{100})
Lowest amount of a substance which kills 100% of test animals under defined conditions.
Note: This value is dependent on the number of organisms used in its assessment.

absorbed dose (of a substance)
Amount (of a substance) taken up by an organism or into organs or tissues of interest.
See *absorption, systemic.*
synonym *internal dose.*

absorbed dose (of radiation)
Energy imparted to a unit mass of matter by ionizing radiation divided by the mass of the absorbing volume.

absorption (in biology)
Penetration of a substance into an organism by various processes, some specialised, some involving expenditure of energy (active transport), some involving a *carrier* system, and others involving passive movement down an electrochemical gradient.
Note: In mammals *absorption* is usually through the respiratory tract, gastrointestinal tract, or skin.

absorption (of radiation)
Phenomenon in which radiation transfers some or all of its energy to matter, which it traverses.

absorption, systemic
Uptake to the blood and transport via the blood of a substance to an organ or *compartment* in the body distant from the site of *absorption.*

absorption coefficient (in biology)
Ratio of the absorbed amount (*uptake*) of a substance to the administered amount (intake).
Note: For exposure by way of the respiratory tract, the absorption coefficient is the ratio of the absorbed amount to the amount of the substance (usually particles) deposited (adsorbed) in the lungs.
synonym *absorption factor.*

acaricide
Substance intended to kill mites, ticks or other Acaridae.

acceptable daily intake (ADI)
Estimate by JECFA of the amount of a food additive, expressed on a body weight basis, that can be ingested daily over a lifetime without appreciable health *risk*.
Note 1: For calculation of ADI, a standard body mass of 60 kg is used.
Note 2: Tolerable daily intake (TDI) is the analogous term used for contaminants.

acceptable residue level of an antibiotic
Acceptable *concentration* of a residue that has been established for an antibiotic found in human or animal foods.

acceptable risk
Probability of suffering disease or injury that is considered to be sufficiently small to be "negligible".

accepted risk
Probability of suffering disease or injury that is accepted by an individual.

accumulation (in biology)
See *bioaccumulation*.

acidosis
Pathological condition in which the hydrogen ion substance concentration of body fluids is above normal and hence the pH of blood falls below the reference interval.
Antonym *alkalosis*.

action level
1. *Concentration* of a substance in air, soil, water or other defined medium at which specified emergency counter-measures, such as the seizure and destruction of contaminated materials, evacuation of the local population or closing down the sources of pollution, are to be taken.
2. *Concentration* of a pollutant in air, soil, water or other defined medium at which some kind of preventive action (not necessarily of an emergency nature) is to be taken.

activation (in biology)
See *bioactivation*.

active metabolite
Metabolite with biological and (or) toxicological activity.

acute

1. Of short duration, in relation to *exposure* or effect.

Note: In experimental *toxicology*, "acute" refers to studies where dosing is either single or limited to one day although the total study duration may extend to 2 weeks.

2. In clinical medicine, sudden and severe, having a rapid onset.

antonym *chronic*.

acute effect

Effect of finite duration occurring rapidly (usually in the first 24 h or up to 14 days) following a single *dose* or short *exposure* to a substance or radiation.

acute exposure

Exposure of short duration.

acute toxicity

1. *Adverse effects* of finite duration occurring within a short time (up to 14 days) after administration of a single *dose* (or *exposure* to a given *concentration*) of a test substance or after multiple doses (exposures), usually within 24 h of a starting point (which may be exposure to the *toxicant*, or loss of reserve capacity, or developmental change, *etc.*).

2. Ability of a substance to cause *adverse effects* within a short time of dosing or *exposure*.

antonym *chronic toxicity*.

added risk

Difference between the *incidence* of an *adverse effect* in a treated group (of organisms or a group of *exposed* humans) and a control group (of the same organisms or the spontaneous incidence in humans).

addiction

Surrender and devotion to the regular use of a medicinal or pleasurable substance for the sake of relief, comfort, stimulation or exhilaration which it affords; often with craving when the drug is absent.

additive effect

Consequence that follows *exposure* to two or more physicochemical agents which act jointly but do not interact: the total effect is the simple sum of the effects of separate exposures to the agents under the same conditions.

adduct

New chemical species AB, each molecular entity of which is formed by direct combination of two separate molecular entities A and B in such a way that there is no change in connectivity of atoms within their moieties A and B.

Note 1: Stoichiometries other than 1:1 are also possible.

Note 2: An intramolecular adduct can be formed when A and B are groups contained within the same molecular entity.

adenocarcinoma

Malignant tumour originating in glandular *epithelium* or forming recognizable glandular structures.

adenoma
Benign tumour occurring in glandular *epithelium* or forming recognizable glandular structures.

adjuvant
1. In pharmacology, a substance added to a *drug* to speed or increase the action of the main component.
2. In immunology, a substance (aluminium hydroxide) or an organism (bovine tuberculosis bacillus) which increases the response to an *antigen*.

adrenergic
See synonym *sympathomimetic*.

advection (in environmental chemistry)
Process of transport of a substance in air or water solely by mass motion.

adverse effect
Change in biochemistry, morphology, physiology, growth, development or lifespan of an organism that results in impairment of functional capacity or impairment of capacity to compensate for additional stress or increase in susceptibility to other environmental influences.

aerobe
Organism, which needs molecular oxygen for respiration and hence for growth and life.

aerobic
Requiring molecular oxygen.

aerodynamic diameter (of a particle)
Diameter of a spherical particle with relative density equal to unity that has the same settling velocity in air as the particle in question.

aerosol
Mixtures of small particles (solid, liquid or a mixed variety) and the *carrier* gas (usually air).
Note 1: Owing to their size, these particles (usually less than $100\,\mu$m and greater than $0.01\,\mu$m in diameter) have a comparatively small settling velocity and hence exhibit some degree of stability in the earth's gravitational field.
Note 2: An aerosol may be characterized by its chemical composition, its radioactivity, the particle size distribution, the electrical charge and the optical properties.

aetiology
1. Science dealing with the cause or origin of disease.
2. In individuals, the cause or origin of disease.

agonist
Substance that binds to cell *receptors* normally responding to naturally occurring substances and which produces a response of its own.
antonym *antagonist*

albuminuria
Presence of albumin, derived from *plasma*, in the urine.

algicide
Substance intended to kill algae.

alkalosis
Pathological condition in which the hydrogen ion substance concentration of body fluids is below normal and hence the pH of blood rises above the reference interval. antonym *acidosis*.

alkylating agent
Substance that introduces an alkyl substituent into a compound.

allele
One of several alternate forms of a *gene* which occur at the same relative position (locus) on homologous *chromosomes* and which become separated during *meiosis* and can be recombined following fusion of *gametes*.

allergen
Antigenic substance capable of producing immediate *hypersensitivity*.

allergy
Symptoms or signs occurring in sensitized individuals following exposure to a previously encountered substance (*allergen*) which would otherwise not cause such *symptoms* or *signs* in non-sensitized individuals. The most common forms of allergy are *rhinitis, urticaria, asthma* and *contact dermatitis*.

allometric
Pertaining to a systematic relationship between growth rates of different parts of an organism and its overall growth rate.

allometric growth
Regular and systematic pattern of growth such that the mass or size of any organ or part of a body can be expressed in relation to the total mass or size of the entire organism according to the *allometric* equation:

$$Y = bx^{\alpha}$$

where Y is the mass of the organ, x the mass of the organism, α the growth coefficient of the organ and b a constant.

allometric scaling
1. Adjustment of data to allow for change in proportion between an organ or organs and other body parts during the growth of an organism.
2. Adjustment of data to allow for differences and make comparisons between species having dissimilar characteristics, for example in size and shape.

allometry (in biology)
Measurement of the rate of growth of a part or parts of an organism relative to the growth of the whole organism.

all-or-none effect
See synonym *quantal effect*.

alveol/us (pulmonary), **-i** pl., **-ar** adj.
Terminal air sac of the lung where gas exchange occurs.

ambient
Surrounding (applied to environmental media such as air, water, sediment or soil).

ambient monitoring
Continuous or repeated measurement of agents in the environment to evaluate ambient exposure and *health* risk by comparison with appropriate reference values based on knowledge of the probable relationship between exposure and resultant adverse health effects.

ambient standard
See synonym *environmental quality standard*.

Ames test
In vitro test for *mutagenicity* using mutant strains of the bacterium *Salmonella typhimurium*, which cannot grow in a given histidine-deficient medium: *mutagens* can cause reverse *mutations* that enable the bacterium to grow on the medium. The test can be carried out in the presence of a given microsomal fraction (S-9) from rat liver (see *microsome*) to allow metabolic transformation of mutagen precursors to active derivatives.

amplification (of genes)
See synonymous term *gene amplification*.

anabolism
Biochemical processes by which smaller molecules are joined to make larger molecules. antonym *catabolism*.

anaerobe
Organism that does not need molecular oxygen for life. Obligate (strict) anaerobes grow only in the absence of oxygen. Facultative anaerobes can grow either in the presence or in the absence of molecular oxygen.
antonym *aerobe*.

anaerobic
Not requiring molecular oxygen.

anaesthetic
Substance that produces loss of feeling or sensation: general anaesthetic produces loss of consciousness; local or regional anaesthetic renders a specific area insensible to pain.

analgesic
Substance that relieves pain, without causing loss of consciousness.

analytic study (in epidemiology)
Study designed to examine associations, commonly putative or hypothesized causal relationships.

anaphylaxis
Severe allergic reaction (see *allergy*) occurring in a person or animal *exposed* to an *antigen* or *hapten* to which they have previously been sensitized.

anaplasia
Loss of normal cell differentiation, a feature characteristic of most *malignancies*.

aneuploid
Cell or organism with missing or extra *chromosomes* or parts of chromosomes.

anoxia
Strictly total absence of oxygen but sometimes used to mean decreased oxygen supply in tissues.

antagonism
Combined effect of two or more factors, which is smaller than the solitary effect of any one of those factors.
Note: In *bioassays*, the term may be used when a specified effect is produced by *exposure* to either of two factors but not by exposure to both together.

anthelmint(h)ic
Substance intended to kill parasitic intestinal worms, such as helminths.
synonym *antihelminth*.

anthracosis (coal miners' pneumoconiosis)
Form of *pneumoconiosis* caused by accumulation of carbon deposits in the lungs due to inhalation of smoke or coal dust.

anthropogenic
1. Caused by or influenced by human activities.
2. Describing a conversion factor used to calculate a *dose* or *concentration* affecting a human that has been derived from data obtained with another species, *e.g.* the rat.

anti-adrenergic
See synonym *sympatholytic*.

antibiotic
Substance produced by, and obtained from, certain living cells (especially bacteria, yeasts and moulds), or an equivalent synthetic substance, which is *biostatic* or *biocidal* at low concentrations to some other form of life, especially pathogenic or noxious organisms.

antibody
Protein molecule produced by the immune system (an *immunoglobulin* molecule), which can bind specifically to the molecule (*antigen* or *hapten*), which induced its synthesis.

anticholinergic
1. adj., Preventing transmission of parasympathetic nerve impulses.
2. n., Substance which prevents transmission of parasympathetic nerve impulses.

anticholinesterase
See synonym *cholinesterase inhibitor*.

anticoagulant
Substance that prevents clotting.

antidote
Substance capable of specifically counteracting or reducing the effect of a potentially *toxic* substance in an organism by a relatively specific chemical or pharmacological action.

antigen
Substance or a structural part of a substance that causes the immune system to produce specific *antibody* or specific cells, which combines with specific binding sites (*epitopes*) on the antibody or cells.

antihelminth
See synonym *anthelmint(h)ic*.

antimetabolite
Substance, structurally similar to a *metabolite*, which competes with it or replaces it, and so prevents or reduces its normal utilization.

antimycotic
Substance used to kill a fungus or to inhibit its growth.
synonym *fungicide*.

antipyretic
Substance that relieves or reduces fever.

antiresistant
Substance used as an additive to a *pesticide* formulation in order to reduce the resistance of insects to the pesticide.

antiserum
Serum containing *antibodies* to a particular *antigen* either because of immunization or after an infectious disease.

aphasia
Loss or impairment of the power of speech or writing, or of the ability to understand written or spoken language or signs, due to a brain injury or disease.

aphicide
Substance intended to kill aphids.

aphid
Common name for a harmful plant parasite in the family Aphididae, some species of which are vectors of plant virus diseases.

aplasia
Lack of development of an organ or tissue, or of the cellular products from an organ or tissue.

apoptosis
Active process of programmed cell death requiring metabolic energy, often characterized by fragmentation of *DNA*, and without associated inflammation.
See also *necrosis*.

arboricide
Substance intended to kill trees and shrubs.

area under the concentration–time curve
See *area under the curve.*

area under the curve (AUC)
Area between a curve and the abscissa, *i.e.*, the area underneath the graph of a function: often, the area under the tissue (*plasma*) *concentration* curve of a substance expressed as a function of time.

area under the moment curve (AUMC)
Area between a curve and the abscissa in a plot of *concentration* $\times$ time versus time.

argyria
Pathological condition characterized by grey-bluish or black pigmentation of tissues (such as skin, retina, mucous membranes and internal organs) caused by the accumulation of metallic silver, due to reduction of a silver compound that has entered the organism during (prolonged) administration or exposure.
synonym *argyrosis.*

argyrosis
See synonym *argyria.*

arrhythmia
Any variation from the normal rhythm of the heartbeat.

artefact
Finding or product of experimental or observational techniques that is not properly associated with the system being studied.

arteriosclerosis
Hardening and thickening of the walls of the arteries.

arthralgia saturnia
Pain in a joint resulting from lead poisoning.

arthritis
Inflammation of a joint, usually accompanied by pain and often by changes in structure.

asbestosis
Form of *pneumoconiosis* caused by inhalation of asbestos fibres.

ascaricide
Substance intended to kill roundworms (Ascaridae).

asphyxia
Condition resulting from insufficient intake of oxygen: symptoms include breathing difficulty, impairment of senses, and, in extreme, convulsions, unconsciousness and death.

asphyxiant
Substance that blocks the transport or use of oxygen by living organisms.

assay
1. Process of quantitative or qualitative analysis of a component of a *sample*.
2. Results of a quantitative or qualitative analysis of a component of a sample.

asthenia
Weakness; lack or loss of strength.

asthma
Chronic respiratory disease characterised by bronchoconstriction, excessive mucus secretion and *oedema* of the pulmonary alveoli, resulting in difficulty in breathing out, wheezing and cough.

astringent
1. adj., Causing contraction, usually locally after topical application.
2. n., Substance causing cells to shrink, thus causing tissue contraction or stoppage of secretions and discharges; such substances may be applied to skin to harden and protect it.

ataxia
Unsteady or irregular manner of walking or movement caused by loss or failure of muscular co-ordination.

atherosclerosis
Pathological condition in which there is thickening, hardening and loss of elasticity of the walls of blood vessels, characterized by a variable combination of changes of the innermost layer consisting of local accumulation of lipids, complex carbohydrates, blood and blood components, fibrous tissue and calcium deposits. In addition, the outer layer becomes thickened and there is fatty degeneration of the middle layer.

atrophy
Wasting away of the body or of an organ or tissue.

auto-immune disease
Pathological condition resulting when an organism produces *antibodies* or specific cells that bind to constituents of its own tissues (*autoantigens*) and cause tissue injury: examples of such disease may include rheumatoid *arthritis, myasthenia* gravis and scleroderma.

autophagosome
Membrane-bound body (secondary *lysosome*) in which parts of the cell are digested.

autopsy
Post-mortem examination of the organs and body tissue to determine cause of death or pathological condition.
synonym *necropsy*.

avicide
Substance intended to kill birds.

axenic animal
See synonym *germ-free animal.*

back-mutation
Process that reverses the effect of a *mutation* that had inactivated a gene; thus it restores the wild phenotype.

bactericide
Substance intended to kill bacteria.

bagassosis
Lung disease caused by the inhalation of dust from sugarcane residues.

Bateman function
Equation expressing the build-up and decay in *concentration* of a substance (usually in *plasma*) based on first-order *uptake* and *elimination* in a *one-compartment model*, having the form

$$C = [fDk_a/V(k_a-k_e)][\exp(-k_et) - \exp(-k_at)]$$

where C is the concentration, D the *dose* of the substance, f the fraction absorbed and V the *volume of distribution*. k_a and k_e are the first-order *rate constants* of uptake and elimination, respectively, and t the time.

B cell
See synonym *B lymphocyte.*

benchmark concentration
Statistical lower confidence limit on the *concentration* that produces a defined *response* (called the *benchmark response* (BMR), usually 5 or 10%) for an *adverse effect* compared to background, defined as 0%.

benchmark dose
Statistical lower confidence limit on the *dose* that produces a defined *response* (called the BMR, usually 5 or 10%) of an *adverse effect* compared to background, defined as 0%.

benchmark guidance value
Biological monitoring guidance value set at the 90th percentile of available *biological monitoring* results collected from a representative *sample* of workplaces with good occupational hygiene practices.

benchmark response
Response, expressed as an excess of background, at which a *benchmark dose* or *benchmark concentration* is set.

benign
1. Of a disease, producing no persisting harmful effects.
2. Tumour that does not invade other tissues (*metastasis*), having lost growth control but not positional control.
antonym *malignant.*

berylliosis
See synonym *beryllium disease*.

beryllium disease
Serious and usually permanent lung damage resulting from chronic inhalation of beryllium.

bilirubin
Orange–yellow pigment, a breakdown product of haem-containing proteins (haemoglobin, myoglobin and *cytochromes*), which circulates in the blood *plasma* bound to albumin or as water-soluble glucuronides, and is excreted in the bile by the liver.

bioaccumulation
Progressive increase in the amount of a substance in an organism or part of an organism that occurs because the rate of intake exceeds the organism's ability to remove the substance from the body.
See also *bioconcentration, biomagnification*.

bioaccumulation potential
Ability of living organisms to concentrate a substance obtained either directly from the environment or indirectly through its food.

bioactivation
Metabolic *conversion* of a *xenobiotic* to a more *toxic* derivative.

bioassay
Procedure for estimating the *concentration* or biological activity of a substance by measuring its effect on a living system compared to a standard system.

bioavailability (general)
Extent of *absorption* of a substance by a living organism compared to a standard system. synonyms *biological availability, physiological availability*.

bioavailability (in pharmacokinetics)
Ratio of the *systemic exposure* from extravascular (ev) exposure to that following intravenous (iv) exposure as described by the equation:

$$F = A_{ev}\, D_{iv}/B_{iv}\, D_{ev}$$

where F is the bioavailability, A and B are the *areas under the plasma concentration time curve* following extravascular and intravenous administration, respectively and D_{ev} and D_{iv} are the administered extravascular and intravenous *doses*.

biochemical (biological) oxygen demand (BOD)
Substance *concentration* of oxygen taken up through the respiratory activity of microorganisms growing on organic compounds present when incubated at a specified temperature (usually 20 °C) for a fixed period (usually 5 days). It is regarded as a measure of that organic *pollution* of water, which can be degraded biologically but includes the oxidation of inorganic material such as sulfide and iron(II). The empirical test used in the

laboratory to determine BOD also measures the oxygen used to oxidize reduced forms of nitrogen unless their oxidation is prevented by an inhibitor such as allyl thiourea.

biocid/e n., **-al** adj.
Substance intended to kill living organisms.

bioconcentration
Process leading to a higher *concentration* of a substance in an organism than in environmental media to which it is *exposed*.
See also *bioaccumulation*.

bioconcentration factor (BCF)
Measure of the tendency for a substance in water to accumulate in organisms, especially fish.
Note 1: The equilibrium *concentration* of a substance in fish can be estimated by multiplying its *concentration* in the surrounding water by its *BCF* in fish.
Note 2: This parameter is an important determinant for human intake of aquatic food by the ingestion route.

biodegradation
Breakdown of a substance catalysed by enzymes *in vitro* or *in vivo*. This may be characterized for purposes of *hazard* assessment as:
1. *Primary*. Alteration of the chemical structure of a substance resulting in loss of a specific property of that substance.
2. *Environmentally acceptable*. Biodegradation to such an extent as to remove undesirable properties of the compound. This often corresponds to primary biodegradation but it depends on the circumstances under which the products are discharged into the environment.
3. *Ultimate*. Complete breakdown of a compound to either fully oxidized or reduced simple molecules (such as carbon dioxide/methane, nitrate/ammonium and water). It should be noted that the products of biodegradation can be more harmful than the substance degraded.

bioelimination
Removal, usually from the aqueous phase, of a test substance in the presence of living organisms by biological processes supplemented by physicochemical reactions.

bioequivalen/ce n., **-t** adj.
Relationship between two preparations of the same *drug* in the same dosage form that have a similar bioavailability.

bioinactivation
Metabolic *conversion* of a *xenobiotic* to a less toxic derivative.

biological assessment of exposure
See *biological monitoring*.

biological cycle
Complete circulatory process through which a substance passes in the biosphere. It may involve transport through the various media (air, water and soil), followed by environmental transformation, and carriage through various ecosystems.

biological effect monitoring (BEM)

Continuous or repeated measurement of early biological effects of *exposure* to a substance to evaluate ambient *exposure* and *health* risk by comparison with appropriate reference values based on knowledge of the probable relationship between ambient exposure and biological effects.

biological exposure indices (BEI)

Guidance value recommended by ACGIH for assessing *biological monitoring* results.

biological half-life

For a substance the time required for the amount of that substance in a biological system to be reduced to one-half of its value by biological processes, when the rate of removal is approximately exponential.

biological half-time ($t_{1/2}$)

See *biological half-life*.

biological monitoring

Continuous or repeated measurement of potentially *toxic substances* or their *metabolites* or biochemical effects in tissues, secreta, excreta, expired air or any combination of these in order to evaluate occupational or environmental *exposure* and *health risk* by comparison with appropriate reference values based on knowledge of the probable relationship between ambient exposure and resultant *adverse* health *effects*.
synonym *biological assessment of exposure*.

biological oxygen demand

See synonym *biochemical oxygen demand*.

biomagnification

Sequence of processes in an ecosystem by which higher *concentrations* are attained in organisms at higher trophic levels (at higher levels in the food web); at its simplest, a process leading to a higher concentration of a substance in an organism than in its food.
synonym *ecological magnification*.

biomarker

Indicator signalling an event or condition in a biological system or *sample* and giving a measure of *exposure*, effect or susceptibility.
Note: Such an indicator may be a measurable chemical, biochemical, physiological, behavioural or other alteration within an organism.

biomarker of effect

Biomarker that, depending upon the magnitude, can be recognized as associated with an established or possible *health* impairment or disease.

biomarker of exposure

Biomarker that relates *exposure* to a *xenobiotic* to the levels of the substance or its *metabolite*, or of the product of an interaction between the substance and some *target* molecule or cell that can be measured in a *compartment* within an organism.

biomarker of susceptibility
Biomarker of an inherent or acquired ability of an organism to respond to *exposure* to a specific substance.

biomass
1. Total amount of biotic material, usually expressed per unit surface area or volume, in a medium such as water.
2. Material produced by the growth of microorganisms, plants or animals.

biomineralization
Complete conversion of organic substances to inorganic derivatives by living organisms, especially microorganisms.

biomonitoring
See synonym *biological monitoring*.

biopsy
Excision of a small piece of living tissue for microscopic or biochemical examination; usually performed to establish a diagnosis.

biosphere
Portion of the planet earth that supports and includes life.

biostatic
Arresting the growth or multiplication of living organisms.

biota
All living organisms as a totality.

biotransformation
Chemical conversion of a substance that is mediated by living organisms or enzyme preparations derived therefrom.

blood–brain barrier
Barrier formed by the blood vessels and supporting tissues of the brain that prevents some substances from entering the brain from the blood.

blood–testis barrier
Membranous barrier separating the blood from the spermatozoa of the seminiferous tubules and consisting of specific junctional complexes between Sertoli cells.

B lymphocyte
Type of *lymphocyte* that synthesizes and secretes *antibodies* in response to the presence of a foreign substance or one identified by it as foreign. The protective effect can be mediated to a certain extent by the antibody alone (contrast *T lymphocyte*). synonym *B cell*.

body burden
Total amount of a substance present in an organism at a given time.

bolus

1. Single *dose* of a substance, originally a large pill.
2. Dose of a substance administered by a single rapid intravenous injection.
3. Concentrated mass of food ready to be swallowed.

brady-
Prefix meaning slow as in bradycardia or bradypnoea.

bradycardia
Abnormal slowness of the heartbeat.
antonym *tachycardia.*

bradypnoea
Abnormally slow breathing.
antonym *tachypnoea.*

breathing zone
Space within a radius of 0.5 m from a person's face.

British anti-Lewisite (BAL)
See synonym *2,3-dimercaptopropan-1-ol.*

bronchoconstriction
Narrowing of the air passages through the bronchi of the lungs.
antonym *bronchodilation.*

bronchodilation
Expansion of the air passages through the bronchi of the lungs.
antonym *bronchoconstriction.*

bronchospasm
Intermittent violent contraction of the air passages of the lungs.

byssinosis
Pneumoconiosis caused by inhalation of dust and associated microbial contaminants and observed in cotton, flax and hemp workers.

cancer
Disease resulting from the development of a *malignancy.*

carboxyhaemoglobin
Compound that is formed between carbon monoxide and haemoglobin in the blood of animals and which is incapable of transporting oxygen.

carcinogen n., **-ic** adj.
Agent (chemical, physical or biological) that is capable of increasing the *incidence* of malignant *neoplasms.*

carcinogen/esis n., **-etic** adj.
Induction, by chemical, physical or biological agents of *malignant neoplasms.*
[21]

carcinogenicity
Process of induction of *malignant neoplasms* by chemical, physical or biological agents.

carcinogenicity test
Long term (*chronic*) test designed to detect any possible carcinogenic effect of a test substance.

carcinoma
Malignant tumour of an epithelial cell.
synonym *epithelioma.*

cardiotoxic
Chemically harmful to the cells of the heart.

carrier protein
1. Protein to which a specific *ligand* or *hapten* is *conjugated.*
2. Unlabelled protein introduced into an assay at relatively high *concentrations* that distributes in a *fractionation* process in the same manner as labelled protein analyte, present in very low concentrations.
3. Protein added to prevent non-specific interaction of reagents with surfaces, *sample* components and each other.
4. Protein found in cell membranes that facilitates transport of a *ligand* across the membrane.

carrier substance
Substance that binds to another substance and transfers it from one site to another.

case control study
Study which starts with the identification of persons with the disease (or other outcome variable) of interest, and a suitable control (comparison, reference) group of persons without the disease. The relationship of an attribute to the disease is examined by comparing the diseased and non-diseased with regard to how frequently the attribute is present or, if quantitative, the levels of the attribute, in the two groups.
synonyms *case comparison study, case compeer study, case history study, case referent study, retrospective study.*

catabolism
1. Reactions involving the oxidation of organic substrates to provide chemically available energy (for example, ATP) and to generate metabolic intermediates.
2. Generally, process of breakdown of complex molecules into simpler ones, often providing biologically available energy.
antonym *anabolism.*

catatonia
Schizophrenia marked by excessive, and sometimes violent, motor activity and excitement, or by generalised inhibition.

cathartic
See synonym *laxative.*

ceiling value (CV)
Airborne *concentration* of a potentially *toxic substance*, which should never be exceeded in a worker's breathing zone.

cell line
Defined unique population of cells obtained by culture from a primary source through numerous generations.
See also *transformed cell line*.

cell-mediated hypersensitivity
State in which an individual reacts with allergic effects caused by the reaction of *antigen*-specific *T-lymphocytes* following exposure to a certain substance (*allergen*) after having been *exposed* previously to the same substance or chemical group.

cell-mediated immunity
Immune response mediated by *antigen*-specific *T-lymphocytes*.

cell strain
Cells having specific properties or markers derived from a primary culture or *cell line*.

certified reference material
Reference material provided by a certifying body such as a National Standards Organization or Metrological Laboratory or by an international body that confirms its purity and analytical values by technically valid procedures and provides a certificate detailing the relevant information.

chain of custody
Sequence of responsibility for a substance from the manufacturer to the distributor, to the user, or to the person(s) ultimately responsible for *waste* disposal. This term is also used in controlled transmission of *samples* from collection to analysis, especially of samples of materials used for medico-legal or forensic purposes.

chemical oxygen demand (COD)
Substance *concentration* of available oxygen (derived from a chemical oxidizing agent) required to oxidize the organic (and inorganic) matter in *waste* water.

chemical safety
Practical certainty that there will be no *exposure* of organisms to toxic amounts of any substance or group of substances: this implies attaining an acceptably low *risk* of exposure to potentially toxic substances.

chemical species (of an element)
Specific form of an element defined as to isotopic composition, electronic or oxidation state and (or) complex or molecular structure.

chemophobia
Irrational fear of chemicals.

chemosis
Chemically induced swelling around the eye caused by *oedema* of the conjunctiva.

chemosterilizer
Substance used to sterilize mites, insects, rodents or other animals.

chloracne
Acne-like eruption caused by *exposure* to certain chlorinated organic substances such as polychlorinated biphenyls or 2,3,7,8-tetrachlorodibenzo-*p*-dioxin.

cholinomimetic
See synonym *parasympathomimetic.*

cholinesterase inhibitor
Substance that inhibits the action of acetylcholinesterase (EC 3.1.1.7) and related enzymes that catalyse the hydrolysis of choline esters: such a substance causes hyperactivity in *parasympathetic* nerves.

chromatid
Either of two filaments joined at the centromere which make up a *chromosome.*

chromatin
Stainable complex of *DNA* and proteins present in the nucleus of a *eukaryotic* cell.

chromosomal aberration
Abnormality of *chromosome* number or structure.

chromosome
Self-replicating structure consisting of *DNA* complexed with various proteins and involved in the storage and transmission of genetic information; the physical structure that contains the *genes.*

chronic
Long-term, (in relation to *exposure* or effect).
1. In experimental toxicology, chronic refers to mammalian studies lasting considerably more than 90 days or to studies occupying a large part of the life-time of an organism.
2. In clinical medicine, long established or long lasting.
antonym *acute.*

chronic effect
Consequence that develops slowly and (or) has a long-lasting course: may be applied to an effect that develops rapidly and is long lasting,
antonym *acute effect.*
synonym *long-term effect.*

chronic exposure
Continued *exposures* occurring over an extended period of time; or a significant fraction of the test species' or of the group of individuals', or of the population's life-time.
antonym *acute exposure.*
synonym *long-term exposure.*

chronic toxicity
1. *Adverse effects* following *chronic exposure*.
2. Effects that persist over a long period of time whether or not they occur immediately upon *exposure* or are delayed.
antonym *acute toxicity*.

chronic toxicity test
Study in which organisms are observed during the greater part of the life span and in which exposure to the test agent takes place over the whole observation time or a substantial part thereof.
antonym *acute toxicity test*.
synonym *long-term test*.

chronotoxicology
Study of the influence of biological rhythms on the *toxicity* of substances.

cirrhosis
1. Liver disease defined by histological examination and characterized by increased fibrous tissue, abnormal physiological changes, such as loss of functional liver cells and increased resistance to blood flow through the liver (portal *hypertension*).
2. Interstitial *fibrosis* of an organ.

clastogen
Agent causing *chromosome* breakage and (or) consequent gain, loss or rearrangement of pieces of chromosomes.

clastogenesis
Occurrence of chromosomal breaks and (or) consequent gain, loss or rearrangement of pieces of *chromosomes*.

clearance (in toxicology)
1. Volume of blood or *plasma* or mass of an organ effectively cleared of a substance by *elimination* (*metabolism* and *excretion*) divided by time of elimination.
Note: Total clearance is the sum of the clearances of each eliminating organ or tissue for that component.
2. (in *pulmonary toxicology*) Volume or mass of lung cleared divided by time of *elimination*; used qualitatively to describe removal of any inhaled substance that deposits on the lining surface of the lung.
3. (in *renal* toxicology) Quantification of the removal of a substance by the kidneys by the processes of filtration and secretion; clearance is calculated by relating the rate of renal excretion to the *plasma concentration*.

clon/e n., **-al** adj.
1. Population of genetically identical cells or organisms having a common ancestor.
2. To produce such a population.
3. *Recombinant DNA* molecules all carrying the same inserted sequence.

clonic
Pertaining to alternate muscular contraction and relaxation in rapid succession.

cocarcinogen
Chemical, physical or biological factor that intensifies the effect of a *carcinogen*.

Codex Alimentarius
Collection of internationally adopted food standards drawn up by the Codex Alimentarius Commission, the principal body implementing the joint FAO/WHO Food Standards Programme.

cohort
Component of the population born during a particular period and identified by period of birth so that its characteristics (such as causes of death and numbers still living) can be ascertained as it enters successive time and age periods. The term "cohort" has broadened to describe any designated group of persons followed or traced over a period of time, as in the term *cohort study* (*prospective study*).

cohort analysis
Tabulation and analysis of *morbidity* or *mortality* rates in relationship to the ages of a specific group of people (cohort), identified by their birth period, and followed as they pass through different ages during part or all of their life span. In certain circumstances, such as studies of migrant populations, cohort analysis may be performed according to duration of residence in a country rather than year of birth, in order to relate *health* or mortality experience to duration of *exposure*.

cohort study
Analytic method of epidemiological study in which subsets of a defined population can be identified who are, have been or in the future may be *exposed* or not exposed, or exposed in different degrees, to a factor or factors hypothesized to influence the prob-ability of occurrence of a given disease or other outcome. The main feature of the method is observation of a large population for a prolonged period (years), with com-parison of *incidence rates* of the given disease in groups that differ in *exposure* levels. synonyms *concurrent study, follow-up study, incidence study, longitudinal study, prospective study.*

cometabolism
Process by which a normally non-biodegradable substance is biodegraded only in the presence of an additional carbon source.

comparative risk
See synonym *relative excess risk.*

compartment
Conceptualized part of the body (organs, tissues, cells or fluids) considered as an independent system for purposes of modelling and assessment of *distribution* and *clearance* of a substance.

compartmental analysis
Mathematical process leading to a model of transport of a substance in terms of compartments and rate constants, usually taking the form

$$C = Ae^{-\alpha t} + Be^{-\beta t}...$$

where each exponential term represents one compartment. *C* is the substance *concen-tration*; *A*, *B*, ... the proportionality constants; α, β, ... the rate constants and *t* the time.

compensation
Adaptation of an organism to changing conditions of the environment (especially chemical) is accompanied by the emergence of stresses in biochemical systems that exceed the limits of normal (*homeostatic*) mechanisms. Compensation is a temporary concealed pathology, which later on can be manifested in the form of explicit pathological changes (decompensation).
synonym *pseudoadaptation.*

competent authority
In the context of European Communities Directive 79/831/EEC, the sixth amendment to the European Community's Directive 67/548/EEC relating to the classification, packaging and labelling of dangerous substances, official government organization or group receiving and evaluating notifications of new substances.

concentration
1. Any one of a group of three quantities characterizing the composition of a mixture and defined as one of mass, amount of substance (chemical amount) or number divided by volume, giving, respectively, mass, amount (of substance) or number concentration.
2. Short form for amount (of substance) concentration (substance concentration in clinical chemistry).

concentration–effect curve
Graph of the relation between *exposure concentration* and the magnitude of the resultant biological change.
synonym *exposure–effect curve.*

concentration–effect relationship
Association between *exposure concentration* and the resultant magnitude of the continuously graded change produced, either in an individual or in a population.

concentration–response curve
Graph of the relation between *exposure concentration* and the proportion of individuals in a population responding with a defined effect.

concentration–response relationship
Association between *exposure concentration* and the incidence of a defined effect in an exposed poopulation.

confounding
1. Situation in which the effects of two processes are not distinguishable from one another: the distortion of the apparent effect of an *exposure* on *risk* brought about by the association of other factors that can influence the outcome.
2. Relationship between the effects of two or more causal factors as observed in a set of data, such that it is not logically possible to separate the contribution that any single causal factor has made to an effect.
3. Situation in which a measure of the effect of an exposure on risk is distorted because of the association of *exposure* with other factor(s) that influence the outcome under study.

confounding variable
Changing factor that can cause or prevent the outcome of interest, is not an intermediate variable, and is associated with the factor under investigation.
synonym *confounder.*

congener
One of two or more substances related to each other by origin, structure or function.

conjugate
1. Molecular species produced in living organisms by covalently linking two chemical moities from different sources.
Example; A conjugate of a *xenobiotic* with some group such as glutathione, sulfate or glucuronic acid, to make it soluble in water or *compartmentalized* within the cell.
See also *phase II reaction.*
2. Material produced by attaching two or more substances together, *e.g.* a *conjugate* of an antibody with a fluorochrome, or an enzyme.

conjunctiva
Mucous membrane that covers the eyeball and lines the under-surface of the eyelid.

conjunctivitis
Inflammation of the *conjunctiva.*

conservative assessment of risk
Assessment of *risk*, which assumes the worst possible case scenario and therefore gives the highest possible value for risk: risk management decisions based on this value will maximize safety.

construct validity
Extent to which a measurement corresponds to theoretical concepts (constructs) concerning the phenomenon under study; for example, if on theoretical grounds, the phenomenon should change with age, a measurement with construct validity would reflect such a change.

contact dermatitis
Inflammatory condition of the skin resulting from dermal *exposure* to an *allergen* (sensitizer) or an irritating (corrosive, defatting) substance.

contraindication
Any condition that renders some particular line of treatment improper or undesirable.

control group
Selected subjects of study, identified as a rule before a study is done, which comprises humans, animals or other species who do not have the disease, intervention, procedure or whatever is being studied, but in all other respects are as nearly identical to the test group as possible.
synonym *comparison group.*

control, matched
Control (individual or group or case) selected to be similar to a study individual or group, or case, in specific characteristics: some commonly used matching variables are age, sex, race and socio-economic status.

convection (as applied to air and water motion)
Vertical motion of air or water, induced by the expansion of the air or water heated by the earth's surface or by human activity, and its resulting buoyancy.

corrosive
Causing a surface-destructive effect on contact; in *toxicology*, this normally means causing visible destruction of the skin, eyes, or the lining of the respiratory tract or the gastro-intestinal tract.

count mean diameter
Mean of the diameters of all particles in a population.
See also *mass mean diameter.*

count median diameter
Calculated diameter in a population of particles in a gas or liquid phase above which there are as many particles with larger diameters as there are particles below it with smaller diameters.
See also *mass median diameter.*

crackles
See synonym *crepitations.*

crepitations
Abnormal respiratory sounds heard on auscultation of the chest, produced by passage of air through passages that contain secretion or exudate or that are constricted by spasm or a thickening of their walls; also referred to as *rhonchi.*
Note: Auscultation is the process of listening for sounds within the body by ear unassisted or using a stethoscope.
synonyms *crackles, râles.*

critical concentration (for a cell or an organ)
Concentration of a substance at and above, which adverse functional changes, reversible or irreversible, occur in a cell or an organ.

critical dose
Dose of a substance at and above, which adverse functional changes, reversible or irreversible, occur in a cell or an organ.

critical effect
For *deterministic effects*, the first *adverse effect* that appears when the *threshold* (*critical*) *concentration* or dose is reached in the *critical organ*: adverse effects with no defined threshold concentration are regarded as critical.

critical end point
Toxic effect used by the USEPA as the basis for a *reference dose* (RfD).

critical group
Part of a *target* population most in need of protection because it is most *susceptible* to a given *toxicant*.

critical organ (in toxicology)
Organ that attains the *critical concentration* of a substance and exhibits the *critical effect* under specified circumstances of *exposure* and for a given population.

critical-organ concentration (of a substance)
Mean *concentration* of a substance in the *critical organ* at the time the substance reaches its *critical concentration* in the most sensitive type of cell in the organ.

critical period (of development)
Stage of development of an organism that is of particular importance in the life cycle if the normal full development of some anatomical, physiological, metabolic or psychological structure or function is to be attained.

critical study
Investigation yielding the no observed *adverse-effect* level that is used by the USEPA as the basis of the RfD.
synonym *pivotal study*.

cross-sectional study (of disease *prevalence* and associations)
Study that examines the relationship between diseases (or other *health*-related characteristics) and other variables of interest as they exist in a defined population at one particular time.
Note: Disease *prevalence* rather than *incidence* is normally recorded in a cross-sectional study and the temporal sequence of cause and effect cannot necessarily be determined.
synonym *disease frequency survey, prevalence study*.

cumulative effect
Overall change, which occurs after repeated *doses* of a substance or radiation.

cumulative incidence
Number or proportion of individuals in a group who experience the onset of a *health*-related event during a specified time interval.
Note: This interval is generally the same for all members of the group, but, as in lifetime *incidence*, it may vary from person to person without reference to age.
synonym *incidence proportion*.

cumulative incidence rate
Proportion of the *cumulative incidence* to the total population.

cumulative incidence ratio
Value obtained by dividing the *cumulative incidence rate* in the *exposed* population by the cumulative incidence rate in the unexposed population.

cumulative-median lethal dose
Estimate of the total administered amount of a substance that is associated with the death of half a population of animals when the substance is administered repeatedly in doses that are generally fractions of the *median lethal dose.*

cutaneous
Pertaining to the skin.
synonym *dermal.*

cyanogenic
Compounds able to produce cyanide.
Examples: Cyanogenic glycosides such as amygdalin in peach and apricot stones.

cyanosis
Bluish colouration, especially of the skin and mucous membranes and fingernail beds, caused by abnormally large amounts of reduced haemoglobin in the blood vessels as a result of deficient oxygenation.

cytochromes
Conjugated proteins containing haem as the *prosthetic group* and associated with electron transport and with redox processes.

cytochrome P-420
Inactive derivative of cytochrome P-450 found in microsomal (see *microsome*) preparations.

cytochrome P-448
Obsolete term for *cytochrome P-450 I*, A1, and A2, one of the major families of the cytochromes P-450 haemoproteins.
Note: During the mono-oxygenation of certain substances, often a detoxification process, these *iso*-enzymes may produce intermediates that initiate *mutations,* chemical *carcinogenesis, immunotoxic* reactions and other forms of chemical *toxicity.*

cytochrome P-450
Member of a superfamily of haem-containing mono-oxygenase enzymes involved in *xenobiotic metabolism,* cholesterol biosynthesis and steroidogenesis, in eukaryotic organisms found mainly in the endoplasmic reticulum and inner mitochondrial membrane of cells. "P-450" refers to a feature in the carbon monoxide absorption difference spectrum at 450 nm caused by a thiolate ligand in the fifth position.

cytogenetics
Branch of genetics that correlates the structure and number of *chromosomes* as seen in isolated cells with variation in *genotype* and *phenotype.*

cytoplasm
Fundamental substance or matrix of the cell (within the *plasma* membrane), which surrounds the nucleus, endoplasmic reticulum, mitochondria and other organelles.

cytotoxic
Causing damage to cell structure or function.

death rate
Estimate of the proportion of a population that dies during a specified period. The numerator is the number of persons dying during the period; the denominator is the size of the population, usually estimated as the mid-year population. The death rate in a population is generally calculated by the formula:
10^n (number of deaths during a specified period)/(number of persons at *risk* of dying during the period).
Note: This rate is an estimate of the person-time death rate, the death rate per 10^n person-years: usually $n = 3$. If the rate is low, it is also a good estimate of the *cumulative death rate*.
synonym *crude death rate*.

defoliant
Substance used for removal of leaves by its *toxic* action on living plants.

denaturation
1. Addition of methanol or acetone to alcohol to make it unfit for drinking.
2. Change in molecular structure of proteins so that they cannot function normally, often caused by splitting of hydrogen bonds following *exposure* to reactive substances or heat.

dermal
Pertaining to the skin.
synonym *cutaneous*.

dermal irritation
Skin reaction resulting from a single or multiple *exposure* to a physical or chemical entity at the same site, characterized by the presence of inflammation; it may result in cell death.

dermatitis
Inflammation of the skin: contact dermatitis is due to local *exposure* and may be caused by irritation, allergy or infection.

desensitization
Suppression of sensitivity of an organism to an allergen to which the organism has been *exposed* previously.

desquamation
Shedding of an outer layer of skin in scales or shreds.

deterministic effect, deterministic process
Phenomenon committed to a particular outcome determined by fundamental physical principles.
See also *stochastic effect*.

detoxification
1. Process, or processes, of chemical modification, which make a *toxic* molecule less toxic.
2. Treatment of patients suffering from poisoning in such a way as to promote physiological processes that reduce the probability or severity of *adverse effects*.

developmental toxicity

Adverse effects on the developing organism (including structural abnormality, altered growth, or functional deficiency or death) resulting from *exposure* prior to conception (in either parent), during prenatal development or post-natally up to the time of sexual maturation.

diploid

Chromosome state in which the chromosomes are present in homologous pairs.

Note: Normal human somatic (non-reproductive) cells are diploid (they have 46 chromosomes), whereas reproductive cells, with 23 chromosomes, are haploid.

disease

Literally, dis-ease, lack of ease; pathological condition that presents a group of symptoms peculiar to it and which establishes the condition as an abnormal entity different from other normal or pathological body states.

disposition

1. Natural tendency shown by an individual or group of individuals, including any tendency to acquisition of specific diseases, often due to hereditary factors.
2. Total of the processes of *absorption* of a chemical into the circulatory systems, *distribution* throughout the body, *biotransformation* and *excretion*.

distributed source

See synonym *area source*.

diuresis

Excretion of urine, especially in excess.

diuretic

Agent that increases urine production.

synonym *micturitic*.

dominant half-life

Half-life of a fraction of a substance in a specific organ or *compartment* if it defines approximately the overall *clearance* rate for that substance at a specific time point.

dosage

Dose divided by product of mass of organism and time of dose.

Note: Often expressed mg (kg body weight)$^{-1}$ day^{-1} and may be used as a synonym for dose.

dose (of a substance)

Total amount of a substance administered to, taken up, or absorbed by an organism, organ or tissue.

dose (of radiation)

Energy or amount of photons absorbed by an irradiated object during a specified *exposure* time divided by area or volume.

dose–effect

Relation between *dose* and the magnitude of a measured biological change.

dose–effect curve
Graph of the relation between *dose* and the magnitude of the biological change produced measured in appropriate units.

dose–effect relationship
Association between *dose* and the resulting magnitude of a continuously graded change, either in an individual or in a population.

dose–response curve
Graph of the relation between *dose* and the proportion of individuals in a population responding with a defined biological effect.

dose–response relationship
Association between *dose* and the *incidence* of a defined biological effect in an *exposed* population usually expressed as percentage.

Draize test
Evaluation of materials for their potential to cause dermal or ocular irritation and corrosion following local *exposure*; generally using the rabbit model (almost exclusively the New Zealand White) although other animal species have been used.

drug
Any substance that when absorbed into a living organism may modify one or more of its functions.
Note: The term is generally accepted for a substance taken for a therapeutic purpose, but is also commonly used for abused substances.
synonyms *medicine, pharmaceutical.*

dysfunction
Abnormal, impaired or incomplete functioning of an organism, organ, tissue or cell.

dysplasia
Abnormal development of an organ or tissue identified by morphological examination.

dyspnoea
Difficult or laboured breathing.

ecogenetics
Study of the influence of hereditary factors on the effects of *xenobiotics* on individual organisms.

ecology
Branch of biology that studies the interactions between living organisms and all factors (including other organisms) in their environment: such interactions encompass environmental factors that determine the distributions of living organisms.

ecosystem
Grouping of organisms (microorganisms, plants and animals) interacting together, with and through their physical and chemical environments, to form a functional entity.

ecotoxicology
Study of the *toxic* effects of chemical and physical agents on all living organisms, especially on populations and communities within defined *ecosystems*; it includes transfer pathways of these agents and their interactions with the environment.

ectoparasiticide
Substance intended to kill parasites living on the exterior of the host.

eczema
Acute or chronic skin inflammation with *erythema*, papules, vesicles, pustules, scales, crusts or scabs, alone or in combination, of varied *aetiology*.

edema
See synonym *oedema*.

effective concentration (EC)
Concentration of a substance that causes a defined magnitude of *response* in a given system.
Note: EC50 is the median concentration that causes 50% of maximal response.

effective dose (ED)
Dose of a substance that causes a defined magnitude of *response* in a given system.
Note: ED50 is the median dose that causes 50% of maximal response.

elimination (in toxicology)
Disappearance of a substance from an organism or a part thereof, by processes of *metabolism, secretion* or *excretion*.
See also *clearance*.

elimination half-life or half-time
Period taken for the *plasma concentration* of a substance to decrease by half.

elimination rate
Differential with respect to time of the *concentration* or amount of a substance in the body, or a part thereof, resulting from *elimination*.

embryo
1. Stage in the developing mammal at which the characteristic organs and organ systems are being formed: for humans, this involves the stages of development from the second to the eighth week (inclusive post-conception).
2. In birds, the stage of development from the fertilization of the ovum up to hatching.
3. In plants, the stage of development within the seed.

embryotoxicity
1. Production by a substance of *toxic* effects in progeny in the first period of pregnancy between conception and the foetal stage.
2. Any *toxic* effect on the conceptus as a result of prenatal *exposure* during the embryonic stages of development: these effects may include malformations and variations, malfunctions, altered growth, prenatal death and altered post-natal function.

emission standard
Quantitative limit on the *emission* or discharge of a substance from a source, usually expressed in terms of a time-weighted average (TWA) *concentration* or a *CV*.

endemic
Present in a community or among a group of people; said of a disease prevailing continually in a region.

endocrine
Pertaining to hormones or to the glands that secrete hormones directly into the bloodstream.

endoplasmic reticulum
Intracellular complex of membranes in which proteins and lipids, as well as molecules for export, are synthesized and in which the *biotransformation* reactions of the mono-oxygenase enzyme systems occur.
Note: May be isolated as microsomes following cell fractionation procedures.

endothelial
Pertaining to the layer of flat cells lining the inner surface of blood and lymphatic vessels, and the surface lining of serous and synovial membranes.

endothelium
Layer of flattened epithelial cells lining the heart, blood vessels and lymphatic vessels.

enteritis
Intestinal inflammation.

enterohepatic circulation
Cyclical process involving intestinal re-*absorption* of a substance that has been excreted through the bile, followed by transfer back to the liver, making it available for biliary *excretion* again.

environmental exposure level (EEL)
Level (*concentration* or amount or a time integral of either) of a substance to which an organism or other component of the environment is *exposed* in its natural surroundings.

environmental fate
Destiny of a chemical or biological *pollutant* after release into the natural environment.

environmental health impact assessment
Estimate of the *adverse effects* to *health* or *risks* likely to follow from a proposed or expected environmental change or development.

environmental health criteria documents
Critical publications of IPCS containing reviews of methodologies and existing knowledge – expressed, if possible, in quantitative terms – of selected substances (or groups of substances) on identifiable, immediate and long-term effects on human *health* and welfare.

environmental impact assessment (EIA)
Appraisal of the possible environmental consequences of a past, ongoing, or planned action, resulting in the production of an environmental impact statement (EIS) or "finding of no significant impact (FONSI)".

environmental impact statement (EIS)
Report resulting from an *EIA*.

environmental monitoring
Continuous or repeated measurement of agents in the environment to evaluate environmental *exposure* and possible damage by comparison with appropriate reference values based on knowledge of the probable relationship between ambient exposure and resultant *adverse effects*.

environmental protection
1. Actions taken to prevent or minimize *adverse effects* to the natural environment.
2. Complex of measures including monitoring of environmental *pollution*, development and practice of environmental protection principles (legal, technical and hygienic), including *risk assessment, risk management and risk communication*.

environmental quality objective (EQO)
Overall state to be aimed for in a particular aspect of the natural environment, for example, "water in an estuary such that shellfish populations survive in good *health*".
Note: Unlike an environmental quality standard (EQS), the EQO is usually expressed in qualitative and not quantitative terms.

environmental quality standard (EQS)
Amount *concentration* or mass concentration of a substance that should not be exceeded in an environmental system, often expressed as a *TWA* measurement over a defined period.
synonym *ambient standard*.

epidemiology
Study of the distribution and determinants of *health*-related states or events in specified populations and the application of this study to control of health problems.

epigen/esis n., -etic adj.
Changes in an organism brought about by alterations in the expression of genetic information without any change in the *genome* itself: the *genotype* is unaffected by such a change but the *phenotype* is altered.

epithelioma
Any tumour derived from *epithelium*.

epithelium
Sheet of one or more layers of cells covering the internal and external surfaces of the body and hollow organs.

epitope
Any part of a molecule that acts as an antigenic determinant: a macromolecule can contain many different epitopes each capable of stimulating production of a different specific *antibody*.

erythema
Redness of the skin produced by congestion of the capillaries.

eschar
Slough or dry scab on an area of skin that has been burnt.

estimated daily intake (EDI)
Prediction of the daily *intake* of a residue of a potentially harmful agent based on the most realistic estimation of the residue levels in food and the best available food consumption data for a specific population: residue levels are estimated taking into account known uses of the agent, the range of contaminated commodities, the proportion of a commodity treated and the quantity of home-grown or imported commodities.
Note: The EDI is expressed in mg residue per person.

estimated exposure concentration (EEC)
Measured or calculated amount or mass *concentration* of a substance to which an organism is likely to be *exposed*, considering *exposure* by all sources and routes.

estimated exposure dose (EED)
Measured or calculated *dose* of a substance to which an organism is likely to be *exposed*, considering *exposure* by all sources and routes.

estimated maximum daily intake (EMDI)
Prediction of the maximum daily intake of a residue of a potentially harmful agent based on assumptions of average food consumption per person and maximum residues in the edible portion of a commodity, corrected for the reduction or increase in residues resulting from preparation, cooking or commercial processing.
Note: The EMDI is expressed in mg residue per person.

etiology
See *aetiology*.

eukaryote
Cell or organism with the genetic material packed in a membrane-surrounded structurally discrete nucleus and with well-developed cell organelles.
Note: The term includes all organisms except archaebacteria, eubacteria and cyanobacteria (until recently classified as cyanophyta or blue-green algae).
antonym *prokaryote*.

European Inventory of Existing Chemical Substances (EINECS)
List of all substances supplied either singly or as components in preparations to persons in a Member State of the European Community on any occasion between 1 January 1971 and 18 September 1981.

eutrophic
Describes a body of water with a high *concentration* of nutrient salts and a high or excessive rate of biological production.

eutrophication
Adverse change in the chemical and biological status of a body of water following depletion of the oxygen content caused by decay of organic matter resulting from high primary production as a result of enhanced input of nutrients.

excess lifetime risk
Additional or excess *risk* incurred over the lifetime of an individual by *exposure* to a *toxic* substance.

excess rate
See synonym *rate difference*.

excipient
Any more or less inert substance added to a *drug* to give suitable consistency or form to the drug.

excretion
Discharge or *elimination* of an absorbed or *endogenous* substance, or of a *waste* product, and (or) its *metabolites*, through some tissue of the body and its appearance in urine, faeces or other products normally leaving the body.
Note: Excretion does not include the passing of a substance through the intestines without absorption.
See also *clearance, elimination*.

excretion rate
Amount of substance and (or) its *metabolites* that is excreted divided by time of excretion.

exogenous
Resulting from causes or derived from materials external to an organism.
antonym *endogenous*.

exogenous substance
See preferred synonym *xenobiotic*.

explant
Living tissue removed from its normal environment and transferred to an artificial medium for growth.

exponential decay
Variation of a quantity according to the law

$$A = A0e^{-\lambda t}$$

where A and A_0 are the values of the quantity being considered at time t and zero respectively, and λ is an appropriate constant.

exposure
1. *Concentration*, amount or intensity of a particular physical or chemical agent or environmental agent that reaches the *target* population, organism, organ, tissue or cell, usually expressed in numerical terms of concentration, duration, and frequency (for chemical agents and microorganisms) or intensity (for physical agents).
2. Process by which a substance becomes available for *absorption* by the *target* population, organism, organ, tissue or cell, by any route.

3. For X- or gamma radiation in air, the sum of the electrical charges of all the ions of one sign produced when all electrons liberated by photons in a suitably small element of volume of air completely stopped, divided by the mass of the air in the volume element.

exposure assessment
Process of measuring or estimating *concentration* (or intensity), duration and frequency of *exposures* to an agent present in the environment or, if estimating hypothetical exposures, that might arise from the release of a substance, or radionuclide, into the environment.

exposure–effect curve
See *concentration–effect curve*.

exposure limit
General term defining an administrative substance *concentration* or intensity of *exposure* that should not be exceeded.

exposure ratio
In a *case control study*, value obtained by dividing the rate at which persons in the case group are *exposed* to a *risk* factor (or to a protective factor) by the *rate* at which persons in the control group are exposed to the risk factor (or to the protective factor) of interest.

exposure–response relationship
See *concentration–response relationship*, *dose–response relationship*.

exposure test
Determination of the level, *concentration* or *uptake* of a potentially *toxic* compound and (or) its *metabolite*(s) in biological *samples* from an organism (blood, urine, hair, *etc.*) and the interpretation of the results to estimate the absorbed dose or degree of environmental *pollution*; or the measuring of biochemical effects, usually not direct *adverse effects* of the substance, and relating them to the quantity of substance absorbed, or to its concentration in the environment.

extracellular space
Volume within a tissue, outside cells and excluding vascular and lymphatic space.

extracellular volume
Volume of fluid outside the cells but within the outer surface of an organism.

extra risk
Probability that an agent produces an observed *response*, as distinguished from the probability that the response is caused by a spontaneous event unrelated to the agent.

extraneous residue limit (ERL)
Refers to a pesticide residue or contaminant arising from environmental sources (including former agricultural uses) other than the use of a pesticide or contaminant substance directly or indirectly on the commodity. It is the maximum *concentration* of a pesticide residue or contaminant that is recommended by the *Codex Alimentarius Commission* to be legally permitted or recognized as acceptable in or on food, agricultural commodity or animal feed.
Note: The mass content is expressed in milligrams of pesticide residue or contaminant per kilogram of commodity.

fecundity
1. Ability to produce offspring frequently and in large numbers.
2. In demography, the physiological ability to reproduce.

feromone
See synonym *pheromone*.

fertility
Ability to conceive and to produce offspring: for litter-bearing species the number of offspring per litter is used as a measure of fertility.
Note: Reduced fertility is sometimes referred to as subfertility.

fetotoxicity
Toxicity to the *fetus*.

fetus (often incorrectly foetus)
Young mammal within the uterus of the mother from the visible completion of characteristic organogenesis until birth.
Note: In humans, this period is usually defined as from the third month after fertilisation until birth (prior to this, the young mammal is referred to as an embryo).

fibrosis
Abnormal formation of fibrous tissue.

first-order process
1. Chemical reaction where the rate is directly proportional to the *concentration* of reactant.
2. Any process changing at a constant fractional rate.

first-pass effect
Biotransformation and, in some cases, *elimination* of a substance in the liver after *absorption* from the intestine and before it reaches the *systemic* circulation.

first-pass metabolism
See *first-pass effect*.

fixed-dose procedure
Acute *toxicity* test in which a substance is tested initially at a small number (3 or 4) predefined *doses* to identify which produces evident toxicity without lethality: the test may be repeated at one or more higher or lower defined discriminating doses to satisfy the criteria.

fluorosis
Adverse effects of fluoride, as in dental or skeletal fluorosis.

foci (singular focus)
Small groups of cells distinguishable, in appearance or histochemically, from the surrounding tissue: indicative of an early stage of a lesion that may lead to the formation of a neoplastic nodule.

foetus
See preferred form *fetus*.

follow-up study
Investigation in which individuals or populations, selected on the basis of whether they have been *exposed* to *risk*, have received a specified preventive or therapeutic procedure, or possess a certain characteristic, are followed to assess the outcome of *exposure*, the procedure, or effect of the characteristic, for example, occurrence of disease.
synonym *cohort study*.

food additive
Any substance not normally consumed as a food by itself and not normally used as a typical ingredient of the food, whether or not it has nutritive value, the intentional addition of which to food for a technological (including organoleptic) purpose in the manufacture, processing, preparation, treatment, packing, packaging, transport or holding of such food results, or may be reasonably expected to result (directly or indirectly) in it or its byproducts becoming a component of or otherwise affecting the characteristics of such foods.
Note: The term does not include "contaminants" or substances added to food for maintaining or improving nutritional qualities.

food chain
Sequence of transfer of matter and energy in the form of food from organism to organism in ascending or descending *trophic levels*.

food intolerance
Physiologically based reproducible, unpleasant (adverse) reaction to a specific food or food ingredient that is not immunologically based.

food web
Network of *food chains*.

foreign substance
See preferred synonym *xenobiotic*.

frame-shift mutation
Point *mutation* involving either the deletion or insertion of one or two nucleotides in a gene: by the frame-shift mutation, the normal reading frame used when decoding nucleotide triplets in the gene is altered.

fumigant
Substance that is vaporized in order to kill or repel pests.

fungicide
Substance intended to kill fungi.

gamete
Reproductive cell (either sperm or egg) containing a haploid set of *chromosomes*.

gametocide
Substance intended to kill *gametes*.

gastrointestinal
Pertaining or communicating with the stomach and intestine.

gavage
Administration of materials directly into the stomach by oesophageal intubation.

gene amplification
Production of extra copies of a chromosomal sequence found either as intra- or extra-chromosomal *DNA*; with respect to a plasmid, it refers to the increase in the number of plasmid copies per cell induced by a specific treatment of transformed cells.

genetic polymorphism
Existence of inter-individual differences in *DNA* sequences coding for one specific *gene* giving rise to different physical and (or) metabolic traits.

genetic toxicology
Study of substances that can produce adverse heritable changes.

genome
Complete set of chromosomal and extrachromosomal *genes* of an organism, a cell, an organelle, or a virus, *i.e.* the complete *DNA* component of an organism.

genomics
1. Science of using *DNA*- and *RNA*-based technologies to demonstrate alterations in gene expression.
2. (in toxicology) Method providing information on the consequences for gene expression of interactions of the organism with environmental stress, *xenobiotics, etc.*

genotoxic

Capable of causing a heritable change to the structure of *DNA* thereby producing a *mutation*.

genotype
Genetic constitution of an organism as revealed by genetic or molecular analysis; the complete set of genes possessed by a particular organism, cell, organelle or virus.

germ-free animal
Animal grown under sterile conditions in the period of post-natal development: such animals are usually obtained by Caesarean operation and kept in special sterile boxes in which there are no viable microorganisms (sterile air, food and water are supplied). synonym *axenic animal.*

glomerular
Pertaining to a tuft or cluster, as of a plexus of capillary blood vessels or nerve fibres, especially referring to the capillaries of the glomeruli of the kidney.

glomerulus
Tuft or a cluster, as of a plexus of capillary blood vessels or nerve fibres, *e.g.* capillaries of the filtration apparatus of the kidney.

glomerular filtration rate
Volume of ultrafiltrate formed in the kidney tubules from the blood passing through the glomerular capillaries divided by time of filtration.

gnotobiont
See synonym *gnotobiote*.

gnotobiota
Specifically and entirely known microfauna and microflora of a specially reared laboratory animal.

gnotobiot/e n., **-ic** adj.
Specially reared laboratory animal whose microflora and microfauna are specifically known in their entirety.

gonadotropic
Pertaining to effects on sex glands and on the systems that regulate them.

good laboratory practice principles (GLP)
Fundamental rules incorporated in national regulations concerned with the process of effective organization and the conditions under which laboratory studies are properly planned, performed, monitored, recorded and reported.

graded effect
Consequence that can be measured on a graded scale of intensity or severity and its magnitude related directly to the *dose* or *concentration* of the substance producing it.
antonym *all-or-none effect, quantal effect, stochastic effect*.

granuloma
Granular growth or *tumour*, usually of lymphoid and epithelial cells.

guideline value
Quantitative measure (a *concentration* or a number) of a constituent of an environmental medium that ensures aesthetically pleasing air, water, or food and does not result in a significant *risk* to the user.

haematoma
Localized accumulation of blood, usually clotted, in an organ, space or tissue, due to a failure of the wall of a blood vessel.

haematuria
Presence of blood in the urine.

haemodialysis
Use of an artificial kidney to remove *toxic* compounds from the blood by passing it through a tube of semipermeable membrane.
Note: The tube is bathed in a dialysing solution to restore the normal chemical composition of the blood while permitting diffusion of toxic substances from the blood.

haemoglobinuria
Presence of free haemoglobin in the urine.

haemolysin
Substance that damages the membrane of erythrocytes causing the release of haemoglobin.

haemolysis
Release of haemoglobin from erythrocytes, and its appearance in the *plasma*.

haemoperfusion
Passing blood through a column of charcoal or adsorbent resin for the removal of *drugs* or *toxins*.

haemosiderin
Iron-containing pigment that is formed from haemoglobin released during the disintegration of red blood cells and that accumulates in individuals who have ingested excess iron.

half-life ($t_{1/2}$)
Time required for the *concentration* of a reactant in a given reaction to reach a value that is the arithmetic mean of its initial and final (equilibrium) values. For a reactant that is entirely consumed it is the time taken for the reactant concentration to fall to one half its initial value.
Note: The half-life of a reaction has meaning only in special cases:
1. For a first-order reaction, the half-life of the reactant may be called the half-life of the reaction.
2. For a reaction involving more than one reactant, with the *concentrations* of the reactants in their stoichiometric ratios, the half-life of each reactant is the same, and may be called the half-life of the reaction.
If the concentrations of reactants are not in their stoichiometric ratios, there are different half lives for different reactants, and one cannot speak of the half-life of the reaction.
synonym *half time*.

half-time ($t_{1/2}$)
See synonym *half-life*

haploid (monoploid)
State in which a cell contains only one set of *chromosomes*.

hapten
Low-molecular-weight molecule that contains an antigenic determinant (*epitope*) that may bind to a specific *antibody* but which is not itself antigenic unless complexed with an antigenic carrier such as a protein or cell; once bound it can cause the *sensitization* of *lymphocytes*, possibly leading to *allergy* or cell-mediated *hypersensitivity*.

hazard
Set of inherent properties of a substance, mixture of substances or a process involving substances that, under production, usage or disposal conditions, make it capable of causing *adverse effects* to organisms or the environment, depending on the degree of *exposure*; in other words, it is a source of danger.
See also *risk*.

hazard assessment
Determination of factors controlling the likely effects of a *hazard* such as the *dose–effect* and *dose–response relationships*, variations in *target* susceptibility, and mechanism of *toxicity*.

hazard evaluation
Establishment of a qualitative or quantitative relationship between *hazard* and benefit, involving the complex process of determining the significance of the identified hazard and balancing this against identifiable benefit.
Note: This may subsequently be developed into a *risk* evaluation.

hazard identification
Determination of substances of concern, their *adverse effects*, *target* populations, and conditions of *exposure*, taking into account *toxicity* data and knowledge of effects on human *health*, other organisms and their environment.

hazard quotient (HQ)
Ratio of *toxicant exposure* (estimated or measured) to a reference value regarded as corresponding to a threshold of *toxicity*: if the total HQ from all toxicants to a target exceeds unity, the combination of toxicants may produce (will produce under assumptions of additivity) an *adverse effect*.

health
1. State of complete physical, mental and social well-being, and not merely the absence of disease or infirmity.
2. State of dynamic balance in which an individual's or a group's capacity to cope with the circumstances of living is at an optimal level.
3. State characterized by anatomical, physiological and psychological integrity, ability to perform personally valued family, work and community roles; ability to deal with physical, biological, psychological and social stress; a feeling of wellbeing; and freedom from the *risk* of disease and untimely death.
4. In ecology, a sustainable steady state in which humans and other living organisms can coexist indefinitely.

health-based exposure limit
Maximum *concentration* or intensity of *exposure* that can be tolerated without significant effect (based on only scientific and not economic evidence concerning exposure levels and associated *health* effects).

health hazard
Any factor or *exposure* that may adversely affect *health*.

health surveillance
Periodic medico-physiological examinations of *exposed* workers with the objective of protecting *health* and preventing occupationally related disease.

healthy worker effect
Epidemiological phenomenon observed initially in studies of occupational diseases: workers usually exhibit lower overall disease and death rates than the general population, due to the fact that the old, severely ill and disabled are ordinarily excluded from employment. Death rates in the general population may be inappropriate for comparison, if this effect is not taken into account.

hepatic
Pertaining to the liver.

hepatotoxic
Poisonous to liver cells.

Henry's law constant
At constant temperature and pressure, the ratio of the partial pressure of a gas above a liquid to its molal solubility in the liquid and therefore a measure of its partition between the gas phase and the solute phase.

herbicide
Substance intended to kill plants.

histology
Study (usually microscopic) of the anatomy of tissues and their cellular and subcellular structure.

histopathology
Microscopic pathological study of the anatomy and cell structure of tissues in disease to reveal abnormal or adverse structural changes.

homeostasis
Normal, internal stability in an organism maintained by co-ordinated responses of the organ systems that automatically compensate for environmental changes.

homo\logy
Degree of identity existing between the nucleotide sequences of two related but not complementary *DNA* or *RNA* molecules.
Note 1: 70 % homology means that on the average 70 out of every 100 nucleotides are identical in a given sequence.
Note 2: The same term is used in comparing the amino acid sequences of related proteins.

hormesis
Stimulatory effect of small doses of a potentially *toxic* substance that is inhibitory in larger doses.

hormone
Substance formed in one organ or part of the body and carried in the blood to another organ or part where it selectively alters functional activity.

human ecology
Interrelationship between humans and the entire environment – physical, biological, socio-economic and cultural, including the interrelationships between individual humans or groups of humans and other human groups or groups of other species.

human equivalent dose
Human *dose* of an agent that is believed to induce the same magnitude of a *toxic* effect that the known animal dose has induced.

hygiene
Science of *health* and its preservation.

hyper-
Prefix meaning above or excessive: when used with the suffix "-aemia" refers to blood and with the suffix "-uria" refers to urine, for example "hyperbilirubinaemia".

hyperaemia
Excessive amount of blood in any part of the body.

hyperalimentation
Ingestion or administration of nutrients in excess of optimal amounts.

hyperbilirubinaemia
Excessive *concentration* of bilirubin in the blood.

hypercalcaemia
Excessive *concentration* of calcium in the blood.

hyperglycaemia
Excessive *concentration* of glucose in the blood.

hyperkalaemia
Excessive *concentration* of potassium in the blood.

hypernatraemia
Excessive *concentration* of sodium in the blood.

hyperparathyroidism
Abnormally increased parathyroid gland activity that affects, and is affected by, *plasma* calcium *concentration*.

hyperplasia
Abnormal multiplication or increase in the number of normal cells in a tissue or organ.

hyper-reactivity
Term used to describe the responses of (effects on) an individual to (of) an agent when they are qualitatively those expected, but quantitatively increased.

hypersensitivity
State in which an individual reacts with *allergic* effects following *exposure* to a certain substance (*allergen*) after having been *exposed* previously to the same substance.

hypersusceptibility
Excessive reaction following *exposure* to a given amount or *concentration* of a substance as compared with the large majority of other *exposed* subjects.

hypertension
Persistently high blood pressure in the arteries or in a circuit, for example pulmonary hypertension or hepatic portal hypertension.

hypertrophy
Excessive growth in bulk of a tissue or organ through increase in size but not in number of the constituent cells.

hypervitaminosis
Condition resulting from the ingestion of an excess of one or more vitamins.

hypo-
Prefix meaning under, deficient: when used with the suffix "-aemia" refers to blood and with the suffix "-uria" refers to urine, for example "hypocalcaemia".

hypocalcaemia
Abnormally low calcium *concentration* in the blood.

hypokalaemia
Abnormally low potassium *concentration* in the blood.

hyponatraemia
Abnormally low sodium *concentration* in the blood.

hypovolaemic
Pertaining to an abnormally decreased volume of circulating fluid (*plasma*) in the body.

hypoxaemia
Deficient oxygenation of the blood.

hypoxia
1. Abnormally low oxygen content or tension.
2. Deficiency of oxygen in the inspired air, in blood or in tissues, short of anoxia.

iatrogenic
Any adverse condition resulting from medical treatment.

icterus
Excess of bile pigment in the blood and consequent deposition and retention of bile pigment in the skin and the sclera.

idiosyncrasy
Genetically based unusually high sensitivity of an organism to the effect of certain substances.

immediately-dangerous-to-life-or-health-concentration (IDLHC)
According to the US NIOSH, the maximum *exposure concentration* from which one could escape within 30 min without any escape-impairing symptoms or any irreversible *health* effects.

immune complex
Product of an antigen–antibody reaction that may also contain components of the complement system.

immune response
Selective reaction of the body to substances that are foreign to it, or that the *immune system* identifies as foreign, shown by the production of antibodies and *antibody-*bearing cells or by a cell-mediated *hypersensitivity* reaction.

immunochemistry
Study of biochemical and molecular aspects of immunology, especially the nature of *antibodies*, *antigens* and their interactions.

immunogen
See synonym *antigen*.

immunoglobulin
Family of closely related glycoproteins capable of acting as antibodies and present in *plasma* and tissue fluids; *immunoglobulin* E is the source of *antibody* in many *hypersensitivity* (*allergic*) reactions.

immunoglobulin E-mediated hypersensitivity
State in which an individual reacts with allergic effects caused fundamentally by the reaction of *antigen*-specific *immunoglobulin* E following *exposure* to a certain substance (*allergen*) after having been *exposed* previously to the same substance.

immunopotentiation
Enhancement of the capacity of the *immune system* to produce an effective response.

immunosuppression
Reduction in the functional capacity of the *immune response*; may be due to:
1. Inhibition of the normal response of the immune system to an antigen.
2. Prevention, by chemical or biological means, of the production of an *antibody* to an *antigen* by inhibition of the processes of transcription, translation or formation of tertiary structure.

immunosurveillance
Mechanisms by which the *immune system* is able to recognize and destroy *malignant* cells before the formation of an overt *tumour*.

immunotoxic
Poisonous to the *immune system*.

incidence
Number of occurrences of illness commencing, or of persons falling ill, during a given period in a specific population: usually expressed as a rate.
Note: When expressed as a rate, it is the number of ill persons divided by the average number of persons in the specified population during a defined period, or alternatively divided by the estimated number of persons at the midpoint of that period.

incidence rate (epidemiology)
Measure of the frequency at which new events occur in a population.
Note: This is the value obtained by dividing the number of new events that occur in a defined period by the population at *risk* of experiencing the event during this period, sometimes expressed as person-time.

indirect exposure
1. *Exposure* to a substance in a medium or vehicle other than the one originally receiving the substance.
2. Exposure of people to a substance by contact with a person directly *exposed*.

individual protective device (IPD)
Device for individual use for protection of the whole body, eyes, respiratory pathways or skin of workers against hazardous and harmful production factors.
synonyms *personal protective device* (PPD), *personal protective equipment* (PPE)

individual risk
Probability that an individual person will experience an *adverse effect*.

inducer
Substance that causes induction.

induction
Increase in the rate of synthesis of an enzyme in response to the action of an inducer or environmental conditions.
Note: Often the inducer is the substrate of the induced enzyme or a structurally similar substance (gratuitous inducer) that is not metabolized.

induction period
Time from the onset of *exposure* to the appearance of signs of disease.
synonym *latent period.*

inhibitory concentration (IC)
Concentration of a substance that causes a defined inhibition of a given system.
Note: IC50 is the median concentration that causes 50% inhibition.

inhibitory dose (ID)
Dose of a substance that causes a defined inhibition of a given system.
Note: ID50 is the median dose that causes 50% inhibition.

initiator
1. Agent that induces a change in a *chromosome* or *gene* that leads to the induction of *tumours* after a second agent, called a *promoter*, is administered to the tissue.
2. Substance that starts a chain reaction.
Note: An initiator is consumed in a chain reaction, in contrast to a catalyst.

insecticide
Substance intended to kill insects.

internal dose
See preferred synonym *absorbed dose.*

interstitial fluid
Aqueous solution filling the narrow spaces between cells.

intervention study
Epidemiological investigation designed to test a hypothesized cause–effect relationship by intentional change of a supposed causal factor in a population.

intoxication
1. Poisoning: pathological process with clinical signs and symptoms caused by a substance of exogenous or *endogenous* origin.
2. Drunkenness following consumption of beverages containing ethanol or other compounds affecting the central nervous system.

intrinsic clearance
Volume of *plasma* or blood from which a substance is completely removed in a period of time under unstressed conditions.

in vitro
In glass, referring to a study in the laboratory usually involving isolated organ, tissue, cell or biochemical systems.
antonym *in vivo*.

in vivo
In the living body, referring to a study performed on a living organism.
antonym *in vitro*.

ionizing radiation
Any radiation consisting of directly or indirectly ionizing particles or a mixture of both or photons with energy higher than the energy of photons of ultraviolet light or a mixture of both such particles and photons.

irritant
1. n., Substance that causes inflammation following immediate, prolonged or repeated contact with skin, mucous membrane or other biological material.
Note: A substance capable of causing inflammation on first contact is called a primary irritant.
2. adj., Causing inflammation following immediate, prolonged or repeated contact with skin, mucous membrane or other tissues.

ischaemia
Local deficiency of blood supply and hence oxygen to an organ or tissue owing to constriction of the blood vessels or to obstruction.

itai-itai disease
Illness observed in Japan possibly resulting from the ingestion of cadmium-contaminated rice: damage occurred to the *renal* and skeleto-articular systems, the latter being very painful ("itai" means pain in Japanese).

jaundice
Pathological condition characterized by deposition of bile pigment in the skin and mucous membranes, including the conjunctivae, resulting in yellow appearance of the patient or animal.

lachrymator
See *lacrimator*.

lacrimator
Substance that irritates the eyes and causes the production of tears or increases the flow of tears.

larvicide
Substance intended to kill larvae.

laryngospasm
Reflex spasmodic closure of the sphincter of the *larynx*, particularly the glottic sphincter.

larynx
Main organ of voice production, the part of the respiratory tract between the pharynx and the trachea.

latent effect
See synonym *delayed effect*.

latent period
1. Delay between *exposure* to a harmful substance and the manifestations of a disease or other *adverse effects*.
2. Period from disease initiation to disease detection.

lavage
Irrigation or washing out of a hollow organ or cavity such as the stomach, intestine or the lungs.

laxative
Substance that causes evacuation of the intestinal contents.
synonyms *cathartic*, *purgative*.

lesion
1. Area of pathologically altered tissue.
2. Injury or wound.
3. Infected patch of skin.

lethal
Deadly; fatal; causing death.

lethal concentration (LC)
Concentration of a substance in an environmental medium that causes death following a certain period of *exposure*.

lethal dose (LD)
Amount of a substance or physical agent (*e.g.* radiation) that causes death when taken into the body.

lethal synthesis
Metabolic formation of a highly *toxic* compound often leading to death of affected cells.

leukaemia
Progressive, *malignant* disease of the blood-forming organs, characterized by distorted proliferation and development of leucocytes and their precursors in the bone marrow and blood.

leukopenia
Reduced *concentration* of leukocytes in the blood.

lgK_{ow}
See synonym *lgP_{ow}*.

lgP_{ow}

Logarithm to the base 10 of the partition coefficient of a substance between octan-1-ol and water.

Note: This is used as an empirical measure for lipophilicity in calculating bioaccumulation, fish toxicity, membrane adsorption and penetration, *etc.*

synonym *lgK_{ow}*.

limacide

Substance intended to kill mollusca including the gastropod mollusc, *Limax*.

limit test

Acute *toxicity* test in which, if no ill-effects occur at a pre-selected maximum dose, no further testing at greater *exposure* levels is required.

limit value (LV)

Limit *concentration* at or below which Member States of the European Community must set their *EQS* and *emission standard* for a particular substance according to community directives.

linearized multistage model

Sequence of steps in which (a) a *multistage model* is fitted to *tumour incidence* data; (b) the maximum linear term consistent with the data is calculated; (c) the low-*dose* slope of the *dose–response* function is equated to the coefficient of the maximum linear term and (d) the resulting slope is then equated to the upper bound of *potency*.

lipophilic/ adj., -ity n.

Having an affinity for fat and high lipid solubility.

Note: This is a physicochemical property, which describes a partitioning equilibrium of solute molecules between water and an immiscible organic solvent, favouring the latter, and which correlates with bioaccumulation.

synonym *hydrophobicity*.

antonym *hydrophilicity, lipophobicity*.

lipophobic/adj., -ity n.

Having a low affinity for fat and a high affinity for water.

synonym *hydrophilicity*.

antonym *hydrophobicity, lipophilicity*.

liposome

1. Originally a lipid droplet in the endoplasmic reticulum of a fatty liver.
2. Now an artificially formed lipid droplet, small enough to form a relatively stable suspension in aqueous media and with potential use in *drug* delivery.

local effect

Change occurring at the site of contact between an organism and a *toxicant*.

log-normal distribution

Distribution function $F(y)$, in which the logarithm of a quantity is normally distributed, *i.e.*

$$F(y) = f_{gauss}(\ln y)$$

where $f_{gauss}(x)$ is a Gaussian *distribution*.

log-normal transformation
Transformation of data with a logarithmic function that results in a normal *distribution*.

lowest-observed-adverse-effect level (LOAEL)
Lowest *concentration* or amount of a substance (*dose*), found by experiment or observation, which causes an *adverse effect* on morphology, functional capacity, growth, development or life span of a *target* organism distinguishable from normal (control) organisms of the same species and strain under defined conditions of *exposure*.

lowest-observed-effect level (LOEL)
Lowest *concentration* or amount of a substance (*dose*), found by experiment or observation, that causes any alteration in morphology, functional capacity, growth, development or life span of *target* organisms distinguishable from normal (control) organisms of the same species and strain under the same defined conditions of *exposure*.

lymphocyte
Animal cell that interacts with a foreign substance or organism, or one which it identifies as foreign, and initiates an immune response against the substance or organism. Note: There are two main groups of lymphocytes, B- and T lymphocytes.

lymphoma
General term comprising *tumours* and conditions allied to tumours arising from some or all of the cells of lymphoid tissue.

lysimeter
Laboratory column of selected representative soil or a protected monolith of undisturbed field soil with which it is possible to *sample* and monitor the movement of water and substances.

lysosome
Membrane-bound cytoplasmic organelle containing hydrolytic enzymes.

macrophage
Large (10–20 μm diameter) amoeboid and phagocytic cell found in many tissues, especially in areas of inflammation, derived from blood monocytes and playing an important role in host defence mechanisms.

macroscopic (gross) pathology
Study of changes associated with disease that are visible to the naked eye without the need for a microscope.

Mad Hatter syndrome
See synonym *mercurialism*.

mainstream smoke (tobacco smoking)
Smoke that is inhaled.

malaise
Vague feeling of bodily discomfort.

malignancy
Population of cells showing both uncontrolled growth and a tendency to invade and destroy other tissues.
Note: A malignancy is life-threatening.

malignant
1. Tending to become progressively worse and to result in death if not treated.
2. In cancer, cells showing both uncontrolled growth and a tendency to invade and destroy other tissues.
antonym *benign*.

mania
Emotional disorder (mental illness) characterized by an expansive and elated state (euphoria), rapid speech, flight of ideas, decreased need for sleep, distractability, grandiosity, poor judgement and increased motor activity.

margin of exposure (MOE)
Ratio of the no-observed-adverse-effect level (*NOAEL*) to the theoretical or EED or EEC.

margin of safety (MOS)
See synonym *margin of exposure.*

mass mean diameter
Diameter of a spherical particle with a mass equal to the mean mass of all the particles in a population.

mass median diameter
Diameter of a spherical particle with the median mass of all the particles in a population.

maximum allowable (admissible, acceptable) concentration (MAC)
Regulatory value defining the *concentration* that if inhaled daily (in the case of work people for 8 h with a working week of 40 h, in the case of the general population 24 h) does not, in the present state of knowledge, appear capable of causing appreciable harm, however long delayed during the working life or during subsequent life or in subsequent generations.

maximum contaminant level (MCL)
Under the Safe Drinking Water Act (USA), primary MCL is a regulatory *concentration* for drinking water, which takes into account both *adverse effects* (including sensitive populations) and technological feasibility (including natural background levels): secondary MCL is a regulatory concentration based on "welfare", such as taste and staining, rather than *health*, but also takes into account technical feasibility. MCL goals (MCLG) under the Safe Drinking Water Act do not consider feasibility and are zero for all human and animal *carcinogens*.

maximum exposure limit (MEL)
Occupational *exposure* limit (OEL) legally defined in the U.K. under COSHH as the maximum *concentration* of an airborne substance, averaged over a reference period,

to which employees may be *exposed* by inhalation under any circumstances, and set on the advice of the HSC Advisory Committee on *Toxic* Substances.

maximum permissible concentration (MPC)
See synonym *maximum allowable concentration*.

maximum permissible daily dose
Maximum daily dose of substance whose penetration into a human body during a lifetime will not cause diseases or *health hazards* that can be detected by current investigation methods and will not adversely affect future generations.

maximum permissible level (MPL)
Level, usually a combination of time and *concentration*, beyond which any *exposure* of humans to a chemical or physical agent in their immediate environment is unsafe.

maximum residue limit for pesticide residues (MRL)
Maximum contents of a *pesticide* residue (expressed as $mg\,kg^{-1}$ fresh weight) recommended by the *Codex Alimentarius Commission* to be legally permitted in or on food commodities and animal feeds.
Note: MRLs are based on data obtained following *good agricultural practice* and foods derived from commodities that comply with the respective MRLs are intended to be toxicologically acceptable.

maximum residue limit for veterinary drugs (MRL)
Maximum contents of a *drug* residue (expressed as $mg\,kg^{-1}$ or $\mu g\,kg^{-1}$ fresh weight) recommended by the *Codex Alimentarius Commission* to be legally permitted or recognized as acceptable in or on food commodities and animal feeds.
Note: The MRLs based on the type and amount of residue considered to be without any toxicological *hazard* for human *health* as expressed by the ADI or on the basis of a temporary ADI that uses an additional uncertainty factor (UF). It also takes into account other relevant public health *risks* as well as food technological aspects.

maximum tolerable concentration (MTC)
Highest *concentration* of a substance in an environmental medium that does not cause death of test organisms or species (denoted by LC_0).

maximum tolerable dose (MTD)
Highest amount of a substance that, when introduced into the body, does not kill test animals (denoted by LD_0).

maximum tolerable exposure level (MTEL)
Maximum amount (*dose*) or *concentration* of a substance to which an organism can be *exposed* without leading to an *adverse effect* after prolonged *exposure* time.

maximum tolerated dose (MTD)
High *dose* used in *chronic toxicity* testing that is expected on the basis of an adequate *subchronic* study to produce limited *toxicity* when administered for the duration of the test period.
Note: It should not induce:
(a) overt toxicity, for example appreciable death of cells or organ dysfunction, or

(b) *toxic* manifestations that are predicted materially to reduce the life span of the animals except as the result of neoplastic development or

(c) 10% or greater retardation of body weight gain as compared with control animals.

Note: In some studies, toxicity that could interfere with a carcinogenic effect is specifically excluded from consideration.

mean residence time (MRT) (in pharmacokinetics)

Average time a *drug* molecule remains in the body or an organ after rapid intravenous injection.

Note 1: Like clearance, its value is independent of dose.

Note 2: After an intravenous bolus:

$$t_r = A_m/A$$

where t_r is the MRT, A the area under the plasma concentration–time curve and A_m the area under the moment curve.

Note 3: For a drug with one-compartment distribution characteristics, MRT equals the reciprocal of the elimination rate constant.

median effective concentration (EC)

Statistically derived *concentration* of a substance in an environmental medium expected to produce a certain effect in test organisms in a given population under a defined set of conditions.

Note: EC_n refers to the median concentration that is effective in $n\%$ of the test population.

median effective dose (ED$_{50}$)

Statistically derived *dose* of a chemical or physical agent (radiation) expected to produce a certain effect in test organisms in a given population or to produce a half-maximal effect in a biological system under a defined set of conditions.

Note: ED_n refers to the median dose that is effective in $n\%$ of the test population.

median lethal concentration (LC$_{50}$)

Statistically derived *concentration* of a substance in an environmental medium expected to kill 50% of organisms in a given population under a defined set of conditions.

median lethal dose (LD$_{50}$)

Statistically derived *dose* of a chemical or physical agent (radiation) expected to kill 50% of organisms in a given population under a defined set of conditions.

median lethal time (TL$_{50}$)

Statistically derived average time interval during which 50% of a given population may be expected to die following *acute* administration of a chemical or physical agent (radiation) at a given *concentration* under a defined set of conditions.

median narcotic concentration (NC$_{50}$)

Statistically derived *concentration* of a substance in an environmental medium expected to cause *narcotic* conditions in 50% of a given population under a defined set of conditions.

median narcotic dose (ND_{50})
Statistically derived dose of a substance expected to cause *narcotic* conditions in 50% of test animals under a defined set of conditions.

meiosis
1. Process of "reductive" cell division, occurring in the production of *gametes*, by means of which each daughter nucleus receives half the number of *chromosomes* characteristic of the somatic cells of the species.
2. See *miosis*.

mercurialism
Chronic poisoning caused by the excessive use of mercury, by breathing its vapour, or by *exposure* in mining or smelting processes.
synonym *Mad Hatter syndrome*.

mesocosm
see *microcosm*.

mesothelioma
Malignant tumour of the mesothelium of the pleura, pericardium or peritoneum, that may be caused by *exposure* to asbestos fibres and some other fibres.

metabolic activation
Biotransformation of a substance to a more biologically active derivative.
synonym *bioactivation*.

metabolic half-life, metabolic half-time
Time required for one half of the quantity of a substance in the body to be metabolized.
Note: This definition assumes that the final quantity in the body is zero. See the definition of *half-life*.

metabolic model
Analysis and theoretical reconstruction of the way in which the body deals with a specific substance, showing the proportion of the intake that is absorbed, the proportion that is stored and in what tissues, the rate of breakdown in the body and the subsequent fate of the metabolic products, and the rate at which it is eliminated (see *elimination*) by different organs as unchanged substance or *metabolites*.

metabolic transformation
Biotransformation of a substance that takes place within a living organism.

metabolism
Sum total of all physical and chemical processes that take place within an organism; in a narrower sense, the physical and chemical changes that take place in a substance within an organism.
Note: It includes the *uptake* and distribution within the body of a substance, the changes (*biotransformation*) undergone by such a substance, and the *elimination* of the substance and of its metabolites.

metabolite
Intermediate or product resulting from *metabolism*.

metabonomics
Evaluation of tissues and biological fluids for changes in *metabolite* levels that follow *exposure* to a given substance, in order to determine the metabolic processes involved and to evaluate the disruption in intermediary metabolic processes that results from exposure to that substance.

metaplasia
Abnormal transformation of an adult, fully differentiated tissue of one kind into a differentiated tissue of another kind.

metastasis
1. Movement of bacteria or body cells, especially *cancer* cells, from one part of the body to another, resulting in change in location of a disease or of its symptoms from one part of the body to another.
2. Growth of pathogenic microorganisms or of abnormal cells distant from the site of their origin in the body.

methaemoglobinaemia
Presence of methaemoglobin (oxidized haemoglobin) in the blood in greater than normal proportion.

methaemoglobin-forming substance
Substance capable of oxidising directly or indirectly the iron(II) in haemoglobin to iron(III) to form methaemoglobin, a derivative of haemoglobin that cannot transport oxygen.

microalbuminuria
Chronic presence of albumin in slight excess in urine.

microcosm
Artificial test system that simulates major characteristics of the natural environment for the purposes of ecotoxicological assessment.
Note: Such a system would commonly have a terrestrial phase, with substrate, plants and herbivores, and an aquatic phase, with vertebrates, invertebrates and plankton. The term "*mesocosm*" implies a more complex and larger system than the term "*microcosm*" but the distinction is not clearly defined.
synonym *experimental model ecosystem.*

microsome
Artefactual spherical particle, not present in the living cell, derived from pieces of the endoplasmic reticulum present in homogenates of tissues or cells.
Note: microsomes sediment from such homogenates when centrifuged at $100,000g$ and higher: the microsomal fraction obtained in this way is often used as a source of mono-oxygenase enzymes.

micturitic
See synonym *diuretic.*

midstream sampling
Taking an *aliquot* of a flowing liquid, such as urine, avoiding initial and terminal flow periods, which are likely to be unrepresentative.

Minamata disease
Neurological disease caused by ingestion of methylmercury-contaminated fish, first seen at Minamata Bay in Japan.

mineralization
Complete conversion of organic substances to inorganic derivatives.

minimum lethal concentration (LC_{min})
Lowest *concentration* of a *toxic* substance in an environmental medium that kills individual organisms or test species under a defined set of conditions.

minimum lethal dose (LD_{min})
Lowest amount of a substance that, when introduced into the body, may cause death to individual species of test animals under a defined set of conditions.

miosis
Abnormal contraction of the pupil of the eye to less than 2 mm.
Alternative spelling (obsolete): meiosis.

miscible
Liquid substances capable of mixing without separation into two phases; refers to liquid mixtures.

mitochondri/on (pl -a)
Eukaryote cytoplasmic organelle that is bounded by an outer membrane and an inner membrane; the inner membrane has folds called cristae that are the centre of ATP synthesis in oxidative phosphorylation in the animal cell and supplement ATP synthesis by the chloroplasts in photosynthetic cells.
Note: The mitochondrial matrix within the inner membrane contains ribosomes, many oxidative enzymes, and a circular *DNA* molecule that carries the genetic information for a number of these enzymes.

mitogen
Substance that induces *lymphocyte* transformation or, more generally, *mitosis* and cell proliferation.

mitosis
Process by which a cell nucleus divides into two daughter nuclei, each having the same genetic complement as the parent cell: nuclear division is usually followed by cell division.

mixed function oxidase
See synonym *mono-oxygenase*.

modifying factor (MF)
See *uncertainty factor*.

molluscicide
Substance intended to kill molluscs.
synonym *limacide*.

monitoring
Continuous or repeated observation, measurement and evaluation of *health* and (or) environmental or technical data for defined purposes, according to prearranged schedules in space and time, using comparable methods for sensing and data collection.
Note: Evaluation requires comparison with appropriate reference values based on knowledge of the probable relationship between ambient *exposure* and *adverse effects*.

monoclonal
Pertaining to a specific protein from a single clone of cells, all molecules of this protein being the same.

monoclonal antibody
Antibody produced by cloned cells derived from a single *lymphocyte*.

mono-oxygenase
Enzyme that catalyses reactions between an organic compound and molecular oxygen in which one atom of the oxygen molecule is incorporated into the organic compound and one atom is reduced to water; involved in the *metabolism* of many natural and foreign compounds giving both unreactive products and products of different or increased *toxicity* from that of the parent compound.
Note: Such enzymes are the main catalysts of phase 1 reactions in the metabolism of *xenobiotics* by the endoplasmic reticulum or by preparations of *microsomes*. synonym *mixed function oxidase*.

Monte Carlo study
Simulation and analysis of a sequence of events using random numbers to generate possible outcomes in an iterative process.

morbidity
Any departure, subjective or objective, from a state of physiological or psychological well-being: in this sense, "sickness", "illness" and "morbid condition" are similarly defined and synonymous.

morbidity rate
Term (to be avoided) used loosely to refer to *incidence* or *prevalence* rates of disease.

morbidity survey
Method for the estimation of the *prevalence* and (or) *incidence* of a disease or diseases in a population.

mortality
Death as studied in a given population or subpopulation.
Note: The word mortality is often used incorrectly instead of mortality rate.

mortality rate
See synonym *death rate*.

mortality study
Investigation dealing with death rates or proportion of deaths attributed to specific causes as a measure of response.

mucociliary transport
Process of removal of particles from the bronchi of the lungs in a mucus stream moved by cilia, thus contributing to *uptake* from the gastrointestinal tract.

multicompartment model
Product of a *compartmental analysis* requiring more than two *compartments*.

multigeneration study
1. *Toxicity* test in which two to three generations of the test organism are *exposed* to the substance being assessed.
2. Toxicity test in which only one generation is exposed and effects on subsequent generations are assessed.

multipotent
Of a cell, capable of giving rise to several different kinds of structure or types of cell.

multistage model
Dose–response model for cancer death estimation of the form

$$P = 1 - \exp[-(q_o + q_1 d_1 + q_2 d_2 + \cdots + q_k d_k)]$$

where P is the probability of cancer death from a continuous *dose* rate, di, of group (or stage) i, the q's are constants and k the number of dose groups (or, if less than the number of dose groups, k is the number of biological stages believed to be required in the *carcinogenesis* process). With the *multistage model*, it is assumed that cancer is initiated by cell mutations in a finite series of steps.

murine
Of or belonging to the family of rats and mice (Muridae).

mutagen
Agent that can induce heritable changes (*mutations*) of the *genotype* in a cell as a consequence of alterations or loss of genetic material.

mutagenesis
Introduction of heritable changes (*mutations*) of the genotype in a cell as a consequence of alterations or loss of *genes* or *chromosomes* (or parts thereof).

mutagenicity
Ability of a physical, chemical or biological agent to induce heritable changes (*mutations*) in the genotype in a cell as a consequence of alterations or loss of *genes* or *chromosomes* (or parts thereof).

mutation
Any relatively stable heritable change in genetic material that may be a chemical transformation of an individual *gene* (gene or point *mutation*), altering its function, or a rearrangement, gain or loss of part of a *chromosome*, that may be microscopically visible (chromosomal mutation).
Note: Mutation can be either germinal, and inherited by subsequent generations, or somatic and passed through cell lineage by cell division.

myasthenia
Muscular weakness.

mycotoxin
Toxin produced by a fungus.

mydriasis
Extreme dilation of the pupil of the eye, either as a result of normal physiological response or in response to a chemical *exposure*.

myelosuppression
Reduction of bone marrow activity leading to a lower *concentration* of platelets, red cells and white cells in the blood.

narcotic
1. Non-specific usage – an agent that produces insensibility or stupor.
2. Specific usage – an opioid, any natural or synthetic *drug* that has morphine-like actions.

natriuretic
Substance increasing the rate of excretion of sodium ion in the urine.

necropsy
See synonym *autopsy*.

necrosis
Sum of morphological changes resulting from cell death by lysis and (or) enzymatic degradation, usually affecting groups of cells in a tissue.
See also *apoptosis*.

negligible risk
1. Probability of *adverse effects* occurring that can reasonably be described as trivial.
2. Probability of adverse effects occurring that is so low that it cannot be reduced appreciably by increased regulation or investment of resources.

nematocide
Substance intended to kill nematodes.

neonat/e n., **-al** adj.
Infant during the first 4 weeks of post-natal life.
Note: For statistical purposes some scientists have defined the period as the first 7 days of post-natal life.

neoplas/ia, -m
New and abnormal formation of tissue as a *tumour* or growth by cell proliferation that is faster than normal and continues after the initial stimulus (i) that initiated the proliferation has ceased.

nephritis
Inflammation of the kidney, leading to kidney failure, usually accompanied by *proteinuria*, *haematuria*, *oedema* and *hypertension*.

nephrotoxic
Chemically harmful to the cells of the kidney.

neural
Pertaining to a nerve or to the nerves.

neuron(e)
Nerve cell, the morphological and functional unit of the central and peripheral nervous systems.

neuropathy
Any disease of the central or peripheral nervous system.

neurotoxic/ adj., **-ity** n.
Able to produce chemically an *adverse effect* on the nervous system: such effects may be subdivided into two types.
1. Central nervous system effects (including transient effects on mood or perform- ance and pre-senile dementia such as Alzheimer's disease).
2. Peripheral nervous system effects (such as the inhibitory effects of organophos- phorus compounds on synaptic transmission).

nitrification
Sequential oxidation of ammonium salts to nitrite and nitrate by microorganisms.

no-acceptable-daily-intake-allocated
This expression is applicable to a substance for which the available information is not sufficient to establish its safety, or when the specifications for identity and purity are not adequate, or when the available data show that the substance is hazardous and should not be used.
Note: The basis for the use of the expression should be determined before action is taken; in the first two cases above, not being able to allocate an ADI does not mean that the substance is unsafe.

***n*-octanol-water partition coefficient**
See synonym *octanol-water partition coefficient*.

no-effect-level (NEL)
Maximum dose (of a substance) that produces no detectable changes under defined conditions of *exposure*.
Note: This term tends to be substituted by *no-observed-adverse-effect level* (NOAEL) or *no-observed-effect level* (NOEL).

no-effect-dose (NED)
Amount of a substance that has no effect on the organism.
Note: It is lower than the threshold of harmful effect and is estimated while establishing the threshold of harmful effect.
synonym *subthreshold dose*.

no-observed-adverse-effect level (NOAEL)
Greatest *concentration* or amount of a substance, found by experiment or observation, which causes no detectable adverse alteration of morphology, functional capacity, growth, development or life span of the *target* organism under defined conditions of *exposure*.

no-observed-effect level (NOEL)
Greatest *concentration* or amount of a substance, found by experiment or observation, that causes no alterations of morphology, functional capacity, growth, development or life span of *target* organisms distinguishable from those observed in normal (control) organisms of the same species and strain under the same defined conditions of *exposure*.

no-response level (NRL)
Maximum *dose* of a substance at which no specified response is observed in a defined population and under defined conditions of *exposure*.

nosocomial
Associated with a hospital or infirmary, especially used of diseases that may result from treatment in such an institution.

noxious substance
See synonym *harmful substance*.

nuisance threshold
Lowest *concentration* of an air pollutant that can be considered objectionable.

nystagmus
Involuntary, rapid, rhythmic movement (horizontal, vertical, rotary and mixed) of the eyeball, usually caused by a disorder of the labyrynth of the inner ear or a malfunction of the central nervous system.

occupational exposure limit (OEL)
Regulatory level of *exposure* to substances, intensities of radiation, *etc.* or other conditions, specified appropriately in relevant government legislation or related codes of practice.

occupational exposure standard (OES)
1. Level of *exposure* to substances, intensities of radiation, *etc. or* other conditions considered to represent specified good practice and a realistic criterion for the control of exposure by appropriate plant design, engineering controls, and, if necessary, the addition and use of personal protective clothing.
2. In the U.K. *health*-based exposure limit defined under COSHH regulations as the *concentration* of any airborne substance, averaged over a reference period, at which, according to current knowledge, there is no evidence that it is likely to be injurious to employees, if they are *exposed* by inhalation, day after day, to that concentration, and set on the advice of the HSE Advisory Committee on *Toxic* Substances.

occupational hygiene
Identification, assessment and control of physicochemical and biological factors in the workplace that may affect the *health* or well-being of those at work and in the surrounding community.

octanol-water partition coefficient (P_{ow}, K_{ow})
Measure of lipophilicity by determination of the equilibrium distribution between octan-1-ol and water, as used in pharmacological studies and in the assessment of environmental fate and transport of organic chemicals.

ocular
Pertaining to the eye.

odds
Ratio of the probability of occurrence of an event to that of non-occurrence, or the ratio of the probability that something is so, to the probability that it is not so.

odds ratio
Quotient obtained by dividing one set of odds by another. The term "odds" or "odds ratio" is defined differently according to the situation under discussion. Consider the following notation for the distribution of a binary *exposure* and a disease in a population or a *sample*.

	Exposed	*Nonexposed*
Disease	*a*	*b*
No disease	*c*	*d*

The odds ratio (cross-product ratio) is *ad/bc*.
synonyms *cross-product ratio, relative odds*
Note 1: The *exposure*-odds ratio for a set of case control data is the ratio of the odds in favour of exposure among the cases (*a/b*) to the odds in favour of exposure among non-cases (*c/d*). This reduces to *ad/(bc)*. With incident cases, unbiased subject selection, and a "rare" disease (say, under 2% cumulative *incidence* rate over the study period), *ad/bc* is an approximate estimate of the *risk* ratio. With incident cases, unbiased subject selection, and density sampling of controls, *ad/bc* is an estimate of the ratio of the person-time incidence rates (force of morbidity) in the *exposed* and unexposed. No rarity assumption is required for this.
Note 2: The disease-odds (rate-odds) ratio for a cohort or cross section is the ratio of the odds in favour of disease among the exposed population (*a/c*) to the odds in favour of disease among the unexposed (*b/d*). This reduces to *ad/bc* and hence is equal to the exposure-odds ratio for the cohort or cross section.
Note 3: The *prevalence*-odds ratio refers to an odds ratio derived cross sectionally, as, for example, an odds ratio derived from studies of prevalent (rather than incident) cases.
Note 4: The *risk*-odds ratio is the ratio of the odds in favour of getting disease, if exposed, to the odds in favour of getting disease if not *exposed*. The odds ratio derived from a cohort study is an estimate of this.

odour threshold
In principle, the lowest *concentration* of an odorant that can be detected by a human being.
Note: In practice, a panel of "sniffers" is used, and the threshold taken as the concentration at which 50% of the panel can detect the odorant (although some workers have also used 100% thresholds).

oedema
Presence of abnormally large amounts of fluid in intercellular spaces of body tissues.

olfactometer
Apparatus for testing the power of the sense of smell.

oliguria
Excretion of a diminished amount of urine in relation to fluid intake.

oncogene
Gene that can cause neoplastic (see *neoplasia*) transformation of a cell; oncogenes are slightly changed equivalents of normal genes known as proto-oncogenes.

oncogenesis
Production or causation of *tumours*.

one-compartment model
Kinetic model, where the whole body is thought of as a single *compartment* in which the substance distributes rapidly, achieving an *equilibrium* between blood and tissue immediately.

one-hit model
Dose–response model of the form

$$P = 1 - e^{-bd}$$

where P is the probability of cancer death from a continuous *dose* rate, d, and b a constant.

onycholysis
Loosening or detachment of the nail from the nailbed following some destructive process.

oogenesis
Process of formation of the ovum (plural ova), the female *gamete*.

operon
Complete unit of *gene* expression and regulation, including structural genes, regulator gene(s) and control elements in *DNA* recognized by regulator gene product(s).

ophthalmic
Pertaining to the eye.

organ dose
Amount of a substance or physical agent (radiation) absorbed by an organ.

organelle
Microstructure or separated compartment within a cell that has a specialized function, for example ribosome, peroxisome, lysosome, Golgi apparatus, mitochondrion, nucleolus and nucleus.

organic carbon partition coefficient (K_{oc})
Measure of the tendency for organic substances to be adsorbed by soil and sediment, expressed as:

$$K_{oc} = \frac{(\text{mg substance adsorbed})/(\text{kg organic carbon})}{(\text{mg substance dissolved})/(\text{litre of solution})}$$

The K_{oc} is substance-specific and is largely independent of soil properties.

organoleptic
Involving an organ, especially a sense organ as of taste, smell or sight.

osteo-
Prefix meaning pertaining to bone.

osteodystrophy
Abnormal development of bone.

osteogenesis
Formation or development of bone.

osteoporosis
Significant decrease in bone mass with increased porosity and increased tendency to fracture.

ovicide
Substance intended to kill eggs.

palpitation
1. Unduly rapid or throbbing heartbeat that is noted by a patient; it may be regular or irregular.
2. Undue awareness by a patient of a heartbeat that is otherwise normal.

paraesthesia
Abnormal sensation, as burning or prickling.

paralysis
Loss or impairment of motor function.

para-occupational exposure
1. *Exposure* of a worker's family to substances carried from the workplace to the home.
2. Exposure of visitors to substances in the workplace.

parasympatholytic
Producing effects resembling those caused by interruption of the parasympathetic nerve; also called anticholinergic.

parasympathomimetic
Producing effects resembling those caused by stimulation of the parasympathetic nervous system.
synonym *cholinomimetic*.

parenteral dosage
Method of introducing substances into an organism avoiding the gastrointestinal tract (subcutaneously, intravenously, intramuscularly, *etc.*).

paresis
Slight or incomplete paralysis.

particulate matter (in atmospheric chemistry)
1. General term used to describe airborne solid or liquid particles of all sizes.
Note: The term *aerosol* is recommended to describe airborne particulate matter.

2. Particles in air, usually of a defined size and specified as PM_n where n is the maximum *aerodynamic diameter* in μm of at least 50% of the particles.

partition coefficient
Ratio of the *distribution* of a substance between two phases when the heterogeneous system of two phases is in *equilibrium*.
Note 1: The ratio of *concentrations* (or, strictly speaking, activities) of the same molecular species in the two phases is constant at constant temperature.
Note 2: The octanol/water partition coefficient is often used as a measure of the *BCF* for modelling purposes.
Note 3: This term is in common usage in toxicology but is not recommended by IUPAC for use in chemistry and should not be used as a synonym for partition constant, partition ratio or distribution ratio.

partition ratio (K_D)
Ratio of the *concentration* of a substance in a single definite form, A, in the extract to its *concentration* in the same form in the other phase at equilibrium, *e.g.* for an aqueous/organic system:

$$K_D(A) = [A]^{org}/[A]^{aq}$$

passive smoking
Inhalation of sidestream smoke by people who do not smoke themselves.
See also *sidestream smoke*.

percutaneous
Through the skin following application on the skin.

perfusion (in physiology)
1. Act of pouring over or through, especially the passage of a fluid through the vessels of a specific organ.
2. Liquid poured over or through an organ or tissue.

perinatal
Relating to the period shortly before and after birth, usually from the 20th–29th week of gestation to 1–4 weeks after birth.

peritoneal dialysis
Method of artificial *detoxication* in which a *toxic* substance from the body is transferred into liquid that is instilled into the peritoneum.
Note: Effectively this represents the employment of the peritoneum surrounding the abdominal cavity as a dialysing membrane for the purpose of removing *waste* products or toxins accumulated as a result of *renal* failure.

permissible-exposure-limit (PEL)
Recommendation by US OSHA for a *TWA concentration* that must not be exceeded during any 8-h work shift of a 40 h working week.

peroxisome
Organelle, similar to a lysosome, characterized by its content of catalase
(EC 1.11.1.6), peroxidase (EC 1.11.1.7) and other oxidative enzymes.

persistence
Attribute of a substance that describes the length of time that the substance remains
in a particular environment before it is physically removed or chemically or biologically transformed.

personal monitoring
Type of environmental monitoring in which an individual's *exposure* to a substance
is measured and evaluated.
Note: This is normally carried out using a personal *sampler*.

personal protective device (PPD)
see synonym *personal protective equipment.*
alternative synonym *individual protective device.*

personal protective equipment (PPE)
Equipment (clothing, gloves, hard hat, respirator and so on) worn by an individual
to prevent *exposure* to a potentially *toxic* substance.
synonyms *individual protective device, personal protective device.*

personal sampler
Compact, portable instrument for individual air sampling, measuring or both, the
content of a harmful substance in the respiration zone of a working person.
synonym *individual monitor.*

pest
Organism that may harm public *health*, that attacks food and other materials essential
to mankind, or otherwise affects human beings adversely.

pesticide
A substance intended to kill pests.
Note: In common usage, any substance used for controlling, preventing or destroying
      animal, microbiological or plant pests.

pesticide residue
Pesticide residue is any substance or mixture of substances in food for man or
animals resulting from the use of a pesticide and includes any specified derivatives,
such as degradation and conversion products, metabolites, reaction products and
impurities considered to be of toxicological significance.

phagocytosis
Process by which particulate material is endocytosed by a cell.
See also e*ndocytosis, pinocytosis.*

pharmacodynamics
Process of interaction of pharmacologically active substances with *target* sites in
living systems, and the biochemical and physiological consequences leading to
therapeutic or *adverse effects.*

pharmacogenetics
Study of the influence of genetic factors on the effects of *drugs* on individual organisms.

pharmacokinetics
1. Process of the *uptake* of *drugs* by the body, the *biotransformation* they undergo, the *distribution* of the *drugs* and their metabolites in the tissues, and the *elimination* of the drugs and their metabolites from the body.
2. Study of such processes.

pharmacology
Science of the use and effects of *drugs*: may be subdivided into *pharmacokinetics* and *pharmacodynamics* defined above.

pharynx
Throat, the part of the digestive tract between the oesophagus below and the mouth and nasal cavities above and in front.

phase I reaction (of *biotransformation*)
Enzymic modification of a substance by oxidation, reduction, hydrolysis, hydration, dehydrochlorination or other reactions catalysed by enzymes of the cytosol, of the endoplasmic reticulum (microsomal enzymes) or of other cell organelles.
See also *cytochrome P450*.

phase II reaction (of *biotransformation*)
Binding of a substance, or its *metabolites* from a phase I reaction, with *endogenous* molecules (*conjugation*), making more water-soluble derivatives that may be excreted in the urine or bile.

phase III reaction (of *biotransformation*)
Further *metabolism* of *conjugated metabolites* produced by *phase II reactions*.

phenotype
Observable structural and functional characteristics of an organism determined by its *genotype* and modulated by its environment.

pheromone
Substance used in olfactory communication between organisms of the same species eliciting a change in sexual or social behaviour.
synonyms *ectohormone, fermone*.

photo-irritation
Inflammation of the skin caused *exposure* to light, especially that due to *metabolites* formed in the skin by photolysis.

photo-oxidant
Substance able to cause oxidation when *exposed* to light of the appropriate wavelength.

photophobia
Abnormal visual intolerance of light.

photosensitization
Allergic reaction (see *allergy*) due to a metabolite formed by the influence of light.

phototoxicity
Adverse effects produced by *exposure* to light energy, especially those produced in the skin.

physiologically based pharmacokinetic modelling (PBPK)
Mathematical modelling of kinetic behaviour of a substance, based on measured physiological parameters.
synonym *toxicologically based pharmacokinetic modelling*.

piscicide
Substance intended to kill fish.

pivotal study
See synonym *critical study*.

plasma (in biology)
1. Fluid component of blood in which the blood cells and platelets are suspended.
   synonym *blood plasma*.
2. Fluid component of semen produced by the accessory glands, the seminal vesicles, the prostate, and the bulbo-urethral glands.
3. Cell substance outside the nucleus, *i.e.* the cytoplasm.

plasma half-life
see synonym *elimination half-life*.

plasmapheresis
Removal of blood from the body and centrifuging it to obtain *plasma* and packed red blood cells: the blood cells are resuspended in a physiologically compatible solution (usually type-specific fresh frozen plasma or albumin) and returned to the donor or injected into a patient who requires blood cells rather than whole blood.

plasmid
Autonomous self-replicating extra-chromosomal circular *DNA* molecule.

pleura
Lining of the lung.

ploidy
Term indicating the number of sets of *chromosomes* present in an organism.

plumbism
Chronic poisoning caused by absorption of lead or lead salts.
synonym *saturnism*.

pneumoconiosis
Usually fibrosis of the lungs that develops owing to (prolonged) inhalation of inorganic or organic dusts.
Cause-specific types of pneumoconiosis:
1. Anthracosis: from coal dust
2. Asbestosis: from asbestos dust
3. Byssinosis: from cotton dust
4. Siderosis: from iron dust
5. Silicosis: from silica dust
6. Stannosis: from tin dust

pneumonitis
Inflammation of the lung.

po
Per os – Latin for by mouth.

point mutation
Reaction that changes a single base pair in *DNA*.

point source
Single *emission* source in a defined location.

poison (in toxicology)
Substance that, taken into or formed within the organism, impairs the *health* of the organism and may kill it.

poison-bearing
Containing a *poison*.

poisoning
Morbid condition produced by a *poison*.
synonym *intoxication*.

pollutant
Any undesirable solid, liquid or gaseous matter in a solid, liquid or gaseous environmental medium.
Note 1: "Undesirability" is often *concentration*-dependent, low concentrations of most substances being tolerable or even essential in many cases.
Note 2: A primary pollutant is one emitted into the atmosphere, water, sediments or soil from an identifiable source.
Note 3: A secondary pollutant is a pollutant formed by chemical reaction in the atmosphere, water, sediments or soil.

pollution
Introduction of *pollutants* into a solid, liquid or gaseous environmental medium, the presence of pollutants in a solid, liquid or gaseous environmental medium, or any undesirable modification of the composition of a solid, liquid or gaseous environmental medium.

polyclonal antibody
Antibody produced by a number of different cell types.

polydipsia
Chronic excessive thirst.

polymorphism (polymorphia) in metabolism
Interindividual variations in *metabolism* of endo- and *exogenous* compounds due to genetic influences, leading to enhanced side effects or *toxicity* of *drugs* (for example, poor vs. fast metabolizers) or to different clinical effects (metabolism of steroid hormones).

polyuria
Excessive production and discharge of urine.

population critical concentration (PCC)
Concentration of a substance in the critical organ at which a specified percentage of the *exposed* population has reached the individual critical organ concentration.
Note: The percentage is indicated by PCC-10 for 10%, PCC-50 for 50%, *etc.* (similar to the use of the term LD_{50}).

population effect
Absolute number or *incidence* rate of cases occurring in a group of people.

population risk
See synonym *societal risk*.

porphyria
Disturbance of porphyrin *metabolism* characterized by increased formation, accumulation, and excretion of porphyrins and their precursors.

posology
Study of *dose* in relation to the physiological factors that may influence response such as age of the *exposed* organisms.

potency (in toxicology)
Expression of relative *toxicity* of an agent as compared to a given or implied standard or reference.

potentiation
Dependent action in which a substance or physical agent at a *concentration* or *dose* that does not itself have an *adverse effect* enhances the harm done by another substance or physical agent.

practical certainty (of safety)
Numerically specified low *risk* of *exposure* to a potentially *toxic* substance (for example, 1 in 1000) or socially acceptable low risk of *adverse effects* from such an exposure applied to decision making in regard to chemical safety.

precursor
Substance from which another, usually more biologically active, substance is formed.

preneoplastic
Before the formation of a *tumour*.

prevalence
Number of instances of existing cases of a given disease or other condition in a given population at a designated time; sometimes used to mean *prevalence rate*. When used without qualification, the term usually refers to the situation at a specified point in time (point prevalence).

prevalence rate (ratio)
Total number of individuals who have an attribute or disease at a particular time (or during a particular period) divided by the population at *risk* of having the attribute or disease at this point in time or midway through the period.

primary pollutant
See *pollutant*.

primary protection standard
Accepted maximum level of a *pollutant* (or its indicator) in the *target* organism, or some part thereof, or an accepted maximum intake of a pollutant or nuisance into the target under specified circumstances.

probit
Probability unit obtained by adding 5 to the normal deviates of a standardized normal distribution of results from a dose response study: addition of 5 removes the complication of handling negative values.
Note: A plot of probit against the logarithm of *dose* or *concentration* gives a linear plot if the distribution of response is a logarithmic normal one. Estimates of the LD_{50} and ED_{50} (or LC_{50} and EC_{50}) can be obtained from this plot.

procarcinogen
Substance that has to be metabolized before it becomes a carcinogen.

prodrug
Precursor converted to an active form of a *drug* within the body.

prokaryote
Unicellular organism, characterized by the absence of a membrane-enclosed nucleus. Prokaryotes include bacteria, blue-green algae and mycoplasmas.

promoter (in oncology)
Agent that induces *cancer* when administered to an animal or human being who has been *exposed* to a cancer *initiator*.

prophage
Latent state of a phage *genome* in a lysogenic bacterium.

proportional mortality rate (ratio)
Proportion of observed deaths from a specified condition in a defined population divided by the proportion of deaths expected from this condition in a standard population, expressed either on an age-specific basis or after age adjustment.

prospective cohort study
See *cohort study*.

proteinuria
Excretion of excessive amounts of protein (derived from blood *plasma* or kidney tubules) in the urine.

proteome
Complete set of proteins encoded by the *genome*.

proteomics
Global analysis of gene expression using a variety of techniques to identify and characterize proteins.
Note: It can be used to study changes caused by *exposure* to chemicals and to determine if changes in mRNA expression correlate with changes in protein expression: the analysis may also show changes in post-translational modification, which cannot be distinguished by mRNA analysis alone.

pseudoadaptation

Apparent adaptation of an organism to changing conditions of the environment (especially chemical) associated with stresses in biochemical systems that exceed the limits of normal (homeostatic) mechanisms.

Note: Essentially there is a temporary concealed pathology that later on can be manifested in the form of explicit pathological changes sometimes referred to as "decompensation".

psychosis

Any major mental disorder characterized by derangement of the personality and loss of contact with reality.

psychotropic

Exerting an effect upon the mind and capable of modifying mental activity.

public health impact assessment

Applying *risk* assessment to a specific target population of known size, giving as the end product a quantitative statement about the number of people likely to be affected in a particular population.

pulmonary

Pertaining to the lung(s).

purgative

see synonym *laxative*.

pyrexia

Condition in which the temperature of a human being or mammal is above normal.

pyrogen

Any substance that produces fever.

quantal effect

Condition that can be expressed only as "occurring" or "not occurring", such as death or occurrence of a tumour.

antonym *graded effect*.

synonym *all-or-none effect*.

quantitative structure–activity relationship (QSAR)

Quantitative structure-biological activity model derived using *regression analysis* and containing as parameters physicochemical constants, indicator variables or theoretically calculated values.

Note: The term is extended by some authors to include chemical reactivity, *i.e.* activity and reactivity are regarded as synonyms. The extension is discouraged.

quantitative structure–*metabolism* relationship (QSMR)
Quantitative association between the physicochemical and (or) the structural properties of a substance and its metabolic behaviour.

rate (in epidemiology)
Measure of the frequency with which an event occurs in a defined population in a specified period of time.
Note 1: Most such rates are ratios, calculated by dividing a numerator, *e.g.* the number of deaths, or newly occurring cases of a disease in a given period, by a denominator, *e.g.* the average population during that period.
Note 2: Some rates are proportions, *i.e.* the numerator is contained within the denominator.

rate difference (RD)
Absolute difference between two *rates*.
Note 1: For example, the difference in *incidence* rate between a population group *exposed* to a causal factor and a population group not exposed to the factor.
Note 2: In comparisons of exposed and unexposed groups, the term *"excess rate"* may be used as a synonym for RD.

rate ratio (RR) (in epidemiology)
Value obtained by dividing the *rate* in an *exposed* population by the rate in an unexposed population.

ratticide
Substance intended to kill rats.

reactive oxygen species (ROS)
Intermediates in the reduction of molecular O2 to water.
Note: Examples are superoxide anion $O_2^-\bullet$. hydrogen peroxide H_2O_2, and hydroxyl radical $HO\bullet$.

readily biodegradable
Arbitrary classification of substances that have passed certain specified screening tests for ultimate biodegradability (see *biodegradation*); these tests are so stringent that such compounds will be rapidly and completely biodegraded in a wide variety of *aerobic* environments.

reasonable maximum exposure (RME)
Highest *exposure* that is reasonably expected to occur.
Note: Typically the 95% upper confidence limit of the *toxicant* distribution is used: if only a few data points (6–10) are available, the maximum detected *concentration* is used.

receptor
Molecular structure in or on a cell that specifically recognizes and binds to a compound and acts as a physiological signal transducer or mediator of an effect.

recovery
1. Process leading to partial or complete restoration of a cell, tissue, organ or organism following its damage from *exposure* to a harmful substance or agent.
2. Term used in analytical and preparative chemistry to denote the fraction of the total quantity of a substance recoverable following a chemical procedure.

reference concentration
Term used for an estimate of air *exposure concentration* to the human population (including sensitive subgroups) that is likely to be without appreciable *risk* of deleterious effects during a lifetime.

reference distribution
Statistical distribution of reference values.

reference dose (RfD)
Term used for an estimate (with uncertainty spanning perhaps an order of magnitude) of a daily *exposure* to the human population (including sensitive subgroups) that is likely to be without appreciable *risk* of deleterious effects during a lifetime.

reference group
See synonym *reference sample group*.

reference individual
Person selected with the use of defined criteria for comparative purposes in a clinical study.

reference interval
Area between and including two reference limits, for example 2.5 and 97.5%.

reference limit
Boundary value defined so that a stated fraction of the reference values is less than or exceeds that boundary value with a stated probability.

reference material
Substance for which one or more properties are sufficiently well established to be used for the calibration of an apparatus, the assessment of a measurement method, or for assigning values to other substances.
synonyms *calibration material, standard material*.

reference population
Group of all reference individuals used to establish criteria against which a population that is being studied can be compared.

reference sample group
Selected reference individuals, statistically adequate numerically to represent the reference population.

reference value
According to IFCC, measured value of a property in a reference individual or *sample* from a reference individual.

regulatory dose
Term used by the USEPA to describe the expected dose resulting from human *exposure* to a substance at the level at which it is regulated in the environment.

relative excess risk (RER)

Measure that can be used in comparison of adverse reactions to drugs, or other *exposures*, based solely on the component of *risk* due to the *exposure* or drug under investigation, removing the *risk* due to background *exposure* experienced by all in the population. The *relative excess risk, R*, is given by

$$R = (R_1 - R_0)/(R_2 - R_0)$$

where R_1 is the *rate* in the population, R_2 is the rate in the comparison population, and R_0 is the rate in the general population.

Note: *Rate* is used here as in epidemiology.

relative odds

See synonym *odds ratio*.

relative risk

1. Ratio of the *risk* of disease or death among the *exposed* to the risk among the unexposed.

synonym *risk ratio*.

2. Ratio of the *cumulative incidence rate* in the exposed to the cumulative incidence rate in the unexposed.

synonym *rate ratio*.

renal

Pertaining to the kidneys.

repellent

Substance used mainly to repel blood sucking insects in order to protect man and animals.

Note: This term may also be used for substances used to repel mammals, birds, rodents, mites, plant pests, *etc.*

reproductive toxicant

Substance or preparation that produces non-heritable harmful effects on the progeny and (or) an impairment of male and female reproductive function or capacity.

reproductive toxicology

Study of the *adverse effects* of substances on the embryo, fetus, neonate and prepubertal mammal and the adult reproductive and neuro-endocrine systems.

reserve capacity

Physiological or biochemical capacity that may be available to maintain homeostasis when the body or an organism is *exposed* to an environmental change.

reservoir (in biology)

Storage *compartment* from which a substance may be released with subsequent biological effects.

residence time

See *mean residence time*.

residual risk

Health risk remaining after risk reduction actions are implemented.

residual time
See *mean residence time*.

resistance (in toxicology)
Ability to withstand the effect of various factors including potentially *toxic* substances.

resorptive effect
Action of a substance after its resorption from the gut into the blood.

respirable dust, respirable particles
Mass fraction of dust (particles) that penetrates to the unciliated airways of the lung (the alveolar region).
Note: This fraction is represented by a cumulative log-normal curve having a median aerodynamic diameter of 4.25 μm and a standard deviation of 1.5 (values for humans).

response
Proportion of an *exposed* population with a defined effect or the proportion of a group of individuals that demonstrates a defined effect in a given time at a given dose rate.

retention
1. Amount of a substance that is left from the total absorbed after a certain time following *exposure*.
2. Holding back within the body or within an organ, tissue or cell of matter that is normally eliminated.

retrospective study
Research design used to test aetiological hypotheses in which inferences about *exposure* to the putative causal factor(s) are derived from data relating to characteristics of the persons or organisms under study or to events or experiences in their past.
Note: The essential feature is that some of the persons under study have the disease or other outcome condition of interest, and their characteristics and past experiences are compared with those of other, unaffected persons. Persons who differ in the severity of the disease may also be compared.

reverse transcription
Process by which an *RNA* molecule is used as a template to make a single-stranded *DNA* copy.

rhabdomyolysis
Acute, fulminating, potentially lethal disease of skeletal muscle that causes disintegration of striated muscle fibres as evidenced by myoglobin in the blood and urine.

rhinitis
Inflammation of the nasal mucosa.

rhonch/us (pl **-i**)
Harsh crepitation in the throat, often resembling snoring.

risk
1. Probability of *adverse effects* caused under specified circumstances by an agent in an organism, a population or an ecological system.
2. Expected frequency of occurrence of a harmful event arising from such an *exposure*.

risk assessment
Identification and quantification of the *risk* resulting from a specific use or occurrence of an agent, taking into account possible harmful effects on individuals *exposed* to the agent in the amount and manner proposed and all the possible routes of *exposure*.
Note: Quantification ideally requires the establishment of *dose–effect* and *dose–response* relationships in likely *target* individuals and populations.

risk assessment management process
Global term for the whole process from *hazard* identification to *risk* management.

risk associated with a lifetime exposure
Probability of the occurrence of a specified undesirable event following *exposure* of an individual person from a given population to a specified substance at a defined level for the expected lifetime of the average member of that population.

risk aversion
Term used to describe the tendency of an individual person to avoid *risk*.

risk characterization
Outcome of *hazard* identification and *risk* estimation applied to a specific use of a substance or occurrence of an environmental *health hazard*.
Note: Risk characterization requires quantitative data on the *exposure* of organisms or people at *risk* in the specific situation. The end product is a quantitative statement about the proportion of organisms or people affected in a target population.

risk communication
Interpretation and communication of *risk* assessments in terms that are comprehensible to the general public or to others without specialist knowledge.

risk de minimis
Risk that is negligible and too small to be of societal concern (usually assumed to be a probability below 10–5 or 10–6).
Note 1: This term can also mean "virtually safe".
Note 2: In the USA, this is a legal term used to mean "negligible *risk* to the individual".
synonym *negligible risk*.

risk estimation
Assessment, with or without mathematical modelling, of the probability and nature of effects of *exposure* to a substance based on quantification of *dose–effect* and *dose–response* relationships for that substance and the population(s) and environmental components likely to be *exposed* and on assessment of the levels of potential exposure of people, organisms and environment at *risk*.

risk evaluation
Establishment of a qualitative or quantitative relationship between *risks* and benefits, involving the complex process of determining the significance of the identified *hazards* and estimated risks to those organisms or people concerned with or affected by them.

risk identification
Recognition of a potential *hazard* and definition of the factors required to assess the probability of *exposure* of organisms or people to that hazard and of harm resulting from such exposure.

risk indicator
see synonym *risk marker*.

risk management
Decision-making process involving considerations of political, social, economic and engineering factors with relevant *risk* assessments relating to a potential *hazard* so as to develop, analyse and compare regulatory options and to select the optimal regulatory response for safety from that hazard.
Note: Essentially risk management is the combination of three steps: *risk evaluation*; *emission* and *exposure* control; *risk monitoring*.

risk marker
Attribute that is associated with an increased probability of occurrence of a disease or other specified outcome and that can be used as an indicator of this increased *risk*.
Note: A risk marker is not necessarily a causal factor.
synonym *risk indicator*.

risk monitoring
Process of following up the decisions and actions within *risk management* in order to check whether the aims of reduced *exposure* and risk are achieved.

risk perception
Subjective perception of the gravity or importance of the *risk* based on a person's knowledge of different risks and the moral, economic and political judgement of their implications.

risk phrases
Word groups identifying potential *health* or environmental *hazards* required under CPL Directives (European Community); may be incorporated into *safety* data sheets.

risk ratio
Value obtained by dividing the probability of occurrence of a specific effect in one group by the probability of occurrence of the same effect in another group, or the value obtained by dividing the probability of occurrence of one potentially hazardous event by the probability of occurrence of another.
Note: Calculation of such ratios is used in choosing between options in *risk management*.

risk-specific dose
Amount of *exposure* corresponding to a specified level of *risk*.

rodenticide
Substance intended to kill rodents.

safety
Reciprocal of *risk*: practical certainty that injury will not result from a *hazard* under defined conditions.
1. Safety of a *drug* or other substance in the context of human *health*: the extent to which a substance may be used in the amount necessary for the intended purpose with a minimum risk of adverse health effects.
2. Safety (toxicological): The high probability that injury will not result from *exposure* to a substance under defined conditions of quantity and manner of use, ideally controlled to minimize exposure.

safety factor (SF)
See synonym *uncertainty factor*.

saluretic
See synonym *natriuretic*.

sarcoma
Malignant tumour arising in a connective tissue and composed primarily of *anaplastic* cells resembling supportive tissue (see *anaplasia*).

saturable elimination
Elimination that becomes *concentration*-independent at a concentration at which the elimination process is functioning maximally.

saturnism
Intoxication caused by lead.
synonym *plumbism*.

scotoma
Area of depressed vision within the visual field, surrounded by an area of less depressed or normal vision.

sclerosis
Hardening of an organ or tissue, especially that due to excessive growth of fibrous tissue.

screening
1. Carrying out of a test or tests, examination(s) or procedure(s) in order to expose undetected abnormalities, unrecognized (incipient) diseases, or defects: examples are mass X-rays and cervical smears.
2. Pharmacological or toxicological screening consists of a specified set of procedures to which a series of compounds is subjected to characterize pharmacological and toxicological properties and to establish *dose–effect* and *dose–response* relationships.

screening level
Decision limit or cut-off point at which a *screening* test is regarded as positive.

secondary metabolite
Product of biochemical processes other than the normal metabolic pathways, mostly produced in microorganisms or plants after the phase of active growth and under conditions of nutrient deficiency.

secondhand smoke
see synonym *sidestream smoke*.

second messenger
Intracellular effector substance increasing or decreasing as a result of the stimulation of a *receptor* by an *agonist*, considered as the "first messenger".

secretion
1. Process by which a substance such as a hormone or enzyme produced in a cell is passed through a *plasma membrane* to the outside, for example the intestinal lumen or the blood (internal secretion).
2. Solid, liquid or gaseous material passed from the inside of a cell through a plasma membrane to the outside as a result of cell activity.

sedative
Substance that exerts a soothing or tranquillising effect.

semichronic
see synonym *subchronic*.

sensibilization
see synonym *sensitization*.

sensitivity (in analytical chemistry)
Extent to which a small change in *concentration* of an analyte can cause a large change in the related measurement.

sensitivity (of a screening test)
Extent (usually expressed as a percentage) to which a method gives results that are free from false negatives.
Note 1: The fewer the false negatives, the greater the sensitivity.
Note 2: Quantitatively, sensitivity is the proportion of truly diseased persons in the screened population who are identified as diseased by the screening test.

sensitization
Immune response whereby individuals become *hypersensitive* to substances, pollen, dandruff, or other agents that make them develop a potentially harmful *allergy* when they are subsequently *exposed* to the sensitizing material (*allergen*).

sensory effect level
1. Intensity, where the detection *threshold* level is defined as the lower limit of the perceived intensity range (by convention the lowest *concentration* that can be detected in 50% of the cases in which it is present).
2. Quality, where the recognition threshold level is defined as the lowest concentration at which the sensory effect can be recognized correctly in 50% of the cases.

3. Acceptability and annoyance, where the nuisance threshold level is defined as the *concentration* at which not more than a small proportion of the population, less than 5%, experiences annoyance for a small part of the time, less than 2%.

Note: Since annoyance will be influenced by a number of factors, a nuisance threshold level cannot be set on the basis of concentration alone.

serum

1. Watery proteinaceous portion of the blood that remains after clotting.

synonym *blood serum*.

2. Clear watery fluid especially that moistening the surface of serous membranes or that exuded through *inflammation* of any of these membranes.

short-term effect

See synonym *acute effect*.

short-term exposure limit (STEL)

Fifteen minute *time-weighted average* (TWA) *exposure* recommended by ACGIH which should not be exceeded at any time during a workday, even if the 8-h TWA is within the *threshold limit value-time-weighted average* (TLV-TWA).

side effect

Action of a *drug* other than that desired for beneficial pharmacological effect.

siderosis

1. *Pneumoconiosis* resulting from the inhalation of iron dust.

2. Excess of iron in the urine, blood or tissues, characterized by haemosiderin granules in urine and iron deposits in tissues.

sidestream smoke

Cloud of small particles and gases that is given off from the end of a burning tobacco product (cigarette, pipe and cigar) between puffs and is not directly inhaled by the smoker.

Note: This is the smoke that gives rise to passive inhalation on the part of bystanders.

synonym *secondhand smoke*.

sign

Objective evidence of a disease, deformity or an effect induced by an agent, perceptible to an examining physician.

silicosis

Pneumoconiosis resulting from inhalation of silica dust.

sink

In environmental chemistry, an area or part of the environment in which, or a process by which, one or more *pollutants* is removed from the medium in which it is dispersed.

Note: For example, moist ground acts as a sink for sulfur dioxide in the air.

sister chromatid exchange (SCE)

Reciprocal exchange of *chromatin* between two replicated *chromosomes* that remain attached to each other until anaphase of *mitosis*; used as a measure of *mutagenicity* of substances that produce this effect.

skeletal fluorosis
Osteosclerosis due to fluoride.

slimicide
Substance intended to kill slime-producing organisms (used on paper stock, water cooling systems, paving stones, *etc.*).

slope factor
Value, in inverse *concentration* or *dose* units, derived from the slope of a *dose–response* curve; in practice, limited to *carcinogenic* effects with the curve assumed to be linear at low concentrations or doses.
Note: The product of the slope factor and the *exposure* is taken to reflect the probability of producing the related effect.

societal risk
Total probability of harm to a human population including the probability of *adverse effects* to *health* of descendants and the probability of disruption resulting from loss of services such as industrial plant or loss of material goods and electricity.

solvent abuse
Deliberate inhalation (or drinking) of volatile solvents, in order to become intoxicated.
synonym "*solvent sniffing*".

solvent sniffing
See synonym *solvent abuse*.

somatic
1. Pertaining to the body as opposed to the mind.
2. Pertaining to nonreproductive cells or tissues.
3. Pertaining to the framework of the body as opposed to the viscera.

soporific
Substance producing sleep.

sorption
Non-committal term used instead of *adsorption* or *absorption* when it is difficult to discriminate experimentally between these two processes.

speciation (in chemistry)
Distribution of an element amongst defined *chemical* species in a system.

speciation analysis (in chemistry)
Analytical activities of identifying and (or) measuring the quantities of one or more individual *chemical species* in a sample.

species
1. In biological systematics, group of organisms of common ancestry that are able to reproduce only among themselves and that are usually geographically distinct.
2. See *chemical species*.

species differences in sensitivity
Quantitative or qualitative differences of response to the action(s) of a potentially *toxic* substance on various species of living organisms.

species-specific sensitivity
Quantitative and qualitative features of response to the action(s) of a potentially *toxic* substance that are characteristic for particular species of living organism.

specific death rate
Death rate computed for a subpopulation of individual organisms or people having a specified characteristic or attribute, and named accordingly.
Note: For example, age-specific death rate, the number of deaths of persons of a specified age during a given period of time, divided by the total number of persons of that age in the population during that time.

specificity (of a screening test)
Proportion of truly non-diseased persons who are identified by the screening test.

specific pathogen free (SPF)
Describing an animal removed from its mother under sterile conditions just prior to term and subsequently reared and kept under sterile conditions.

spreader
Agent used in some *pesticide* formulations to extend the even disposition of the active ingredient.

stability half-life (half-time)
Time required for the amount of a substance in a formulation to decrease, for any reason, by one-half (50%).

standard(ized) mortality (morbidity) ratio (SMR)
Ratio of the number of deaths observed in the study group or population to the number of deaths that would be expected if the study population had the same specific rates as the standard population, multiplied by 100.
Note: This ratio is usually expressed as a percentage.

stannosis
Pneumoconiosis resulting from inhalation of tin dust.

steady state (in toxicology)
State of a system in which the conditions do not change in time.

stem cell
Multipotent cell with mitotic potential that may serve as a precursor for many kinds of differentiated cells.

stochastic
Pertaining to or arising from chance and hence obeying the laws of probability.

stochastic effect, stochastic process
Phenomenon pertaining to or arising from chance, and hence obeying the laws of probability.

stratification (in epidemiology)
Process of or result of separating a *sample* into several subsamples according to specified criteria such as age groups, socio-economic status, *etc.*

stratified sample
Subset of a population selected according to some important characteristic.

structure activity relationship (SAR)
Association between specific aspects of molecular structure and defined biological action.
See also *quantitative structure–activity relationship.*

structure–*metabolism* relationship (SMR)
Association between the physicochemical and (or) the structural properties of a substance and its metabolic behaviour.

subacute (effect)
See *subchronic (effect).*

subchronic
Repeated over a short period, usually about 10% of the life span; an imprecise term used to describe *exposures* of intermediate duration.

subchronic effect
Biological change resulting from an environmental alteration lasting about 10% of the lifetime of the test organism.
Note: In practice with experimental animals, such an effect is usually identified as resulting from multiple or continuous *exposures* occurring over 3 months (90 days). Sometimes a subchronic effect is distinguished from a *subacute* effect on the basis of its lasting for a much longer time.

subchronic toxicity test
Animal experiment serving to study the effects produced by the test substance when administered in repeated *doses* (or continually in food, drinking water and air) over a period of up to about 90 days.

subclinical effect
Biological change following *exposure* to an agent known to cause disease either before symptoms of the disease occur or when they are absent.

subthreshold dose
See synonym *non-effective dose.*

sudorific
Substance that causes sweating.

suggested no adverse response level (SNARL)
Maximum *dose* or *concentration* that on current understanding is likely to be tolerated by an *exposed* organism without producing any harm.

Superfund
Federal authority, established by the US Comprehensive Environmental Response, Compensation, and Liability Act (CERCLA) in 1980, to respond directly to releases or threatened releases (such as from dumps) of hazardous substances that may endanger *health* or welfare.

super-threshold dose
See *toxic dose*.

surrogate
Relatively well studied *toxicant* whose properties are assumed to apply to an entire chemically and toxicologically related class; for example, benzo(a)pyrene data may be used as toxicologically equivalent to that for all *carcinogenic* polynuclear aromatic hydrocarbons.

surveillance
Systematic ongoing collection, collation and analysis of data and the timely dissemination of information to those who need to know in order that action can be taken to initiate investigative or control measures.

sympatholytic
1. adj., Blocking transmission of impulses from the adrenergic (sympathetic) postganglionic fibres to effector organs or tissues.
2. n., Agent that blocks transmission of impulses from the adrenergic (sympathetic) postganglionic fibres to effector organs or tissues.
synonym *antiadrenergic*.

sympathomimetic
1. adj., Producing effects resembling those of impulses transmitted by the postganglionic fibres of the sympathetic nervous system.
2. n., Agent that produces effects resembling those of impulses transmitted by the postganglionic fibres of the sympathetic nervous system.
synonym *adrenergic*.

symptom
Any subjective evidence of a disease or an effect induced by a substance as perceived by the affected subject.

symptomatology
General description of all of the signs and symptoms of *exposure* to a *toxicant*
Note: Signs are the overt (observable) responses associated with exposure (such as convulsions, death, *etc.*) whereas symptoms are covert (subjective) responses (such as nausea, headache, *etc.*).

synapse
Functional junction between two neurones, where a nerve impulse is transmitted from one neurone to another.

synaptic transmission
See *synapse*.

syndrome
Set of signs and symptoms occurring together and often characterizing a particular disease-like state.

synergism (in toxicology)
Pharmacological or toxicological interaction in which the combined biological effect of two or more substances is greater than expected on the basis of the simple summation of the *toxicity* of each of the individual substances.

synergistic effect
See *synergism*.

systemic
Relating to the body as a whole.

systemic effect
Consequence that is either of a generalized nature or that occurs at a site distant from the point of entry of a substance.
Note: A systemic effect requires *absorption* and distribution of the substance in the body.

tachy-
Prefix meaning rapid as in *tachycardia* and *tachypnoea*.

tachycardia
Abnormally fast heartbeat.
antonym *bradycardia*.

tachypnoea
Abnormally fast breathing.
antonym *bradypnoea*.

taeniacide
Substance intended to kill tapeworms.

target (in biology)
Any organism, organ, tissue, cell or cell constituent that is subject to the action of an agent.

target population (epidemiology)
1. Collection of individuals, items, measurements, *etc.* about which inferences are required: the term is sometimes used to indicate the population from which a *sample* is drawn and sometimes to denote any reference population about which inferences are needed.
2. Group of persons for whom an intervention is planned.

T cell
see synonym *T lymphocyte*.

temporary acceptable daily intake
Value for the ADI proposed for guidance when data are sufficient to conclude that use of the substance is safe over the relatively short period of time required to generate and evaluate further safety data, but are insufficient to conclude that use of the substance is safe over a lifetime.
Note: A higher-than-normal uncertainty factor is used when establishing a temporary ADI and an expiration date is established by which time appropriate data to resolve the safety issue should be available.

temporary maximum residue limit
Regulatory value established for a specified, limited time when only a temporary acceptable daily intake has been established for the *pesticide* concerned or, with the existence of an agreed ADI, the available residue data are inadequate for firm maximum residue recommendations.

teratogen
Agent that, when administered prenatally (to the mother), induces permanent structural malformations or defects in the offspring.

teratogenicity
Potential to cause or the production of structural malformations or defects in offspring.

tetanic
Pertaining to tetanus, characterized by tonic muscle spasm.

therapeutic index
Ratio between *toxic* and therapeutic doses (the higher the ratio, the greater the safety of the therapeutic dose).

three-dimensional quantitative structure–activity relationship (3D-QSAR)
Quantitative association between the three-dimensional structural properties of a substance and its biological properties.
see *quantitative structure–activity relationship.*

threshold
Dose or *exposure concentration* below which an effect will not occur.

threshold concentration
See *threshold.*

threshold dose
See *threshold.*

threshold limit value®-ceiling (TLV-C)
Concentration of a potentially *toxic* substance that should not be exceeded during any part of the working *exposure.*

threshold limit value-short term exposure limit (TLV-STEL)
Concentration to which it is believed that workers can be *exposed* continuously for a short period of time without suffering from (1) irritation, (2) *chronic* or irreversible tissue damage, or (3) *narcosis* of sufficient degree to increase the likelihood of accidental injury, impair self-rescue or materially reduce work efficiency, and provided that the daily TLV-TWA is not exceeded.
Note: It is not a separate independent *exposure* guideline; rather, it supplements the TLV-TWA limit where there are recognized *acute* effects from a substance whose *toxic* effects are primarily of a chronic nature. TLV-STELs are recommended only where *toxic* effects have been reported from high short-term exposures in either humans or animals.

threshold limit value-time-weighted average (TLV-TWA)
TWA *concentration* for a conventional 8-h workday and a 40-hour workweek, to which it is believed nearly all workers may be repeatedly *exposed*, day after day, without *adverse effect*.

thrombocytopenia
Decrease in the number of blood platelets (thrombocytes).

tidal volume
Quantity of air or test gas that is inhaled and exhaled during one respiratory cycle.

time-weighted-average-exposure (TWAE), or concentration (TWAC)
Concentration in the *exposure* medium at each measured time interval multiplied by that time interval and divided by the total time of observation
Note: For occupational exposure a working shift of 8 h is commonly used as the averaging time.

tinnitus
Continual noise in the ears, such as ringing, buzzing, roaring or clicking.

tissue dose
Amount of a substance or physical agent (radiation) absorbed by a tissue.

tissue/plasma partition coefficient
See *partition ratio*.

T lymphocyte
Animal cell, which possesses specific cell surface *receptors* through which it binds to foreign substances or organisms, or those which it identifies as foreign, and which initiates *immune responses*.

tolerable daily intake (TDI)
Estimate of the amount of a potentially harmful substance (*e.g.* contaminant) in food or drinking water that can be ingested daily over a lifetime without appreciable *health risk*.
Note: *ADI* is normally used for substances not known to be harmful, such as food additives.

tolerable risk
Probability of suffering disease or injury that can, for the time being, be tolerated, taking into account the associated benefits, and assuming that the *risk* is minimized by appropriate control procedures.

tolerable weekly intake (TWI)
Estimate of the amount of a potentially harmful substance (*e.g.* a contaminant) in food or drinking water that can be ingested weekly over a lifetime without appreciable *health risk*.

tolerance
1. Adaptive state characterized by diminished effects of a particular *dose* of a substance: the process leading to tolerance is called "adaptation."

2. In food *toxicology*, dose that an individual can tolerate without showing an effect.
3. Ability to experience *exposure* to potentially harmful amounts of a substance without showing an *adverse effect*.
4. Ability of an organism to survive in the presence of a *toxic* substance: increased tolerance may be acquired by adaptation to constant *exposure*.
5. In immunology, state of specific immunological unresponsiveness.

tonic
1. Characterised by tension, especially muscular tension.
2. Medical preparation that increases or restores normal muscular tension.

topical (in medicine)
Applied directly to the surface of the body.

topical effect
Consequence of the application of a substance to the surface of the body, which occurs at the point of application.

toxic
Able to cause injury to living organisms as a result of physicochemical interaction.

toxicant
See synonym *toxic substance*.

toxic chemical
See synonym *toxic substance*.

toxic dose
Amount of a substance that produces intoxication without lethal outcome.
synonym *super-threshold dose*.

toxicity
1. Capacity to cause injury to a living organism defined with reference to the quantity of substance administered or absorbed, the way in which the substance is administered and distributed in time (single or repeated *doses*), the type and severity of injury, the time needed to produce the injury, the nature of the organism(s) affected and other relevant conditions.
2. *Adverse effects* of a substance on a living organism defined as in 1.
3. Measure of incompatibility of a substance with life: this quantity may be expressed as the reciprocal of the absolute value of *median lethal dose* ($1/LD_{50}$) or *concentration* ($1/LC_{50}$).

toxicity equivalency factor (TEF, f)
Factor used in *risk assessment* to estimate the *toxicity* of a complex mixture, most commonly a mixture of chlorinated dibenzo-p-dioxins, furans, and biphenyls: in this case, TEF is based on relative toxicity to 2,3,7,8-tetrachlorodibenzo-*p*-dioxin for which the TEF = 1.

toxicity equivalent (TEQ)
Contribution of a specified component (or components) to the *toxicity* of a mixture of related substances. The amount-of-substance (or substance *concentration*) of total TEQ is the sum of that for the components B, C ... N. TEQ is most commonly used

in relation to the reference toxicant 2,3,7,8-tetrachlorodibenzo-*p*-dioxin by means of the TEF, f; which is 1 for the reference substance. Hence:

$$n(\text{TEQ}) = \sum_{i=B}^{N} f_i n_i$$

$$n(\text{TEQ}) = \sum_{i=B}^{N} f_i n_i$$

toxic substance
Substance causing injury to living organisms as a result of physicochemical interactions.

$t_{1/2}$
See *half-life, half-time*.

toxicity test
Experimental study of the *adverse effects* of *exposure* of a living organism to a substance for a defined duration under defined conditions.

toxic material
See synonym *toxic substance*.

toxicodynamics
Process of interaction of potentially *toxic substances* with *target* sites, and the biochemical and physiological consequences leading to *adverse effects*.

toxicogenetics
Study of the influence of hereditary factors on the effects of potentially *toxic* substances on individual organisms.

toxicokinetics
Process of the *uptake* of potentially *toxic* substances by the body, the *biotransformation* they undergo, the distribution of the substances and their metabolites in the tissues, and the *elimination* of the substances and their metabolites from the body.

toxicologically based pharmacokinetic modelling (TBPK)
See *physiologically based pharmacokinetic modeling*.

toxicology
Scientific discipline involving the study of the actual or potential danger presented by the harmful effects of substances on living organisms and ecosystems, of the relationship of such harmful effects to *exposure*, and of the mechanisms of action, diagnosis, prevention and treatment of intoxications.

toxicometry
Term sometimes used to indicate a combination of investigative methods and techniques for making a quantitative assessment of *toxicity* and the *hazards* of potentially *toxic* substances.

toxicophobia
Morbid dread of *poisons*.

toxicophoric (toxophoric) group
Structural moiety that upon metabolic activation exerts *toxic* effects: the presence of a toxicophoric group indicates only potential and not necessarily actual *toxicity* of a *drug* or other substances.
synonym *toxogenic group*.

toxicovigilance
Active process of identification, investigation and evaluation of various *toxic* effects in the community with a view to taking measures to reduce or control *exposure*(s) involving the substance(s) which produces these effects.

toxic substance
Material causing injury to living organisms as a result of physicochemical interactions.
synonyms *chemical etiologic agent, poison, toxicant, toxic chemical, toxic material.*

toxification
Metabolic conversion of a potentially *toxic* substance to a product that is more toxic.

toxin
Poisonous substance produced by a biological organism such as a microbe, animal or plant.

toxinology
Scientific discipline involving the study of the chemistry, biochemistry, pharmacology and *toxicology* of *toxins*.

toxogenic group
synonym *toxicophoric group*.

transcription
Process by which the genetic information encoded in a linear sequence of nucleotides in one strand of *DNA* is copied into an exactly complementary sequence of *RNA*.

transcriptomics
Global analysis of gene expression to identify and evaluate changes in synthesis of m*RNA* after chemical *exposure*.

transformation
1. Alteration of a cell by incorporation of foreign genetic material and its subsequent expression in a new *phenotype*.
2. Conversion of cells growing normally to a state of rapid division in culture resembling that of a *tumour*.
3. Chemical modification of substances in the environment.

transformed cell
Cell which has become genetically altered spontaneously or by incorporation of foreign *DNA* to produce a cell with an extended lifetime in culture.

transformed cell line
See *cell line, transformed cell.*

transgenic
Adjective used to describe animals carrying a *gene* introduced by micro-injecting *DNA* into the nucleus of the fertilized egg.

triage
Assessment of sick, wounded and injured persons following a disaster to determine priority needs for efficient use of available medical facilities.

trophic level
Amount of energy in terms of food that an organism needs.
Note: Organisms not needing organic food, such as plants, are said to be on a low trophic level, whereas predator species needing food of high energy content are said to be on a high trophic level. The trophic level indicates the level of the organism in the food chain.

tubular reabsorption
Transfer of solutes from the *renal* tubule lumen to the tubular epithelial cell and normally from there to the peritubular fluid.

tumorigenic
Able to cause *tumours.*

tumour
1. Any abnormal swelling or growth of tissue, whether benign or malignant.
2. An abnormal growth, in rate and structure, that arises from normal tissue, but serves no physiological function.
synonym *neoplasm.*

tumour progression
Sequence of changes by which a *benign tumour* develops from the initial lesion to a *malignant* stage.

turnover time
See synonym *mean life.*

two-compartment model
Product of *compartmental analysis* requiring two *compartments.*
See *compartmental modelling, multicompartment analysis.*

ulcer
Defect, often associated with *inflammation*, occurring locally or at the surface of an organ or tissue owing to sloughing of necrotic (see *necrosis*) tissue.

ultrafine particles
Particles in air of *aerodynamic diameters* < 0.1 μm ($PM_{0.1}$).

uncertainty factor (UF)
1. In assay methodology, confidence interval or fiducial limit used to assess the probable precision of an estimate.

2. In *toxicology*, value used in extrapolation from experimental animals to man (assuming that man may be more sensitive) or from selected individuals to the general population. For example, a value applied to the NOEL or NOAEL to derive an ADI or TDI.
Note: The NOEL or NOAEL is divided by the value to calculate the ADI or TDI.
See *modifying factor, safety factor.*

unit risk

Upper-bound excess lifetime *cancer risk* estimated to result from continuous *exposure* to an agent at a *concentration* of $1\ \mu g\ L^{-1}$ in water, or $1\ \mu g\ m^{-3}$ in air.
Note: The interpretation of unit risk is as follows: if unit risk $= 1.5 \times 10^{-6}\ \mu g\,L^{-1}$, 1.5 excess tumors are expected to develop per 1,000,000 people if exposed daily for a lifetime to $1\ \mu g$ of the chemical in $1\ L$ of drinking water.

upper boundary

Estimate of the plausible upper limit to the true value of a quantity.
Note: This is usually not a statistical confidence limit.

uptake

Entry of a substance into the body, into an organ, into a tissue, into a cell or into the body fluids by passage through a membrane or by other means.

urticaria

Vascular reaction of the skin marked by the transient appearance of smooth, slightly elevated patches (wheals, hives) that are redder or paler than the surrounding skin and often attended by severe itching.

vacuole

Membrane-bound cavity within a cell.

vasoconstriction

Decrease of the calibre of the blood vessels leading to a decreased blood flow.
antonym *vasodilation.*

vasodilation

Increase in the calibre of the blood vessels, leading to an increased blood flow.
antonym *vasoconstriction.*

vehicle

Substance(s) used to formulate active ingredients for administration or use.
Note: In this context, it is a general term for solvents, suspending agents, *etc.*

venom

Animal toxin generally used for self-defence or predation and usually delivered by a bite or sting.

ventilation

1. Process of supplying a building or room with fresh air.
2. Process of exchange of air between the ambient atmosphere and the lungs.
3. In physiology, the amount of air inhaled per day.
4. Oxygenation of blood.

ventricular fibrillation
Irregular heartbeat characterized by uncoordinated contractions of the ventricle.

vermicide
Substance intended to kill worms.

vermifuge
Substance that causes the expulsion of intestinal worms.

vertigo
Dizziness; an illusion of movement as if the external world were revolving around an individual or as if the individual were revolving in space.

vesicant
1. adj., Producing blisters on the skin.
2. n., Substance that causes blisters on the skin.

vesicle
1. Small sac or bladder containing fluid.
2. Blister like elevation on the skin containing serous fluid.

volume of distribution
Apparent (hypothetical) volume of fluid required to contain the total amount of a substance in the body at the same *concentration* as that present in the *plasma* assuming equilibrium has been attained.

waste
Anything that is discarded deliberately or otherwise disposed of on the assumption that it is of no further use to the primary user.

wasting syndrome
Disease marked by weight loss and atrophy of muscular and other connective tissues that is not directly related to a decrease in food and water consumption.

withdrawal effect
Adverse event following withdrawal from a person or animal of a *drug* to which they have been chronically *exposed* or on which they have become dependent.

x-disease
Hyperkeratotic disease in cattle following *exposure* to chlorinated dibenzo-*p*-dioxins, naphthalenes and related compounds.

xenobiotic
Compound with a chemical structure foreign to a given organism.
Note: The term is frequently restricted to manmade compounds.

zoocide
Substance intended to kill animals.

zygote
1. Cell such as a fertilized egg resulting from the fusion of two *gametes*.
2. Cell obtained as a result of complete or partial fusion of cells produced by *meiosis*.

Appendix C: Abbreviations and Acronyms Used in Toxicology

ADI	Acceptable daily intake
AF	Assessment factor
ALARA(P)	As low as reasonably achievable (practicable) In UK, regulations relating to worker exposure In USA, goal of risk management (NRC regulations)
AUC	Area under the concentration–time curve
AUMC	Area under the moment curve
BCF	Bioconcentration factor
BEI	Biological exposure indices (ACGIH)
BEM	Biological effect monitoring
BOD	Biochemical oxygen demand
b.w.	Body weight
CMR	Carcinogenic, mutagenic and reproductive (toxicant)
CoMFA	Comparative molecular field analysis
CV	Ceiling value
Cyt	Cytochrome
DNA	Deoxyribonucleic acid
DNEL	Derived no-effect level
EC	Enzyme classification or effective concentration
EC_n	Median effective concentration to $n\%$ of a population
EDI	Estimated daily intake
ED_n	Median effective dose to $n\%$ of a population
EEC	Estimated exposure concentration
EED	Estimated exposure dose
EEL	Environmental exposure level

EMDI	Estimated maximum daily intake
EQS	Environmental quality standard
GLP	Good laboratory practice
HSG	Health and safety guide (IPCS)
HQ	Hazard quotient
IC	Inhibitory concentration
i.c.	Intracutaneous
i.d.	Intradermal
i.m.	Intramuscular
inhl	By inhalation
i.p.	Intraperitoneal
I-TEF	International toxicity equivalency factor
i.v.	Intravenous
K_M	Michaelis constant
K_{oc}	Organic carbon partition coefficient
K_{OW}	Octanol water partition coefficient
LADD	Lifetime average daily dose
LC_n	Median concentration lethal to $n\%$ of a test population
LC_{50}	See LC_n
LD_n	Median dose lethal to $n\%$ of a test population
LD_{50}	See LD_n
LEL	Lowest effect level, same as LOEL
LOAEL	Lowest observed adverse effect level
LOEL	Lowest observed effect level
LT_n	Median time for death of $n\%$ of a test population
LV	Limit value
MAC	Maximum allowable concentration
MEL	Maximum exposure limit
MF	Modifying factor
MOE	Margin of exposure
MPC	Maximum permissible concentration
MRL	Maximum residue limit

mRNA	Messenger ribonucleic acid
MSDS	Material safety data sheet
MTC	Maximum tolerable concentration
MTD	Maximum tolerable dose, maximum tolerated dose
MTEL	Maximum tolerable exposure level
NADP(H)	Nicotinamide adenine dinucleotide phosphate (reduced)
ND_n	Median dose narcotic to $n\%$ of a population
NEL	No effect level, same as NOEL
NOAEL	No observed adverse effect level
NOEL	No observed effect level
NSC	Normalized sensitivity coefficients
PBT	Persistent, bioaccumulative and toxic
PBPK	Physiologically based pharmacokinetics modelling
PEL	Permissible exposure limit
p.c.	Per cutim (Latin) = through the skin
$PM_{2.5}$	Particles in air of with a maximum aerodynamic diameter of $2.5\,\mu m$
PM_{10}	Particles in air of with a maximum aerodynamic diameter of $10\,\mu m$
PMR	Proportionate mortality rate, ratio
p.o.	Per os (Latin) = by mouth
POW	Octanol water partition coefficient
PPAR	Peroxisome proliferator-activated receptor
PTWI	Provisional tolerable weekly intake
QSAR	Quantitative structure–activity relationship
3D-QSAR	Three-dimensional quantitative structure–activity relationship
QSMR	Quantitative structure–metabolism relationship
RD	Rate difference
RfC	Reference concentration
RfD	Reference dose
RNA	Ribonucleic acid
ROS	Reactive oxygen species
RR	Rate ratio
SAR	Structure–activity relationship

s.c.	Subcutaneous
SCE	Sister chromatid exchange
SMR	Standard mortality ratio
SMR	Structure–metabolism relationship
SNARL	Suggested no-adverse-response level
STEL	Short-term exposure limit
$t_{1/2}$	Half life, half time
TBPK	Toxicologically based pharmacokinetic modeling
TCDD	2,3,7,8-tetrachlorodibenzo-p-dioxin
TDI	Tolerable daily intake
TEF	Toxicity equivalency factor
TEQ	Toxicity equivalent
TL_n	See LT_n
TLV	Threshold limit value (ACGIH)
TMDI	Theoretical maximum daily intake
TWA	Time-weighted average
TWAC	Time-weighted average concentration
TWAE	Time-weighted average exposure
TWI	Tolerable weekly intake
UF	Uncertainty factor
V_{max}	Maximum velocity
vPvB	Very persistent and very bioaccumulative

Appendix D: Abbreviations and Acronyms of Names of International Bodies and Legislation

ACGIH	American Conference of Governmental Industrial Hygienists
ATSDR	Agency for Toxic Substances and Diseases Registry
BCR	Bureau Communautaire de Référence (Bruxelles)
BIBRA	British Industrial Biological Research Association
CCFA	Codex Committee on Food Additives
CCPR	Codex Committee on Pesticide Residues
CDC	Centers for Disease Control and Prevention
CEC	Commission of the European Communities
CERCLA	Comprehensive Environmental Response, Compensation and Liability Act (USA)
CHIP	Classification, Hazard Information and Packaging (UK)
COSHH	Control of Substances Hazardous to Health Regulations (UK)
CPL	Classification, Packaging and Labelling
EC	European Community, European Commission
ECB	European Chemicals Bureau
EEA	European Environmental Agency
EEC	European Economic Community
EINECS	European Inventory of Existing Chemical Substances
ELINCS	European List of New Chemical Substances
EPA	Environmental Protection Agency (USA), same as USEPA
EUROTOX	European Society of Toxicology
EUSES	European Uniform System for Evaluation of Substances
FAO	Food and Agricultural Organization
FDA	Food and Drug Administration (USA)

HSC	Health and Safety Commission (U.K.)
HSE	Health and Safety Executive (U.K.)
IAEA	International Atomic Energy Agency
IARC	International Agency for Research on Cancer
ICH	International Conference for Harmonization
ICRP	International Commission on Radiological Protection
ICSU	International Council of Scientific Unions (since 1998, International Council of Science)
IFCC	International Federation of Clinical Chemists
ILO	International Labour Organization
IPCS	International Programme on Chemical Safety, UNEP, ILO, WHO
IRIS	Integrated Risk Information System (USA)
IRPTC	International Register of Potentially Toxic Chemicals, now UNEP Chemicals
ISO	International Organization for Standardization
IUPAC	International Union of Pure and Applied Chemistry
IUTOX	International Union of Toxicology
JECFA	Joint FAO/WHO Expert Committee on Food Additives
JMPR	Joint FAO/WHO Meeting on Pesticide Residues
NBS	National Bureau of Standards (USA), now NIST
NIH	National Institutes of Health (USA)
NIOSH	National Institute of Occupational Safety and Health (USA)
NIST	National Institute of Standards and Technology (USA), formerly NBS
NRC	National Research Council (USA)
OECD	Organization for Economic Cooperation and Development
OMS	Organization Mondiale de la Santé, same as WHO
OSHA	Occupational Safety and Health Administration (USA)
REACH	Registration, Evaluation, and Authorization of Chemicals (EC)
RSC	Royal Society of Chemistry
SCOPE	Scientific Committee on Problems of the Environment (ICSU)
TOSCA	Toxic Substances Control Act (USA)
UNEP	United Nations Environment Programme
USEPA	United States Environmental Protection Agency, same as EPA
WHO	World Health Organization, same as OMS

Index

Page numbers in italic, e.g. *29*, refer to figures. Page numbers in bold, e.g. **246**, signify entries in tables.